Geometry and Topology

LECTURE NOTES

IN PURE AND APPLIED MATHEMATICS

1. *N. Jacobson,* Exceptional Lie Algebras
2. *L. -Å. Lindahl and F. Poulsen,* Thin Sets in Harmonic Analysis
3. *I. Satake,* Classification Theory of Semi-Simple Algebraic Groups
4. *F. Hirzebruch, W. D. Newmann, and S. S. Koh,* Differentiable Manifolds and Quadratic Forms (out of print)
5. *I. Chavel,* Riemannian Symmetric Spaces of Rank One (out of print)
6. *R. B. Burckel,* Characterization of C(X) Among Its Subalgebras
7. *B. R. McDonald, A. R. Magid, and K. C. Smith,* Ring Theory: Proceedings of the Oklahoma Conference
8. *Y.-T. Siu,* Techniques of Extension on Analytic Objects
9. *S. R. Caradus, W. E. Pfaffenberger, and B. Yood,* Calkin Algebras and Algebras of Operators on Banach Spaces
10. *E. O. Roxin, P.-T. Liu, and R. L. Sternberg,* Differential Games and Control Theory
11. *M. Orzech and C. Small,* The Brauer Group of Commutative Rings
12. *S. Thomeier,* Topology and Its Applications
13. *J. M. Lopez and K. A. Ross,* Sidon Sets
14. *W. W. Comfort and S. Negrepontis,* Continuous Pseudometrics
15. *K. McKennon and J. M. Robertson,* Locally Convex Spaces
16. *M. Carmeli and S. Malin,* Representations of the Rotation and Lorentz Groups: An Introduction
17. *G. B. Seligman,* Rational Methods in Lie Algebras
18. *D. G. de Figueiredo,* Functional Analysis: Proceedings of the Brazilian Mathematical Society Symposium
19. *L. Cesari, R. Kannan, and J. D. Schuur,* Nonlinear Functional Analysis and Differential Equations: Proceedings of the Michigan State University Conference
20. *J. J. Schäffer,* Geometry of Spheres in Normed Spaces
21. *K. Yano and M. Kon,* Anti-Invariant Submanifolds
22. *W. V. Vasconcelos,* The Rings of Dimension Two
23. *R. E. Chandler,* Hausdorff Compactifications
24. *S. P. Franklin and B. V. S. Thomas,* Topology: Proceedings of the Memphis State University Conference
25. *S. K. Jain,* Ring Theory: Proceedings of the Ohio University Conference
26. *B. R. McDonald and R. A. Morris,* Ring Theory II: Proceedings of the Second Oklahoma Conference
27. *R. B. Mura and A. Rhemtulla,* Orderable Groups
28. *J. R. Graef,* Stability of Dynamical Systems: Theory and Applications
29. *H.-C. Wang,* Homogeneous Branch Algebras
30. *E. O. Roxin, P.-T. Liu, and R. L. Sternberg,* Differential Games and Control Theory II
31. *R. D. Porter,* Introduction to Fibre Bundles
32. *M. Altman,* Contractors and Contractor Directions Theory and Applications
33. *J. S. Golan,* Decomposition and Dimension in Module Categories
34. *G. Fairweather,* Finite Element Galerkin Methods for Differential Equations
35. *J. D. Sally,* Numbers of Generators of Ideals in Local Rings
36. *S. S. Miller,* Complex Analysis: Proceedings of the S.U.N.Y. Brockport Conference
37. *R. Gordon,* Representation Theory of Algebras: Proceedings of the Philadelphia Conference
38. *M. Goto and F. D. Grosshans,* Semisimple Lie Algebras
39. *A. I. Arruda, N. C. A. da Costa, and R. Chuaqui,* Mathematical Logic: Proceedings of the First Brazilian Conference

40. *F. Van Oystaeyen,* Ring Theory: Proceedings of the 1977 Antwerp Conference
41. *F. Van Oystaeyen and A. Verschoren,* Reflectors and Localization: Application to Sheaf Theory
42. *M. Satyanarayana,* Positively Ordered Semigroups
43. *D. L. Russell,* Mathematics of Finite-Dimensional Control Systems
44. *P.-T. Liu and E. Roxin,* Differential Games and Control Theory III: Proceedings of the Third Kingston Conference, Part A
45. *A. Geramita and J. Seberry,* Orthogonal Designs: Quadratic Forms and Hadamard Matrices
46. *J. Cigler, V. Losert, and P. Michor,* Banach Modules and Functors on Categories of Banach Spaces
47. *P.-T. Liu and J. G. Sutinen,* Control Theory in Mathematical Economics: Proceedings of the Third Kingston Conference, Part B
48. *C. Byrnes,* Partial Differential Equations and Geometry
49. *G. Klambauer,* Problems and Propositions in Analysis
50. *J. Knopfmacher,* Analytic Arithmetic of Algebraic Function Fields
51. *F. Van Oystaeyen,* Ring Theory: Proceedings of the 1978 Antwerp Conference
52. *B. Kedem,* Binary Time Series
53. *J. Barros-Neto and R. A. Artino,* Hypoelliptic Boundary-Value Problems
54. *R. L. Sternberg, A. J. Kalinowski, and J. S. Papadakis,* Nonlinear Partial Differential Equations in Engineering and Applied Science
55. *B. R. McDonald,* Ring Theory and Algebra III: Proceedings of the Third Oklahoma Conference
56. *J. S. Golan,* Structure Sheaves over a Noncommutative Ring
57. *T. V. Narayana, J. G. Williams, and R. M. Mathsen,* Combinatorics, Representation Theory and Statistical Methods in Groups: YOUNG DAY Proceedings
58. *T. A. Burton,* Modeling and Differential Equations in Biology
59. *K. H. Kim and F. W. Roush,* Introduction to Mathematical Consensus Theory
60. *J. Banas and K. Goebel,* Measures of Noncompactness in Banach Spaces
61. *O. A. Nielson,* Direct Integral Theory
62. *J. E. Smith, G. O. Kenny, and R. N. Ball,* Ordered Groups: Proceedings of the Boise State Conference
63. *J. Cronin,* Mathematics of Cell Electrophysiology
64. *J. W. Brewer,* Power Series Over Commutative Rings
65. *P. K. Kamthan and M. Gupta,* Sequence Spaces and Series
66. *T. G. McLaughlin,* Regressive Sets and the Theory of Isols
67. *T. L. Herdman, S. M. Rankin, III, and H. W. Stech,* Integral and Functional Differential Equations
68. *R. Draper,* Commutative Algebra: Analytic Methods
69. *W. G. McKay and J. Patera,* Tables of Dimensions, Indices, and Branching Rules for Representations of Simple Lie Algebras
70. *R. L. Devaney and Z. H. Nitecki,* Classical Mechanics and Dynamical Systems
71. *J. Van Geel,* Places and Valuations in Noncommutative Ring Theory
72. *C. Faith,* Injective Modules and Injective Quotient Rings
73. *A. Fiacco,* Mathematical Programming with Data Perturbations I
74. *P. Schultz, C. Praeger, and R. Sullivan,* Algebraic Structures and Applications Proceedings of the First Western Australian Conference on Algebra
75. *L. Bican, T. Kepka, and P. Nemec,* Rings, Modules, and Preradicals
76. *D. C. Kay and M. Breen,* Convexity and Related Combinatorial Geometry: Proceedings of the Second University of Oklahoma Conference
77. *P. Fletcher and W. F. Lindgren,* Quasi-Uniform Spaces
78. *C.-C. Yang,* Factorization Theory of Meromorphic Functions
79. *O. Taussky,* Ternary Quadratic Forms and Norms
80. *S. P. Singh and J. H. Burry,* Nonlinear Analysis and Applications
81. *K. B. Hannsgen, T. L. Herdman, H. W. Stech, and R. L. Wheeler,* Volterra and Functional Differential Equations

82. *N. L. Johnson, M. J. Kallaher, and C. T. Long,* Finite Geometries: Proceedings of a Conference in Honor of T. G. Ostrom
83. *G. I. Zapata,* Functional Analysis, Holomorphy, and Approximation Theory
84. *S. Greco and G. Valla,* Commutative Algebra: Proceedings of the Trento Conference
85. *A. V. Fiacco,* Mathematical Programming with Data Perturbations II
86. *J.-B. Hiriart-Urruty, W. Oettli, and J. Stoer,* Optimization: Theory and Algorithms
87. *A. Figa Talamanca and M. A. Picardello,* Harmonic Analysis on Free Groups
88. *M. Harada,* Factor Categories with Applications to Direct Decomposition of Modules
89. *V. I. Istrăţescu,* Strict Convexity and Complex Strict Convexity: Theory and Applications
90. *V. Lakshmikantham,* Trends in Theory and Practice of Nonlinear Differential Equations
91. *H. L. Manocha and J. B. Srivastava,* Algebra and Its Applications
92. *D. V. Chudnovsky and G. V. Chudnovsky,* Classical and Quantum Models and Arithmetic Problems
93. *J. W. Longley,* Least Squares Computations Using Orthogonalization Methods
94. *L. P. de Alcantara,* Mathematical Logic and Formal Systems
95. *C. E. Aull,* Rings of Continuous Functions
96. *R. Chuaqui,* Analysis, Geometry, and Probability
97. *L. Fuchs and L. Salce,* Modules Over Valuation Domains
98. *P. Fischer and W. R. Smith,* Chaos, Fractals, and Dynamics
99. *W. B. Powell and C. Tsinakis,* Ordered Algebraic Structures
100. *G. M. Rassias and T. M. Rassias,* Differential Geometry, Calculus of Variations, and Their Applications
101. *R.-E. Hoffmann and K. H. Hofmann,* Continuous Lattices and Their Applications
102. *J. H. Lightbourne, III, and S. M. Rankin, III,* Physical Mathematics and Nonlinear Partial Differential Equations
103. *C. A. Baker and L. M. Batten,* Finite Geometries
104. *J. W. Brewer, J. W. Bunce, and F. S. Van Vleck,* Linear Systems Over Commutative Rings
105. *C. McCrory and T. Shifrin,* Geometry and Topology: Manifolds, Varieties, and Knots

Other Volumes in Preparation

Geometry and Topology

MANIFOLDS, VARIETIES, AND KNOTS

Edited by

Clint McCrory
Theodore Shifrin
The University of Georgia
Athens, Georgia

MARCEL DEKKER, INC. New York and Basel

ISBN 0-8247-7621-6

MARCEL DEKKER, INC.
270 Madison Avenue, New York, New York 10016

Current Printing (last digit):
10 9 8 7 6 5 4 3 2 1

PRINTED IN THE UNITED STATES OF AMERICA

Preface

This volume is the proceedings of the 1985 Georgia topology conference, which was held at the University of Georgia in Athens from August 5 to August 16, 1985. This conference followed the tradition of the Georgia topology conferences of 1961, 1969, and 1977, in bringing together many of the leading research mathematicians working in topology and allied fields.

Topics of discussion ranged from traditional areas of topology, such as knot theory and the topology of manifolds, to areas such as differential and algebraic geometry. Geometry continues to provide a rich source of both topological examples and tools with which to attack fundamental topological problems. The recent work of Simon Donaldson on four-manifolds is a striking example of the power of geometry, and is also a reminder of the essential unity of mathematics. An application of these ideas is given in the article by Friedman and Morgan.

Knot theory was well represented at the conference. An exciting trend in knot theory is its increasing application to chemistry, a provocative example of which is described in DeWitt Sumners' survey article on DNA research. The tendency of applications to engender new theoretical questions is illustrated by Keith Wolcott's work on the knotting of graphs.

Other topics discussed in this volume are three-manifolds, group actions, and algebraic varieties. The article by Akbulut and King is an exposition of their important work on the topological characterization of real algebraic varieties. A beautiful application of classical geometry to topology is described by Davis in his article on convex cell complexes.

Helpful advice in planning the conference was received from R. H. Bing, R. C. Kirby, J. Morgan, D. Sullivan, and W. Thurston. We are grateful, in addition, to J. C. Cantrell and J. G. Hollingsworth for their patient advice and unwavering assistance in the conference.

Invited addresses were presented by S. Akbulut, J. Cannon, S. Cappell, M. Davis, S. Donaldson, R. Edwards, F. T. Farrell, D. Gabai, W. Goldman, J. Harer, A. Hatcher, W-C. Hsiang, S. Kerckhoff, W. Meeks, W. Pardon, F. Quinn, P. Scott, P. Shalen, S. Weinberger, and J. West.

Financial support for the conference was provided by the National Science Foundation, the University of Georgia Research Foundation, the Franklin College of Arts and Sciences of the University of Georgia, and the Department of Mathematics of the University of Georgia. The papers submitted for this volume were referred by conference participants. The referees have done an excellent job of improving the exposition of the mathematics presented here. Mrs. Dianne Byrd typed the proceedings, and we wish to thank her warmly for her patient and painstaking job. We also thank the staff of Marcel Dekker for their assistance in producing this book.

Clint McCrory
Theodore Shifrin

Contents

Contributors

SELMAN AKBULUT Mathematics Department, Michigan State University, East Lansing, Michigan

DOUGLAS R. ANDERSON Department of Mathematics, Syracuse University, Syracuse, New York

MARK D. BAKER Department of Mathematics, Brown University, Providence, Rhode Island

SYLVAIN CAPPELL Courant Institute of Mathematical Sciences, New York University, New York, New York

MICHAEL W. DAVIS Department of Mathematics, Ohio State University, Columbus, Ohio

S. K. DONALDSON Mathematical Institute, University of Oxford, Oxford, England

ALAN H. DURFEE Department of Mathematics, Mount Holyoke College, South Hadley, Massachusetts

MICHAEL FALK Department of Mathematics, University of Iowa, Iowa City, Iowa

F. T. FARRELL Department of Mathematics, Columbia University, New York, New York

ROBERT FRIEDMAN Department of Mathematics, Columbia University, New York, New York

WILLIAM M. GOLDMAN Department of Mathematics, Massachusetts Institute of Technology, Cambridge, Massachusetts

F. GONZÁLES-ACUÑA Instituto de Mathematicas, Universidad Nacional Autonoma de Mexico, Mexico City, Mexico

JEAN-CLAUDE HAUSMANN Section de Mathêmatiques, University of Geneva, Geneva, Switzerland

DAVID L. JOHNSON Department of Mathematics, Lehigh University, Bethlehem, Pennsylvania

L. E. JONES Mathematics Department, State University of New York, Stony Brook, New York

HENRY KING Mathematics Department, University of Maryland, College Park, Maryland

DARRYL McCULLOUGH Department of Mathematics, University of Oklahoma, Norman, Oklahoma

JOHN J. MILLSON Department of Mathematics, University of California, Los Angeles, California

JOHN W. MORGAN Department of Mathematics, Columbia University, New York, New York

HANS JØRGEN MUNKHOLM Matematisk Institut, Odense Universiteit, Odense, Denmark

ANTHONY PHILLIPS Mathematics Department, State University of New York, Stony Brook, New York

FRANK QUINN Department of Mathematics, Virginia Polytechnic Institute and State University, Blacksburg, Virginia

RICHARD RANDELL Department of Mathematics, University of Iowa, Iowa City, Iowa

LEE RUDOLPH Adamsville, Rhode Island

URI SREBRO Department of Mathematics, Technion-Israel Institute of Technology, Technion, Haifa, Israel

MARK STEINBERGER Department of Mathematics, Northern Illinois University, Dekalb, Illinois

DAVID A. STONE Brooklyn College, City University of New York, Brooklyn, New York

D. W. SUMNERS Department of Mathematics, Florida State University, Tallahassee, Florida

BRONISLAW WAJNRYB Department of Mathematics, Technion-Israel Institute of Technology, Technion, Haifa, Israel

WILBUR WHITTEN Department of Mathematics, University of Southwestern Louisiana, Lafayette, Louisiana

SHMUEL WEINBERGER Department of Mathematics, The University of Chicago, Chicago, Illinois

JAMES WEST Department of Mathematics, Cornell University, Ithaca, New York

KEITH WOLCOTT Department of Mathematics, University of Iowa, Iowa City, Iowa

Geometry and Topology

Introduction to Resolution Towers

SELMAN AKBULUT / Mathematics Department, Michigan State University, East Lansing, Michigan

HENRY KING / Mathematics Department, University of Maryland, College Park, Maryland

In this article we give an overview of the theory of resolution towers. This is a summary of our ongoing work on the topology of real algebraic sets. A real algebraic set is the solution of real polynomial equations $f_i(x_1, \cdots, x_m) = 0$, $i = 1, \cdots, k$ in $\mathbb{R}^m$. We are interested in classifying all possible topological types of such sets. First of all in the case of nonsingular algebraic sets this problem is solved (see [1], [2]):

$$\left\{\begin{matrix}\text{Nonsingular} \\ \text{Algebraic sets}\end{matrix}\right\} = \left\{\begin{matrix}\text{Interiors of smooth} \\ \text{compact manifolds}\end{matrix}\right\}$$

Also according to [1] a set is homeomorphic to an algebraic set if and only if the one point compactification is homeomorphic to an algebraic set. Hence it suffices to understand the compact singular algebraic sets. First we start with the case $m = 1$. For the sake of argument assume that the coefficients of the polynomials are in $\mathbb{Q}$ (it turns out that our methods give us such polynomials, so this is not a serious restriction). Recall that the elements of algebraic subsets of $\mathbb{R}$ are called algebraic numbers, and that an algebraic subset of $\mathbb{R}$ is a finite number of points in $\mathbb{R}$. We could consider this case to be trivial and go on to the case $m > 1$. But already in this case we see a glimpse of a certain internal structure, reflecting the fact that these points lie in an extension field of the rational numbers. For example, if x is an algebraic number such as $a + \sqrt{2}\, b$, $a,b \in \mathbb{Q}$, we can associate to it a set $\mathcal{J} = \{a,b,x^2 = 2\}$ consisting of two rational numbers and a monomial equation. We can say that the realization $|\mathcal{J}|$ of this set is the algebraic number x. Naively, we can consider the number x as being obtained by glueing two rational numbers with a monomial. For the high dimensional algebraic sets we might expect a similar structure. First of all we can consider the set of compact smooth manifolds as a high dimensional analogue of the rationals (recall that there are countable number of them). If we can define 'topological monomial maps' between smooth manifolds, we can conjecture that the objects such as $\mathcal{J} = \{V_i, f_{ji}\}$, where V_i, $i = 0, \cdots, n$ are closed smooth manifolds and $f_{ji} : V_i \to V_j$ are topological monomial maps, should classify real algebraic sets. Then, we can expect that the realization

$$|\mathcal{J}| = \bigcup_{i=o}^{n} V_i/x \sim f_{ji}(x)$$

of $\mathcal{J}$ (i.e., the set obtained by glueing the smooth manifolds V_i, $i = 0, \cdots, n$ together by the maps f_{ji}) should describe the underlying stratified space of the algebraic set. Surprisingly, it turns out that the algebraic sets do possess such a structure. We call these structures $\mathcal{J}$ resolution towers. Historically, mathematicians attempted to understand algebraic sets by stratified spaces which are high dimensional analogues of real numbers. They are too general to characterize algebraic sets. It was clear that the complexification property of the real algebraic sets did impose some restriction on their topological types [12]. It was also suspected that the resolution of singularities theorem [11] should further restrict the possible topological types of stratified spaces that can occur as algebraic sets. We hope that the resolution towers give a definite answer to these expectations. A resolution tower $\mathcal{J}$ has a rich internal structure which imposes many strong restrictions on the topology of its realization $|\mathcal{J}|$. Roughly, the manifolds and the maps of $\mathcal{J}$ reflect the resolution and the complexification properties of the algebraic set $|\mathcal{J}|$, respectively. To be able to define these objects, we need to define topological monomial maps, which brings us to the notion of ticos (acronym for transversally intersection codimension one submanifolds).

1. TICOS

Let $f : M \to N$ be a smooth map between closed smooth manifolds. We would like to find a topological condition which will make f 'look like' a monomial map. One futile attempt would be to use the coordinate charts to define a monomial map. If M,N are not already algebraic manifolds, this definition will not make sense. Since we need coordinates to define monomials and the coordinate charts are no help, we put fixed coordinates directly on M and N as follows:

DEFINITION: A <u>tico</u> $\mathcal{A}$ in M is a finite collection of properly immersed closed smooth codimension one submanifolds of M.

M

One should consider a tico to be the generalization of the coordinate hyperplanes $\{\mathbb{R}^n_i\}^n_{i=1}$ in $\mathbb{R}^n$, where $\mathbb{R}^n_i = \{x \in \mathbb{R}^n \mid x_i = 0\}$. We call the elements of $\mathcal{A}$ the <u>sheets</u>. We say $\mathcal{A}$ is a <u>regular tico</u> if each element is a properly imbedded submanifold. We define the <u>realization</u> of a tico $\mathcal{A}$ to be $|\mathcal{A}| = \cup_{S\in\mathcal{A}} S$. Of course $\{|\mathcal{A}|\}$ is also a tico consisting of a single sheet. For simplicity often we call the pair $(M,\mathcal{A})$ a tico. A tico $\mathcal{A}$ induces a natural stratification of M, where a codimension d stratum is a connected component of the d-fold self intersections of $|\mathcal{A}|$. We say a tico $(M,\mathcal{A})$ is an <u>algebraic tico</u> if M is a nonsingular algebraic set and each sheet of $\mathcal{A}$ is a nonsingular algebraic subset of M. In particular algebraic ticos are regular. Now by using ticos as coordinates we can define topological monomials.

DEFINITION. Let $(M,\mathcal{A})$, $(N,\mathcal{B})$ be a smooth manifold with ticos. A <u>tico map</u> $f : (M,\mathcal{A}) \to (N,\mathcal{B})$ is a smooth map from M to N with the following local property. Pick any $p \in M$, and pick any charts $\psi : (\mathbb{R}^m,0) \to (M,p)$ and $\theta : (\mathbb{R}^n,0) \to (N,f(p))$ such that $\psi^{-1}(|\mathcal{A}|) = \cup^a_{j=1} \mathbb{R}^m_j$, $\theta^{-1}(|\mathcal{B}|) = \cup^b_{i=1} \mathbb{R}^n_i$. Let $f_i(x)$ be the i-th coordinate of $\theta^{-1} \circ f \circ \psi(x)$. Then there are nonnegative integers α_{ij}, $1 \leq i \leq b$, $1 \leq j \leq a$ and smooth functions $\phi_i : \mathbb{R}^m \to \mathbb{R}$ such that:

$$f_i(x) = \prod_{j=1}^{a} x_j^{\alpha_{ij}} \phi_i(x)$$

for all x near 0 and $\phi_i(0) \neq 0$ for $i = 1, \cdots, b$. The following properties can easily be checked [5].

a) Up to permuation of i and j the exponents α_{ij} above depend only on f and on $p \in M$, not on the local charts ψ and θ.

b) $|\mathcal{A}| \supset f^{-1}(|\mathcal{B}|)$. Furthermore if $\mathcal{A}$ is regular then $f^{-1}(|\mathcal{B}|)$ is a union of components of sheets of $\mathcal{A}$.

c) We may pick $\phi_i(x)$ so that $\phi_i^{-1}(0) = \emptyset$.

The property (a) says that the exponents of a tico map is well defined, and (b) says that tico maps have an 'analytic continuation' like property. Because of (a) for every $S \in \mathcal{A}$ and $T \in \mathcal{B}$ we can define a function $\alpha_{ST} : S \to \mathbb{Z}$ as follows. For $p \in S$ if $f(p) \neq T$ let $\alpha_{ST}(p) = 0$, otherwise choose coordinates as above, say $\psi^{-1}(S) = \mathbb{R}^m_j$ and $\theta^{-1}(T) = \mathbb{R}^n_i$ and let $\alpha_{ST}(p) = \alpha_{ij}$. It follows from definitions that α_{ST} is continuous, hence it is locally constant on each component of S; furthermore

$$f^{-1}(T) = \bigcup_{S \in \mathcal{A}} \alpha_{ST}^{-1}(\mathbb{Z}\text{-}o)$$

One can impose some niceness conditions on tico maps to make them more useful, the two most important ones are:

DEFINITION. A tico map $f : (M,\mathcal{A}) \to (N,\mathcal{B})$ is called <u>type N</u> if for every $p \in M$ there are charts $\psi : (U,0) \to (M,p)$, $\theta : (\mathbb{R}^m,0) \to (N,f(p))$, where $0 \in U \subset \mathbb{R}^m$ is an open subset, such that

$$\psi^{-1}(|\mathcal{A}|) = (\bigcup_{j=1}^{a} \mathbb{R}^m) \cap U, \quad \theta^{-1}(|\mathcal{B}|) = \bigcup_{i=1}^{b} \mathbb{R}^n_i$$

and for some

$$c \geq a \quad f_i(x) = \prod_{j=1}^{c} x_j^{\alpha_{ij}}$$

for $x \in U$, where $f_i(x)$ is the i-th coordinate of $\theta^{-1} \circ f \circ \psi(x)$; and the $b \times c$ matrix (α_{ij}) is onto and $m - c \geq n - b$. We say f is <u>submersive</u> if in addition, we have $f_i(x) = x_{i-n+m}$ for all $i > b$ and all $x \in U$.

So basically a type N tico map is a tico map such that under some choice of local coordinates it becomes a pure monomial ($\phi_i(x) \equiv 1$), and its exponent matrix is onto. The later condition means that it submerses the top stratum of $(M,\mathcal{A})$. A submersive tico map is a type N tico map which submerses each stratum of $(M,\mathcal{A})$ to a stratum of $(N,\mathcal{B})$.

The local topological behavior of tico maps is quite restrictive: they are composition of folds and crushes. For example if $M = \mathbb{R}^2$ and $\mathcal{A} = \{\mathbb{R}^2_1, \mathbb{R}^2_2\}$, type N tico maps $f : (M,\mathcal{A}) \to (M,\mathcal{A})$ are of the form $f(x,y) = (x^a y^b, x^c y^d)$ where $ad - bc \neq 0$. The 'topological behavior' of f depends only on the parity of a,b,c,d. The case b = 0 arises naturally in classification of 3-dimensional algebraic sets [8]. They are up to sign compositions of four basic tico maps $I = (x,y)$, $g = (x,y^2)$, $h = (x^2,y)$, $k = (x,xy)$. They can be classified into eight different topological types. We can distinguish them according to how they map the square $[-1,1] \times [-1,1]$ into itself. In particular it is enough to know how the two vertical sides of the square is mapped into itself. In [8] these are symbolically denoted as follows

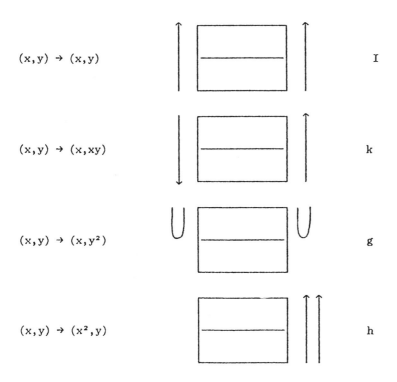

$(x,y) \to (x,y)$

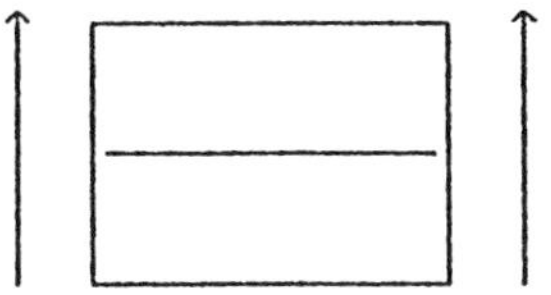

I

$(x,y) \to (x,xy)$

k

$(x,y) \to (x,y^2)$

g

$(x,y) \to (x^2,y)$

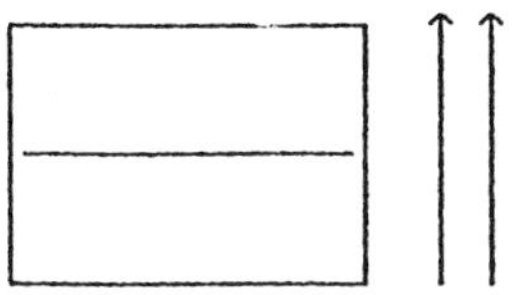

h

$(x,y) \to (x,xy^2)$

$k \cdot g$

$(x,y) \to (x^2,y^2)$

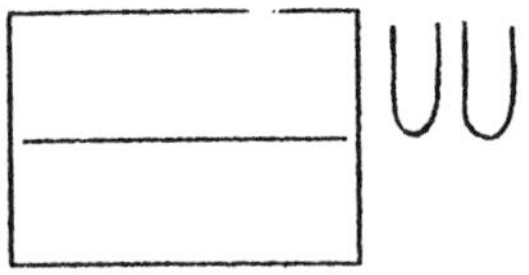

$h \cdot g$

$(x,y) \to (x^2,xy)$

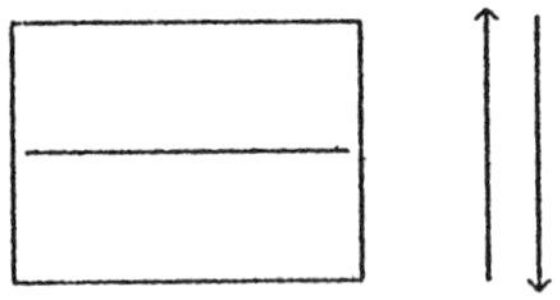

$h \cdot k$

$(x,y) \to (x^2,xy^2)$

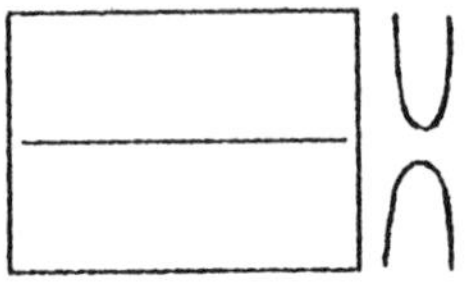

$h \cdot k \cdot g$

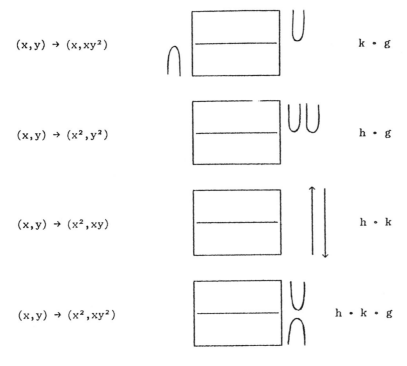

Finally we make the following definition to be able to talk about the germs of tico maps

DEFINITION. Let $(M,\mathcal{A})$, $(N,\mathcal{B})$ be smooth manifolds with ticos and let $\mathcal{C} \subset \mathcal{A}$. We call a map $f : |\mathcal{C}| \to N$ a <u>mico</u> if there is a neighborhood U of $|\mathcal{C}|$ in M and a tico map $g : (U, U \cap \mathcal{A}) \to (N,\mathcal{B})$ such that $f = g|_{|\mathcal{C}|}$.

2. RESOLUTION TOWERS

DEFINITION. A resolution tower $\mathcal{T} = \{V_i, \mathcal{A}_i, p_i\}_{i=0}^{n}$ is a collection of compact smooth manifolds with ticos $(V_i,\mathcal{A}_i)$ $i = 0, \cdots, n$ and a collection of maps $p_i = \{p_{ji}\}_{j=o}^{i-1}$ with $p_{ji} : V_{ji} \to V_j$ such that each $V_{ji} = |\mathcal{A}_{ji}|$ for some $\mathcal{A}_{ji} \subset \mathcal{A}_i$ and

(I) $p_{ji}(V_{ji} \cap V_{ki}) \subset V_{kj}$ for $0 \leq k < j < i \leq n$

(II) $p_{kj} \circ p_{ji}| = p_{ki}|_{V_{ji} \cap V_{ki}}$ for $0 \leq k < j < i \leq n$

(III) $p_{ji}^{-1}(\cup_{k<m} V_{kj}) = \cup_{k<m} V_{ki}) \cap V_{ji}$

(IV) $\mathcal{A}_i = \cup_{j<i} \mathcal{A}_{ji}$ and $\mathcal{A}_{ji} \cap \mathcal{A}_{ki} = \emptyset$ if $j \neq k$

(Notice that this does not rule out $|\mathcal{C}_{ji}| \cap |\mathcal{C}_{ki}| \neq \emptyset$)

There are some extra properties which resolution towers can satisfy, the basic ones are:

R - Each tico $\mathcal{C}_i$ is regular

M - Each p_{ji} is a mico

N - Each p_{ji} is a type N mico

S - Each p_{ji} is a submersive mico

U - $p_{ji}|_S$ is a submersion for every stratum S of $(V_i,\mathcal{C}_i)$ with $S \subset V_{ji} - \cup_{k<j} V_{ki}$

F - Each $(V_i,\mathcal{C}_i)$ is full. This means if S is V_i or any intersection of sheets of $\mathcal{C}_i$, $H_*(S;\mathbb{Z}/2\mathbb{Z})$ is generated by imbedded smooth submanifolds of S.

We say a resolution tower is type R if it satisfies R, is type RM if it satisfies both properties R and M, etc. We define $\partial\mathcal{T} = \{\partial V_i, \partial\mathcal{C}_i, p_i|\}$ where $\partial\mathcal{C}_i = \{\partial S|\ S \in \mathcal{C}_i\}$ and $p_i|$ is the restriction. An <u>algebraic resolution tower</u> is a resolution tower $\mathcal{T}$ such that each $(V_i,\mathcal{C}_i)$ is an algebraic tico and each p_{ji} is an entire rational function. To emphasize the fact that the resolution towers are purely topologically defined objects we sometimes call them <u>topological resolution towers</u>.

Let $\mathscr{T}$ be the set of topological resolution towers and $\mathscr{A}$ be the set of algebraic resolution towers. When we put subscripts R,M,N,S,U,F to $\mathscr{T}$ or $\mathscr{A}$ we mean the set of resolution towers of that type. For example $\mathscr{T}_{RM}$ is the set of resolution towers of type RM. Clearly we have the forgetful inclusion $\mathscr{A} \subset \mathscr{T}$, also $\mathscr{A}_R = \mathscr{A}$ and $\mathscr{T}_S \subset \mathscr{T}_{NU} \subset \mathscr{T}_{MU}$, $\mathscr{A}_S \subset \mathscr{A}_{NU} \subset \mathscr{A}_{MU}$. For $\mathcal{T} \in \mathscr{T}$ we define the <u>realization</u> of $\mathcal{T}$

$$|\mathcal{T}| = \bigcup_{i=o}^{n} V_i/x \sim p_{ji}(x)$$

That is, $|\mathcal{T}|$ is obtained by identifying the points x and $p_{ji}(x)$ in the disjoint union $\cup V_i$. $|\mathcal{T}|$ is a stratified space with strata $\{V_j - \cup_{r<j} V_{rj}\}$. Define $\dim(\mathcal{T})$ = dimension of $|\mathcal{T}|$. For example, the following is a resolution tower of type S : $\mathcal{T} = \{V_i,\mathcal{C}_i,p_i\}_{i=o}^{2}$ where V_2 is the surface of genus 2, $V_1 = S^1$, $V_0 = \{a,b,c\}$, $\mathcal{C}_{12} = \{D\}$, $\mathcal{C}_{02} = \{A,A',B,B',C\}$, and $\mathcal{C}_{01} = \{a',b',c'\}$, where A,A',B,B',C,D are circles on V_2 and a',b',c' are points on V_1 as indicated by the picture

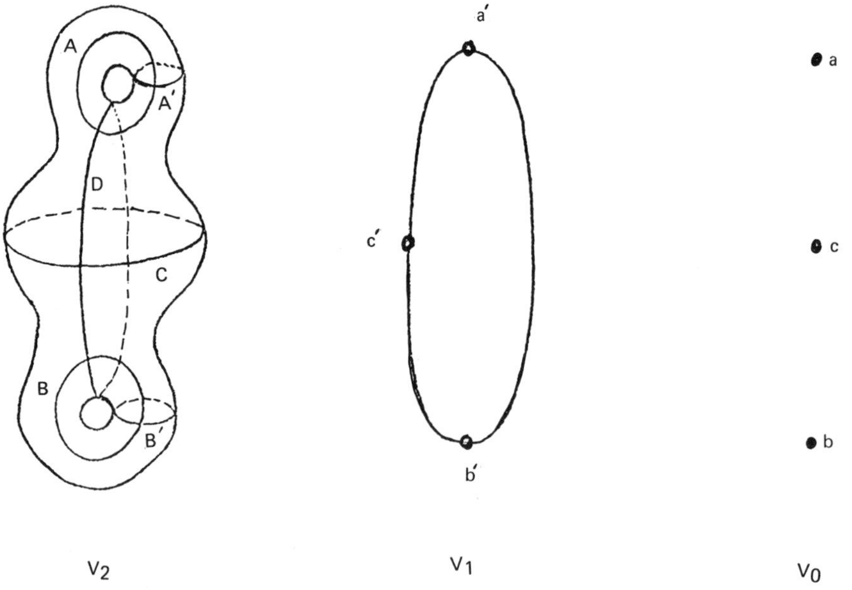

p_{02} collapses A ∪ A' to a, B ∪ B' to b, C to c and p_{01} identifies a',b',c' to a,b,c respectively, and p_{12} folds D onto the arc $\overline{a'c'b'}$ on V_1. Then the realization $|\mathcal{T}|$ of this tower is

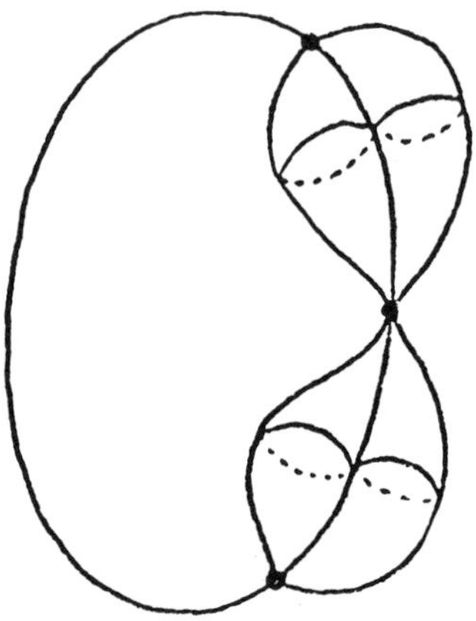

We can usually modify resolution towers to nicer ones without changing their realizations, for example:

PROPOSITION 1. ([5]). $\mathcal{T} \in \mathscr{T}$ then there is $\mathcal{T}' \in \mathscr{T}_R$ with $|\mathcal{T}'| = |\mathcal{T}|$. Furthermore $\mathcal{T}'$ has the same type as $\mathcal{T}$.

PROPOSITION 2. ([5]). If $\mathcal{T} \in \mathscr{T}_M$ (or $\mathscr{A}_M$) then there is $\mathcal{T}' \in \mathscr{T}_{RN}$ (or $\mathscr{A}_{RN}$) with $|\mathcal{T}'| = |\mathcal{T}|$. Furthermore $\mathcal{T}'$ has the same type as $\mathcal{T}$.

Our main theorems relating algebraic sets to resolution towers are the following:

THEOREM 1. ([7]). If $\mathcal{T} \in \mathscr{A}_U$ then $|\mathcal{T}|$ is isomorphic to an algebraic set as a stratified space.

THEOREM 2. ([6]). Any real algebraic set is homeomorphic to $|\mathcal{T}|$ for some $\mathcal{T} \in \mathscr{A}_{FUN}$. This homeomorphism is an isomorphism of stratified sets for some algebraic stratification of the algebraic set.

This theorem turns out to be true for complex algebraic sets where the complex algebraic version $\mathscr{A}^{\mathbb{C}}$ of $\mathscr{A}$ is similarly defined, except in this case we can only take $\mathcal{T} \in \mathscr{A}^{\mathbb{C}}_{UN}$.

THEOREM 3. ([7]). If $\mathcal{T} \in \mathscr{T}_{FS}$ then there exists $\mathcal{T}' \in \mathscr{A}_{FS}$ such that $|\mathcal{T}| = |\mathcal{T}'|$. In particular $|\mathcal{T}|$ is isomorphic, as a stratified set, to an algebraic set (by Theorem 1).

Let us make the convention that if $\mathscr{S}$ is a subset of $\mathscr{T}$ or $\mathscr{A}$ then $|\mathscr{S}| = \{ |\mathcal{T}| \mid \mathcal{T} \in \mathscr{S}\}$. Consider the set $|\mathscr{S}|$ up to P.L. isomorphism, that is two stratified sets in $|\mathscr{S}|$ are equivalent if they have the same isomorphic subdivisions. Also, let Alg denote the set of P.L. isomorphism classes of all compact real algebraic sets. Then we can summarize the above results by the following diagram.

$$
\begin{array}{ccccccc}
\text{Alg} = |\mathscr{A}_{FUN}| & \xleftarrow{\ i\ } & |\mathscr{A}_{FS}| & = & |\mathscr{T}_{FS}| & = & |\mathscr{T}_{RFS}| \\
{\scriptstyle j}\nwarrow \quad \| & & & & \uparrow & & \\
|\mathscr{A}_U| & & & & & & \\
\uparrow & & & & & & \\
|\mathscr{A}_S| & \xrightarrow{\ \ k\ \ } & & & |\mathscr{T}_S| & = & |\mathscr{T}_{RS}|
\end{array}
$$

where i,j,k are induced by inclusions. Hence proving i onto would topologically classify real algebraic sets; it would imply that $\mathrm{Alg} = |\mathcal{T}_{FS}|$. Alternatively proving j and k onto would give the topological classification $\mathrm{Alg} = |\mathcal{T}_S|$. The fact $|\mathcal{A}_U| = \mathrm{Alg} = |\mathcal{A}_{FUN}|$ suggests that getting condition F should be easy to establish, i.e., we should expect to have $|\mathcal{T}_S| = |\mathcal{T}_{FS}|$. This would imply k onto. So an important remaining problem is to refine the proof of Theorem 2 to show the surjectivity of j. We conjecture that this is the case. In low dimensional case i is an isomorphism, for example,

$$\{x \in \mathrm{Alg} \mid \dim X \leq 3\} = \{|\mathcal{T}| \in \mathcal{A}_{FS} \mid \dim \mathcal{T} \leq 3\}$$

By using this one can even combinatorially classify all real algebraic sets of dimension ≤ 3 ([8]). One of the nice properties of resolution towers is that they enjoy many properties of manifolds. For example, we can define the cobordism groups of resolution towers, whereas it would be hard to make sense of cobordisms of stratified spaces. This allows us to talk about the cobordism groups of algebraic sets. In [8] the cobordism group of algebraic sets of dimension ≤ 2 is defined and computed to be $(\mathbb{Z}/2\mathbb{Z})^{15}$ and the pictures of 15 generators are given. Resolution towers naturally generalize the notion of A-spaces of [10]. Finally, to justify the need and the importance of type F resolution towers we refer the reader to [3], [4].

REFERENCES

1. Akbulut, S., and King, H., The Topology of Real Algebraic Sets with Isolated Singularities, Ann. of Math., 113 (1981), 425-446.
2. Akbulut, S., and King, H., The Topology of Real Algebraic Sets, L'Enseignment Math 29 (1983), 221-261.
3. Akbulut, S., and King, H., Submanifolds and Homology of Nonsingular Algebraic Varieties, Amer. Journ. Math (1985).
4. Akbulut, S., and King, H., A Resolution Theorem for Homology Cycles of Real Algebraic Varieties, Invent. Math. 79 (1985), 589-601.
5. Akbulut, S., and King, H., Resolution Towers, M.S.R.I. preprint (1985).
6. Akbulut, S., and King, H., Resolution Tower Structures on Algebraic sets, M.S.R.I. preprint (1985).
7. Akbulut, S., and King, H., Algebraic Structures on Resolution Towers, M.S.R.I. preprint (1985).
8. Akbulut, S., and King, H., The Topological Classification of 3-dimensional Real Algebraic Sets, M.S.R.I. preprint (1985).

9. Akbulut, S., and King, H., The Topology of Resolution Towers, M.S.R.I., preprint (1985).

10. Akbulut, S., and King, H., Real Algebraic Structures on Topological Spaces, Publ. I.H.E.S. 53 (1981), 79-162.

11. Hironaka, H., Resolution of Singularities of an Algebraic Variety Over a Field of Characteristic Zero, Ann. of Math. 79 (1964), 109-326.

12. Sullivan, D., Combinatorial Invariants of Analytic Spaces, Proc. Liverpool Singularities, Springer Notes, 192 (1971), 165-168.

A Geometric Construction of the Boundedly Controlled Whitehead Group

DOUGLAS R. ANDERSON / Department of Mathematics, Syracuse University, Syracuse, New York

HANS JØRGEN MUNKHOLM / Matematisk Institut, Odense Universiteit, Odense, Denmark

In our paper "An Introduction to Boundedly Controlled Simple Homotopy Theory" which appears elsewhere in these Proceedings, we introduced the notions of a space Z with a boundedness control structure (P,C) and the category of finite boundedly controlled (or, simply, bc) CW complexes over Z. This category has objects (X,p) where X is a finite dimensional CW complex and $p : X \to Z$ is a proper map with some additional properties (cf. [2; section 1]) and is denoted by $\underline{\underline{CW}}_f^c/Z$. Since our paper [2] mainly surveys the development of simple homotopy theory in $\underline{\underline{CW}}_f^c/Z$, it contains very few details. It is the purpose of this note to fill in some of those details by describing a construction of $Wh^c(X,p)$, the boundedly controlled Whitehead group of (X,p), and showing that the correspondence $(X,p) \mapsto Wh^c(X,p)$ defines a homotopy functor $Wh^c : \underline{\underline{CW}}_f^c/Z \to \underline{\underline{Ab}}$, where $\underline{\underline{Ab}}$ is the category of abelian groups. This result is Theorem 2.3 below.

The contruction of the functor Wh^c given here follows closely the ideas of Siebenmann [6] and Eckmann [5]. We have chosen this approach for several reasons: it is categorical in nature and exploits the fact that $\underline{\underline{CW}}_f^c$ is a category; it advertises Eckmann's paper [5], which has been largely overlooked by topologists; and it clarifies some of the details of Siebenmann's paper [6] which are omitted in [6] and have puzzled many topologists. An alternative construction of $Wh^c(X,p)$ using the approach of Cohen [4] will be used in [3]. It is equivalent to the construction given here, but is less formal.

Both authors wish to acknowledge with thanks support for this project received from the Scientific Affairs Division of NATO (grant number 670/84). In addition, the first named author was partially supported by

the NSF under grant number MCS-8201776 and the second author, by the
Danish Natural Sciences Research Council. Finally, the first author would
like to thank with pleasure the Matematisk Institut of Odense Universitet
for its hospitality during the period when this manuscript was written.

1. THE FORMAL FRAMEWORK FOR THE CONSTRUCTION OF THE WHITEHEAD GROUP

This section describes the category theoretic ideas that underlie our
construction, from a geometric viewpoint, of the Whitehead group functor.
Our approach is a modification of the one in Eckmann [5] or Siebenmann [6]
and has been chosen to make it apply easily in the present context.

Following [5], we let $\underline{C}$ and $\underline{D}$ be categories, Σ be a family of
morphisms in $\underline{C}$ containing all isomorphisms and closed under compositions,
and $H : \underline{C} \to \underline{D}$ be a functor which is bijective on objects and carries the
morphisms in Σ into isomorphisms. It then follows that H factors uniquely
through the category of fractions $\underline{C}(\Sigma^{-1})$ as indicated in the diagram

$$
\begin{array}{ccc}
\underline{C} & \xrightarrow{\ H\ } & \\
{\scriptstyle Q}\downarrow & \nearrow & \underline{D} \\
\underline{C}(\Sigma^{-1}) & {\scriptstyle H'} &
\end{array}
$$

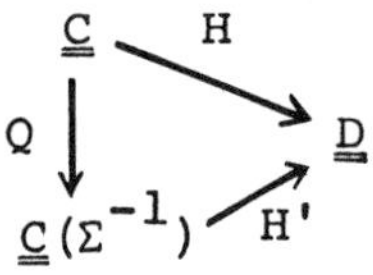

and that H' is bijective on objects.

We shall assume that $\underline{C}$, $\underline{D}$, Σ, and H have the following properties:

1.1. Let $f : X \to Y$ be a morphism in $\underline{C}$ such that $H(f)$ is an isomorphism.
Then any diagram in $\underline{C}$

$$
\begin{array}{ccc}
X & \xrightarrow{\ f\ } & Y \\
{\scriptstyle g}\downarrow & & \downarrow {\scriptstyle g'} \\
Z & \dashrightarrow & W \\
& {\scriptstyle f'} &
\end{array}
$$

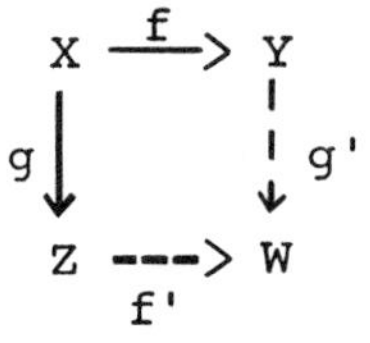

has a pushout as indicated (dotted) such that $H(f')$ is an isomorphism in
$\underline{D}$. Furthermore, if $f \in \Sigma$, so does f'.

1.2. Let $f, g : X \to Y$ be morphisms in $\underline{C}$ such that $H(f) = H(g)$. Then there
exist $s, t \in \Sigma$ such that $sf = tg$.

1.3. Given any morphism $\phi : X \to Y$ in $\underline{D}$, there exists f, s in $\underline{C}$ with $s \in \Sigma$
such that $H(s)\phi = h(f)$.

1.4. For any $X \in |\underline{\underline{C}}|$, there exist $Z_X \in |\underline{\underline{C}}|$ and $r_X, s_X : X \to Z_X$ in Σ such that whenever $f_0, f_1 : X \to Y$ have $H(f_0) = H(f_1)$, then there exists a commutative diagram in $\underline{\underline{C}}$ with $s \in \Sigma$:

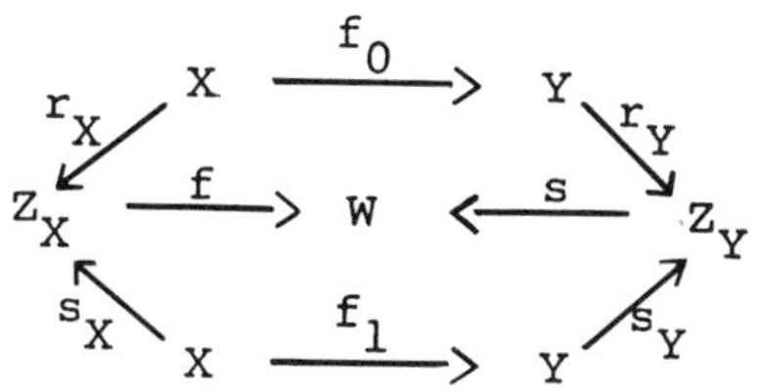

PROPOSITION 1.5. If C, D, Σ, and H satisfy 1.1, 1.3., and 1.4. above, then $H' : C(\Sigma^{-1}) \to D$ is an isomorphism of categories.

Proof: This follows immediately from the arguments given in section 5 of Eckmann [5].

Let $X \in |\underline{\underline{D}}|$ and $\phi_i : X \to Y_i$ $(i = 1,2)$ be an isomorphism in $\underline{\underline{D}}$. We say ϕ_1 and ϕ_2 are <u>simply equivalent</u> if they differ by a composition of morphisms of the forms $H(s)$ and $H(t)^{-1}$ for $s,t \in \Sigma$. We let $E(X)$ denote the equivalence classes of isomorphisms in $\underline{\underline{D}}$ with domain X and assume $E(X)$ is a set for all X.

PROPOSITION 1.6. If $\underline{\underline{C}}$, $\underline{\underline{D}}$, Σ, and H satisfy 1.1 through 1.4 above, then $E(X)$ has an abelian group structure. Furthermore, the correspondence $X \mapsto E(X)$ extends to a functor $E : \underline{\underline{D}} \to \underline{Ab}$, the category of abelian groups.

Proof: The arguments given in sections 3 and 4 of Eckmann [5], combined with Proposition 1.5, easily specialize to prove this proposition.

We note that it follows from property 3 above that any element $x \in E(X)$ has a representative of the form $H(f)$ where $f : X \to Y$ is a morphism in $\underline{\underline{C}}$. Furthermore, if $f_i : X \to Y_i$ $(i = 1,2)$ is a morphism in $\underline{\underline{C}}$ representing $x_i \in E(X)$ $(i = 1,2)$, then $x_1 + x_2$ is represented by $f_2'f_1 = f_1'f_2$ in the pushout diagram

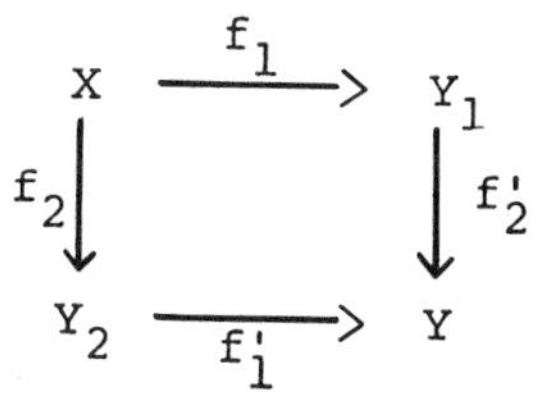

(The pushout exists by property 1.1 above.) Finally, if $g : X \to Z$ is a morphism in C and $f : X \to Y$ is a morphism in C representing $x \in E(X)$, then

$E(g)(x)$ is represented by f' in the pushout diagram

$$\begin{array}{ccc} X & \xrightarrow{f} & Y \\ {\scriptstyle g}\downarrow & & \downarrow{\scriptstyle g'} \\ Z & \xrightarrow[f']{} & W \end{array}$$

2. THE GEOMETRIC DEFINITION OF THE WHITEHEAD GROUP

In this section we describe how the formal machinery of section 1 is used to construct the functor $Wh^c : (\underline{\underline{CW}}^c_f/Z)_{hty} \to \underline{\underline{Ab}}$. Here $(\underline{\underline{CW}}^c_f/Z)_{hty}$ is the category of finite, bc CW complexes and bc homotopy classes of maps. The main results are Theorems 2.2 and 2.4 whose proofs are given in section 4.

The remarks at the end of section 1 will show the construction given here is geometric in the following sense: Each element $x \in Wh^c(X,p)$ can be represented by an inclusion $i : (X,p) \to (Y,q)$ which is a bc homotopy equivalence. Furthermore, if $i_j : (X,p) \to (Y_j,q_j)$ is an inclusion representing x_j $(j = 1,2)$, then there is a pushout diagram

$$\begin{array}{ccc} (X,p) & \xrightarrow{i_1} & (Y_1,q_1) \\ {\scriptstyle i_2}\downarrow & & \downarrow{\scriptstyle k_1} \\ (Y_2,q_2) & \xrightarrow[k_2]{} & (Y_1 \underset{X}{\cup} Y_2,q) \end{array}$$

and $x_1 + x_2$ is represented by $k_1i_1 = k_2i_2$. In this sense the approach taken here is a formalization of the construction of the classical Whitehead group given by Cohen [4].

In the sequel when Z is clear from the context, we shall denote $\underline{\underline{CW}}^c_f/Z$ simply by $\underline{\underline{CW}}^c_f$.

Let $(\underline{\underline{CW}}^c_f)_{inc}$ be the subcategory of $\underline{\underline{CW}}^c_f$ whose objects are finite, be CW complexes and whose morphisms are inclusions $i : (X,p) \to (Y,q)$ (i.e., i is a bc map that maps X isomorphically onto a subcomplex of Y). Following Eckmann [5] or Siebenmann [6], $(\underline{\underline{CW}}^c_f)_{inc}$ plays the role of $\underline{\underline{C}}$ in section 1; $(\underline{\underline{CW}}^c_f)_{hty}$ the role of $\underline{\underline{D}}$; and the functor that assigns to an inclusion its bc homotopy class, the role of H. Finally, the class of all expansions in the sense of the following definition will play the role of Σ.

DEFINITION 2.1. A finite composite of elementary expansions is called an <u>expansion</u>. A morphism $s : (X,p) \to (Y,q)$ in $(\underline{\underline{CW}}^c_f)_{inc}$ is an <u>elementary expansion</u> if there exists a bounded collection of finite subcomplexes of Y, $\{Y_\lambda | \lambda \in \Lambda\}$ (cf. [2; section 1] for definitions) such that

1. $Y = X \cup \cup Y_\lambda$ and $Y_\lambda \cap Y_\mu \subset X$ if $\lambda \neq \mu$; and
2. For any $\lambda \in \Lambda$, the inclusion $X = X \cap Y_\lambda \to Y_\lambda$ is an (usual) expansion of finite complexes.

Notice that for each $\lambda \in \Lambda$, there is a deformation retraction $r_\lambda : Y_\lambda \to X_\lambda$. Since $\{Y_\lambda | \lambda \in \Lambda\}$ is a bounded family of subsets of Y, the map $r : (Y,q) \to (X,p)$ given by setting $r|Y_\lambda = r_\lambda$ $(\lambda \in \Lambda)$ and $r|X = 1$ (or, more precisely s^{-1}) is easily seen to be a bc homotopy inverse for s. Hence, every morphism in Σ is a bc homotopy equivalence and H factors uniquely as indicated in the following diagram

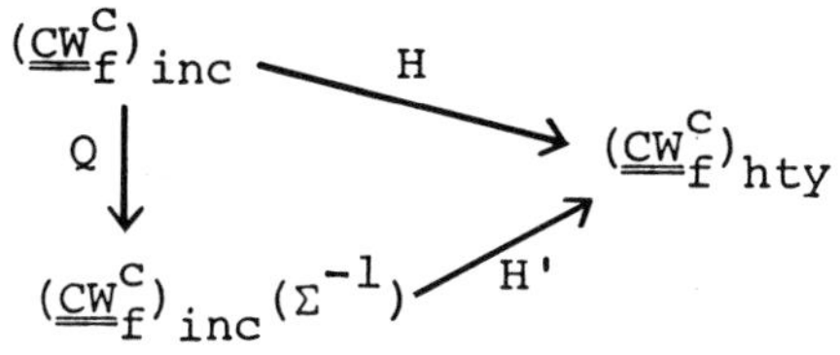

THEOREM 2.2. The functor H' is an isomorphism of categories.

For $(X,p) \in |(\underline{\underline{CW}}^c_f)_{hty}|$, let $Wh^c(X,p)$ be the set of homotopy classes of bc homotopy equivalences with domain (X,p) under the relation that $f_i : (X,p) \to (Y_i,q_i)$ $(i = 1,2)$ are identified whenever there exists a composite $g : (Y_1,q_1) \to (Y_2,q_2)$ of maps in Σ and homotopy inverses of maps in Σ such that f_2 is bc homotopic to gf_1.

THEOREM 2.3. For every $(X,p) \in |(\underline{\underline{CW}}^c_f)_{hty}|$, $Wh^c(X,p)$ is an abelian group. The correspondence $(X,p) \mapsto Wh^c(X,p)$ extends to a functor $Wh^c : (\underline{\underline{CW}}^c_f)_{hty} \to \underline{\underline{Ab}}$.

The proofs of 2.2 and 2.3 proceed by verifying properties 1.1 through 1.4. Since these verifications require some preliminary results, the proofs of 2.1 and 2.2 are deferred to section 4.

3. SOME PRELIMINARY RESULTS

This section contains some preliminary results needed in the proofs of 2.2 and 2.3. It should be no surprise that they concern mostly pushouts and mapping cylinders.

LEMMA 3.1. Let $f : (X,p) \to (Y,q)$ be a map in $\underline{\underline{CW}}{}_f^c$ that is a homeomorphism of spaces. Then f is a bc homeomorphism.

COROLLARY 3.2. A map $f : (X,p) \to (Y,q)$ is bc if and only if $1_X : (X,p) \to (X,qf)$ is a bc homeomorphism.

Proof of 3.1: To show f^{-1} is bc, let $K \in P$ where (P,C) is the boundedness control structure on Z, suppose $y \in q^{-1}K$, and write $y = f(x)$. Then x is in the interior of some cell $e \in X$. Since $(X,p) \in |\underline{\underline{CW}}{}_f^c/Z|$, there exists an integer m, independent of e, and a minimal element $K_e \in P$ such that $p(e) \subset C^m(K_e)$. In particular, $p(x) \in C^m(K_e)$.

Since f is bc, there exists an integer n such that for all $L \in P$, $fp^{-1}(L) \subset q^{-1}(C^n L)$. In particular, $y = f(x) \in q^{-1}(C^{n+m}(K_e))$. Hence $q(y) \in K \cap C^{n+m}(K_e)$. It follows that $K_e \subset C^{\Theta(n+m)}(K)$ where $\Theta : Z_+ \to Z_+$ is the function given by iv) of a boundedness control structure (cf. [2; section 1]). But then $x = f^{-1}(y) \in p^{-1}(C^m(K_e)) \subset p^{-1}(C^{m+\Theta(m+n)}(K_e))$. Hence $f^{-1}q^{-1}(K) \subset p^{-1}(C^d(K))$ for $d = m+\Theta(m+n)$ and f^{-1} is bc. The lemma follows.

Proof of 3.2: It is easily verified that if $f : (X,p) \to (Y,q)$ is bc, then so is $1_X : (X,p) \to (X,qf)$. Hence 1_X is a bc homeomorphism by 3.1.

If $1_X : (X,p) \to (X,qf)$ is a bc homeomorphism, then $f : (X,p) \to (Y,q)$ is the composite $(X,p) \xrightarrow{1_X} (X,qf) \xrightarrow{\bar{f}} (Y,q)$ of the bc map 1_X and the obviously bc map $f : (X,qf) \to (Y,q)$. Hence f is bc and 3.2 follows.

In the sequel when we refer to maps in CW_f^c, we shall often omit the adjective "boundedly controlled" and use it mainly for emphasis.

The next lemma gives a sufficient condition for the existence of a pushout of an angle diagram in $(\underline{\underline{CW}}{}_f^c)_{inc}$

$$\begin{array}{ccc} (X_0,p_0) & \xrightarrow{f_1} & (X_1,p_1) \\ {\scriptstyle f_2}\big\downarrow & & \\ (X_2,p_2) & & \end{array}$$

LEMMA 3.3. Suppose there exists a retraction $r : (X_1,p_1) \to (X_0,p_0)$. Then the diagram above has a pushout in $(\underline{\underline{CW}}^c_f)_{inc}$. In particular, this occurs if f_1 is a homotopy equivalence.

Proof: By 3.1, the identity maps $1_{X_0} : (X_0,p_0) \to (X_0,p_2f_2)$ and $1_{X_1} : (X_1,p_1) \to (X_1,p_2(f_2r))$ map the angle above isomorphically onto the angle of the diagram

$$\begin{array}{ccc} (X_0,p_2f_2) & \xrightarrow{f_1} & (X_1,p_2f_2r) \\ {\scriptstyle f_2}\downarrow & & \vdots\, {\scriptstyle k_1} \\ (X_2,p_2) & \overset{}{\underset{k_2}{\dashrightarrow}} & (X_1 \cup_{X_0} X_2, p) \end{array}$$

Let $p : X_1 \cup_{X_0} X_2 \to Z$ be such that $p|X_1 = p_2f_2r$ and $p|X_2 = p_2$. Then p is continuous and it is easily verified that the above square is a pushout diagram in $(\underline{\underline{CW}}^c_f)_{inc}$. The first part of 3.3 follows.

The last part of 3.3 follows from the proof of the Whitehead Theorem in [1; section 13].

Now let $f : (X,p) \to (Y,q)$ be a map in $\underline{\underline{CW}}^c_f$ and $(A,p|A)$ be a subcomplex of (X,p). Let $\bar{M}_f$ be obtained from $X \times I \coprod Y$ by identifying $(x,1)$ with $f(x)$ for any $x \in X$ and (a,s) with (a,t) for any $a \in A$ and $s,t \in I$. Let $r : \bar{M}_f \to Y$ be the obvious retraction. The pair $(\bar{M}_f,qr)$ is called the <u>mapping cylinder of f reduced modulo A</u> or simply the <u>reduced mapping cylinder</u> of f when A is clear from the context. When $A = \emptyset$, $\bar{M}_f$ is the usual mapping cylinder of f and will be denoted by M_f. The following remarks are obvious:

REMARK 3.4. 1. If $f : (X,p) \to (Y,q)$ is cellular, then $(\bar{M}_f,qr)$ is a finite, bc CW complex.

2. If $i : (X,p) \to (\bar{M}_f,qr)$ is given by $i(x) = (x,0)$, then i is bc and the following diagram in $\underline{\underline{CW}}^c_f$ commutes:

$$\begin{array}{ccc} & & (\bar{M}_f,qr) \\ & {\scriptstyle i}\nearrow & \downarrow{\scriptstyle r} \\ (X,p) & \xrightarrow{f} & (Y,q) \end{array}$$

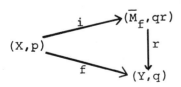

Now let $f : (X,p) \to (Y,q)$ be a cellular map in $\underline{\underline{CW}}_f^c$ and $(A,p|A)$ and $(B,p|B)$ be subcomplexes of (X,p). If $B \neq \emptyset$, let $\overline{M}_{f|B}$ be the mapping cylinder of $f|B$ reduced modulo $A \cap B$; while if $B = \emptyset$, let $\overline{M}_{f|B} = Y$. In either case there is an inclusion $j : (\overline{M}_{f|B}, qr\ ) \to (\overline{M}_f, qr)$ which is clearly in $(\underline{\underline{CW}}_f^c)_{inc}$.

LEMMA 3.5. In the situation described above, $j : (\overline{M}_{f|B}, qr|) \to (\overline{M}_f, qr)$ is in Σ.

Proof: A standard induction argument shows that it suffices to prove 3.5 in the case when $X - (A \cup B) = \{e_\lambda^n | \lambda \in \Lambda\}$ consists entirely of n-cells. In this case, we show j is an elementary expansion in the sense of 2.1.

To do this, notice that $\overline{M}_f = \overline{M}_{f|B} \cup \bigcup e^n \cup \bigcup (e_\lambda^n \times I)$ where both unions run over $\lambda \in \Lambda$. For each λ, let M_λ be the smallest subcomplex of $\overline{M}_f$ containing $e_\lambda^n \times I$. Since the inclusion $M_\lambda \cap \overline{M}_{f|B} \to M_\lambda$ is a classical elementary expansion, it now suffices to show that $\{M_\lambda | \lambda \in \Lambda\}$ is a bounded family of finite subcomplexes of M_f. But this follows by an easy argument using property iv) of a boundedness control structure (cf. [3] or [1; proof of Lemma 7.3]).

COROLLARY 3.6. Let (X,p) be in $\underline{\underline{CW}}_f^c$ and $(B,p|B)$ be a subcomplex. Then the inclusion map $j : (X \times 0 \cup B \times I, p\pi|) \to (X \times I, p\pi)$ is in Σ. Hence, there exists a deformation retraction $r : (X \times I, p\pi) \to (X \times 0 \cup B \times I, p\pi|)$.

Proof: The first sentence follows immediately from 3.5 by taking $f = 1 : (X,p) \to (X,p)$ and $A = \emptyset$. The last sentence follows from the remarks following definition 2.1.

COROLLARY 3.7. (The Boundedly Controlled Homotopy Homotopy Extension Property): Let (X,p) be in $\underline{\underline{CW}}_f^c$ and $(B,p|B)$ be a subcomplex. Then any map $F : (X \times 0 \cup B \times I, p\pi|) \to (Y,q)$ extends to a map $G : (X \times I, p\pi) \to (Y,q)$.

Proof: Set $G = Fr$ where r is the retraction of 3.6.

LEMMA 3.8. Let $f : (X,p) \to (Y,q)$ be a cellular morphism in $\underline{\underline{CW}}_f^c$ and $(A,p|A)$ be a subcomplex of (X,p). If the inclusion $s : (A,p|A) \to (X,p)$ is an expansion, so is the inclusion $(M_{f|A} \cup X, qr|) \to (M_f, qr)$.

Proof: It suffices to prove the lemma in the case when s is an elementary expansion. In this case, let $\{X_\lambda | \lambda \in \Lambda\}$ be a bounded family of finite subcomplexes of X satisfying the conditions of definition 2.1. Since $f : (X,p) \to (Y,q)$ is bc, there exists a bounded family of finite subcomplexes of Y, $\{Y_\lambda | \lambda \in \Lambda\}$ such that for any $\lambda \in \Lambda$, $f(X_\lambda) \subset Y_\lambda$. Let

$f_\lambda = f|X_\lambda : X_\lambda \to Y_\lambda$ and notice that $M_{f_\lambda} \cap (M_{f|A} \cup X) = M_{f_\lambda}|A_\lambda \cup X_\lambda$ where $A_\lambda = A \cap X_\lambda$.

Now consider $\{M_{f_\lambda}|\lambda \in \Lambda\}$ as a family of finite subcomplexes of M_f. It is easy to see that this family is bounded, that $M_f = (M_{f|A} \cup X) \cup \cup M_{f_\lambda}$, and that $M_{f_\lambda} \cap M_{f_\mu} \subset M_{f|A} \cup X$ if $\lambda \neq \mu$. Since for any $\lambda \in \Lambda$, the inclusion $A \cap X_\lambda = A_\lambda \to X_\lambda$ is an ordinary expansion of finite complexes, so is the inclusion $(M_{f|A} \cup X) \cap M_{f_\lambda} = M_{f_\lambda}|A_\lambda \cup X_\lambda \to M_{f_\lambda}$ (cf. [4; Lemma 5.1, p. 16]). It now follows from 2.1 that $(M_{f|A} \cup X, qr|) \to (M_f, qr)$ is an elementary expansion and 3.8 follows.

4. THE PROOFS OF 2.2 AND 2.3.

This section contains the proofs of 2.2 and 2.3. They proceed by verifying that $(\underline{\underline{CW}}_f^C)_{inc}$, $(\underline{\underline{CW}}_f^C)_{hty}$, H, and Σ of section 2 satisfy properties 1.1 through 1.4 of section 1 and then by applying 1.5 and 1.6.

To see that property 1.1 holds, let

$$
\begin{array}{ccc}
(X,p) & \xrightarrow{\ f\ } & (Y,q) \\
g \downarrow & & \downarrow \\
(Z,r) & \xdashrightarrow{\ f'\ } & (W,s)
\end{array}
$$

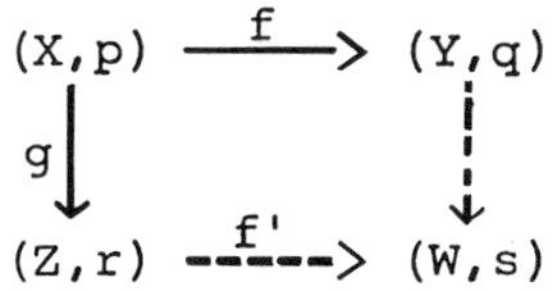

be a diagram in $(\underline{\underline{CW}}_f^C)_{inc}$ such that f is a homotopy equivalence. Then there is a strong deformation retraction R_t $(0 \leq t \leq 1)$ of (Y,q) onto (X,p) by [1; section 13]. It follows from 3.3 that the dotted pushout exists; furthermore, as a space $W = Y \cup_X Z$. It follows easily that R_t extends to a strong deformation retraction of (W,s) onto (Z,r) and that f! is a homotopy equivalence.

Finally, to show that if $f \in \Sigma$ so does f', it suffices to prove this in the case when f is an elementary expansion. But if $\{Y_\lambda|\lambda \in \Lambda\}$ is the bounded family of finite subcomplexes of Y satisfying 1) and 2) of 2.1 showing that $f : (X,p) \to (Y,q)$ is an elementary expansion, the same family thought of as subcomplexes of W shows that f' is also an elementary expansion. Thus 1.1 follows.

The verification of 1.2 is somewhat tedious and is temporarily deferred.

To see that 1.3 holds, let $\phi : (X,p) \to (Y,q)$ in $(\underline{\underline{CW}}_f^C)_{hty}$ be represented by $f : (X,p) \to (Y,q)$ in $\underline{\underline{CW}}_f^C$. It follows from 3.4 and 3.5 that the following diagram homotopy commutes:

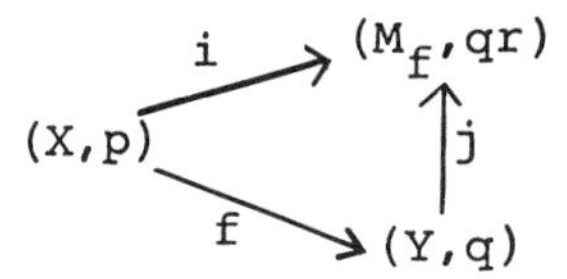

Since i and j are morphisms in $(\underline{\underline{CW}}_f^c)_{inc}$ and $j \in \Sigma$ by 3.5, 1.3 follows.

To prove 1.4, for any $(X,p) \in |(\underline{\underline{CW}}_f^c)_{inc}|$, we let $(X \times I, p\pi)$ and the inclusions $j_i(x) = (x,i)$ $(i = 0,1)$ play the roles of Z_X, r_X, and s_X, respectively, where $\pi : X \times I \to X$ is projection on the first factor. It follows from 3.5 that $j_i \in \Sigma (i = 0,1)$. Now, if $f_0, f_1 : (X,p) \to (Y,q)$ are morphisms in $(\underline{\underline{CW}}_f^c)_{inc}$ which are homotopic, let $F : (X \times I, p\pi) \to (Y \times I, q\pi)$ be a level preserving cellular homotopy and let $(\overline{M}_F, q\pi r)$ be the mapping cylinder of F reduced modulo $X \times \{0,1\}$. Then clearly the diagram

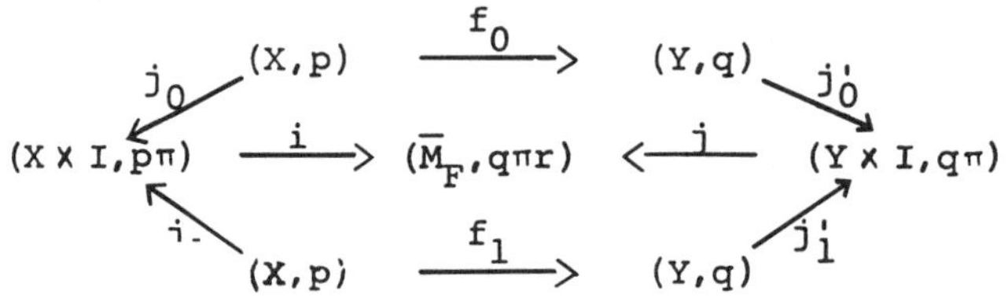

commutes where i and j are the inclusions. Since $j \in \Sigma$ by 3.5, 1.4 follows.

The following proposition asserts that 1.2 holds in the setting of section 2. Proving it will complete the verification of 1.1 to 1.4 and thus the proofs of 2.2 and 2.3.

PROPOSITION 4.1. Let $f_0, f_1 : (X,p) \to (Y,q)$ be morphisms in $(\underline{\underline{CW}}_f^c)_{inc}$ which are homotopic. Then there exist morphisms $s_0, s_1 : (Y,q) \to (W,t)$ in Σ such that $s_0 f_0 = s_1 f_1$.

The proof of this proposition was found after studying an unpublished argument given by Marshall Cohen. The authors would like to thank him for sharing that argument with them.

Proof: Let $F : (X \times I, p\pi) \to (Y \times I, q\pi)$ be a level preserving homotopy from f_0 to f_1 and form the mapping cylinder $(M_F, q\pi r)$. Notice that the usual embedding $(X \times I, p\pi) \to (M_F, q\pi r)$ extends to an embedding

$(X\times[-1,2],p\pi)\to(M_F,q\pi r)$ that maps $X\times[-1,0]$ into M_{f_0} and $X\times[1,2]$ into M_{f_1}. (See Figure 1, below).) Let $\hat{M}_F$ be obtained from M_F by collapsing each seqment $x\times[-1,2]$ to a point $(x\in X)$ and let $W = M_\rho$ be the mapping cylinder of the collapsing map $\rho: M_F\to\hat{M}_F$.

The mapping cylinder M_F. The rays of the mapping cylinder are drawn in. The embedded copy of $X\times[-1,2]$ is the flat strip in the front.

Let s_i be the composite $Y\xrightarrow{\times\{i\}} Y\times I \xrightarrow{j} M_F \xrightarrow{\rho} \hat{M}_F \longrightarrow M_\rho$ $(i = 0,1)$ and notice that s_i is an embedding. Furthermore, by construction, $s_0f_0 = s_1f_1$. In order to complete the proof of 4.1, it is necessary to construct a map $\hat{q}:\hat{M}_F\to Z$ and to show $s_0,s_1 : (Y,q)\to(M_\rho,\hat{q}\hat{r})$ are morphisms in Σ where $\hat{r}: M_\rho\to\hat{M}_F$ is the usual mapping cylinder retraction. The construction of $\hat{q}$ is based on the following lemma:

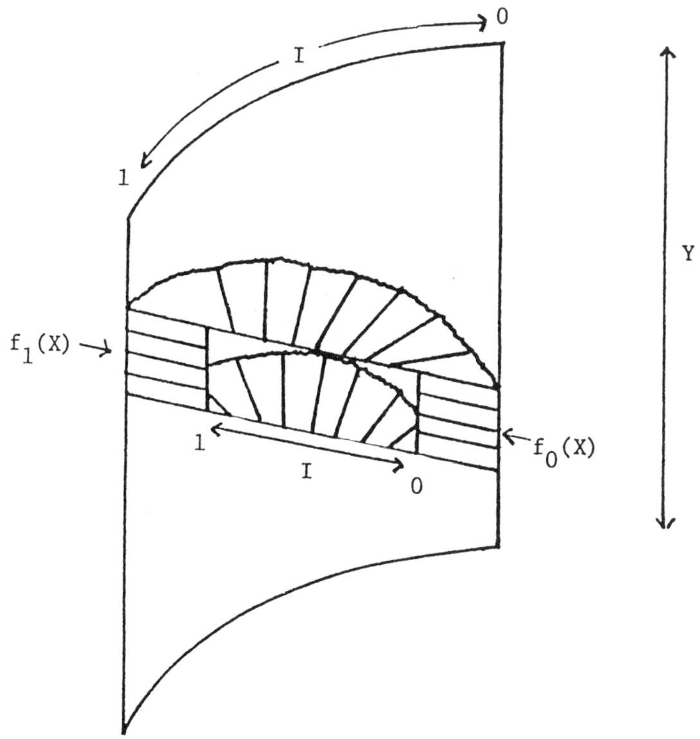

FIGURE 1

LEMMA 4.2. There exists a map $q_1 : M_F \to Z$ such that

1. If $A = M_F$, $Y \times I$, or $Y \times \{i\}$ $(i = 0,1)$, $1_A : (A, q\pi r|A) \to (A, q_1|A)$ is a bc homeomorphism;

2. There exists a map $\hat{q}_1 : X \to Z$ such that $q_1|X \times [-1,2] = \hat{q}_1\pi$ where $\pi : X \times [-1,2] \to X$ is projection on the first factor.

COROLLARY 4.3. The inclusion $(Y \times \{i\}, q_1|Y \times \{i\}) \to (M_F, q_1)$ $(i = 0,1)$ is in Σ.

Proof of the Corollary: By 4.2 the indicated inclusion is homeomorphically equivalent to the composite $(Y \times \{i\}, q\pi r|Y \times \{i\}) \to (Y \times I, q\pi r|Y \times I) \to (M_F, q\pi r)$. Since each of these maps is in Σ, 4.3 follows.

The proof of 4.2 is based on the following lemma.

LEMMA 4.4. Let $(V,g_0) \in |\underline{\underline{CW}}{}^c_f/Z|$ and let $G : (V \times I, g_0\pi) \to (Z, 1_Z)$ be a bc homotopy from $g_0 = G|V \times 0$ to $g_1 = G|V \times 1$. Then (V,g_1) is a finite, bc CW complex and $1_V : (V,g_0) \to (V,g_1)$ is a bc homeomorphism.

Proof: Let n_1 be such that for any cell $e \in V$, there exists a minimal element $K_e \in P$ such that $g_0(e) \subset C^{n_1}K_e$. Let n_2 be such for all $K \in P$, $G(g_0^{-1}(K) \times I) = G((g_0\pi)^{-1}(K)) \subset C^{n_2}K$. It is easy to verify that for any $e \in V$, $g_1(e) \subset C^{n_1+n_2}(K_e)$ and for any $K \in P$, $g_0^{-1}(K) \subset g_1^{-1}(C^{n_2}K)$. It follows from the first of these assertions that (V,g_1) is a finite, bc CW complex and from the second that $1_V : (V,g_0) \to (V,g_1)$ is bc. The last part of the lemma now follows from 3.1.

Proof of 4.2: By the bc homotopy extension theorem, 3.7, there is a bc homotopy $Q : (M_F \times I, q\pi r) \to (Z, 1_Z)$ such that $Q|M_F \times 0 = q\pi r$ and such that for all $x \in X$, $Q|\{x\} \times [-1,2] \times 1$ is constant. Let $q_1 = Q|M_F \times 1$. Then 1) follows immediately from 4.4; while 2) follows from the fact that $q_1|\{x\} \times [-1,2]$ $(x \in X)$ is constant.

It follows immediately from 4.2 that $q_1 : M_F \to Z$ induces a map $\hat{q} : \hat{M}_F \to Z$ such that

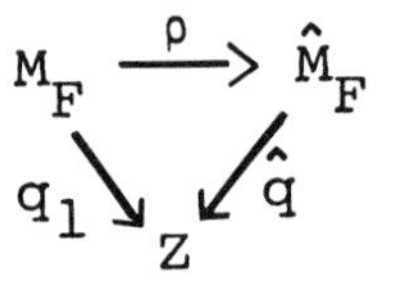

commutes and that $\hat{q}\hat{r}k = q_1$ where $k : M_F \to M_\rho$. Notice also that s_i is the composite

$$(Y,q) \xrightarrow{\times\{i\}} (Y \times I, q\pi) \longrightarrow (M_F, q\pi r) \longrightarrow (M_F, q_1)$$

$$\xrightarrow{\rho} (\hat{M}_F, \hat{q}) \xrightarrow{k} (M_\rho, \hat{q}\hat{r})$$

of bc maps. Since $s_i : (Y,q) \to (s_i(Y), \hat{q}\hat{r}|s_i(Y))$ is a homeomorphism by 3.1 and hence is in Σ, to complete the proof of 4.1, it suffices to prove the following lemma:

LEMMA 4.5. The inclusion $(s_i(Y), \hat{q}\hat{r}|s_i(Y)) \to (M\ ,qr)$ is in Σ for $i = 0,1$.

Proof: To simplify notation, we write $\hat{q}\hat{r}|$ instead of $\hat{q}\hat{r}|A$ when $A \subset M_\rho$. We also let $\rho_1 = \rho| : Y \times 0 \to s_0(Y)$ and $\rho_2 = \rho| : Y \times 0 \cup X \times [-1,2] \to s_0(Y)$. Now consider Figure 2 which depicts the subspace $M_F \cup M_{\rho_2}$ of M_ρ. Notice that $(s_0(Y), \hat{q}\hat{r}|) \to (M_{\rho_1}, \hat{q}\hat{r}|)$ is in Σ by 3.5. In addition, since $\hat{q}\hat{r}|M_F = q_1$,

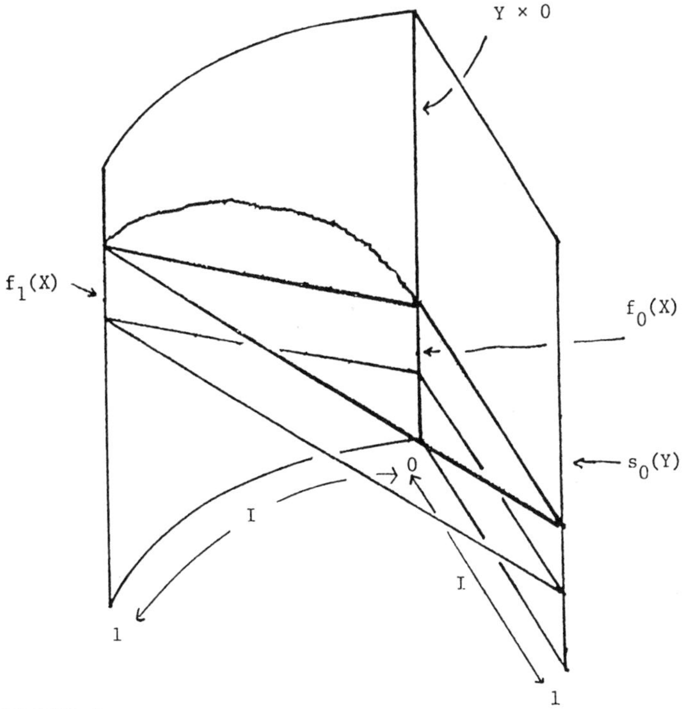

FIGURE 2

$(Y \times 0, \hat{q}\hat{r}|) \to (M_F, \hat{q}\hat{r}|)$ is in Σ by 4.3. A pushout argument using 1.1 now shows that $(M_{\rho_1}, \hat{q}\hat{r}|) \to (M_F \cup M_{\rho_1}, \hat{q}\hat{r}|)$ is in Σ. On the other hand, since $(Y \times 0, \hat{q}\hat{r}|) \to (Y \times 0 \cup X \times [-1,2], \hat{q}\hat{r}|)$ is in Σ, so is $(M_{\rho_1} \cup X \times [-1,2], \hat{q}\hat{r}|) \to (M_{\rho_2}, \hat{q}\hat{r}|)$ by 3.8. A pushout argument using this fact, now shows that $(M_F \cup M_{\rho_1}, \hat{q}\hat{r}|) \to (M_F \cup M_{\rho_2}, \hat{q}\hat{r}|)$ is also in Σ.

The arguments above now show that each map of the composite $(s_0(Y), \hat{q}\hat{r}|) \to (M_{\rho_1}, \hat{q}\hat{r}|) \to (M_F \cup M_{\rho_1}, \hat{q}\hat{r}|) \to (M_F \cup M_{\rho_2}, \hat{q}\hat{r}|)$ is in Σ. Since Σ is closed under finite compositions, $(s_0(Y), \hat{q}\hat{r}|) \to (M_F \cup M_{\rho_2}, \hat{q}\hat{r}|)$ is also in Σ.

We now claim $(M_F \cup M_{\rho_2}, \hat{q}\hat{r}|) \to (M_\rho, \hat{q}\hat{r})$ is in Σ. To see this, note that $\rho : M_F \to \hat{M}_F$ is a cellular isomorphism away from $X \times [-1,2]$. Thus, the part of M outside $M_F \cup M_{\rho_2}$ is isomorphic to $[M_F - (X \times [-1,2] \cup Y \times 0)] \times I$. An argument similar to the proof of 3.5 now establishes the claim. It then follows that $(s_0(Y), \hat{q}\hat{r}|) \to (M_\rho, \hat{q}\hat{r})$ is in Σ.

The proof that $(s_1(Y), \hat{q}\hat{r}|) \to (M_\rho, \hat{q}\hat{r})$ is in Σ is entirely similar and is omitted. This completes the proofs of 4.1, 2.1, and 2.2.

REFERENCES

1. Anderson, D. R., and Munkholm, H. J., The Algebraic Topology of Controlled Spaces, preprint, Odense Universitet, 1984.
2. Anderson, D. R., and Munkholm, H. J., An Introduction to Boundedly Controlled Simple Homotopy Theory, these Proceedings.
3. Anderson, D. R., and Munkholm, H. J., The Simple Homotopy Theory of Boundedly Controlled CW Complexes, preprint (to appear), Odense Universitet.
4. Cohen, M. M., A Course in Simple Homotopy Theory, Springer Verlag, New York, 1973.
5. Eckmann, B., Simple Homotopy Type and Categories of Fractions, Symposia Mathematica (Instituto Nazionale di Alta Matematica), Vol. 5, 1971, 285-299.
6. Siebenmann, L. C., Infinite Simple Homotopy Types, Indag Math., 32 (1970), 479-495.

An Introduction to Boundedly Controlled Simple Homotopy Theory

DOUGLAS R. ANDERSON / Department of Mathematics, Syracuse University, Syracuse, New York

HANS JØRGEN MUNKHOLM / Matematisk Institut, Odense Universiteit, Odense, Denmark

In the last few years, the general area of topology "parameterized over the space Z" or of topology "with control" has undergone an extensive development, particularly, in papers by Chapman [8], Quinn [12,13], and Pedersen [9,10], among others. On examining these papers, one discovers that there are actually two different and distinct approaches to the general area--the work of Chapman and Quinn focuses on topology "with ε-control;" while that of Pedersen is concerned with "bounded topology" (usually over R^k).

In the Chapman-Quinn approach, one starts with a map $p : M \to Z$ from a manifold, say, to a metric space Z, fixes a number $\varepsilon > 0$, and tries to answer some question about M to within the given ε. For example, if $\partial M = \partial_0 M \amalg \partial_1 M$ and $\delta > 0$, then Chapman [8] and Quinn [12] define what it means to say that $(M, \partial_0 M)$ is a (δ,h)-cobordism and investigate the following problem:

PROBLEM. Let $\varepsilon > 0$, under what circumstances does a (δ,h)-cobordism have an ε-product structure?

In Pedersen's work, one is really working in a category in which an object is a pair (M,p) where M is a manifold and $p: M \to R^k$. A morphism $f : (M,p) \to (N,q)$ is a continuous map $f : M \to N$ for which $p - qf : M \to R^k$ is bounded. In this setting, one defines what it means to say that $(M, \partial_0 M, p)$ is a bounded h-cobordism and investigates the following problem:

PROBLEM. Under what circumstances does a bounded h-cobordism have a bounded structure?

In the Pedersen approach, only the existence of some bound is at issue; while in the Chapman-Quinn approach, a specific bound ($\varepsilon > 0$) is set and the problem must be resolved to within the given bound.

We present here a third approach to controlled simple homotopy theory which we call "boundedly controlled." We have chosen this term for two reasons. First, the work presented here is clearly a considerable generalization of Pedersen's work and, thus, has a "bounded" side; and second, the authors conjecture that this work is closely related to that of Chapman and Quinn and, thus, has a "controlled" side. The authors plan to investigate the relationship between this work and that of Chapman-Quinn in the near future.

Some of the foundational material underlying the results presented here has already appeared in preprint form, [1]. We expect that two more preprints, [2,3], will appear shortly giving full details of the simple homotopy theory of boundedly controlled CW complexes and the s-cobordism theorem for boundedly controlled manifolds, respectively. The present note should also be viewed as an introduction to the research monograph, [4], into which we hope that the above mentioned preprints will grow.

Complete details of the construction of the boundedly controlled, geometric Whitehead group of a finite, boundedly controlled CW complex can be found in a companion paper to this one, [5], which also appears in these Proceedings.

Details of the computational results mentioned in section 7 of this note will appear separately, [6].

In an earlier preprint, [7], we outlined one approach to simple homotopy theory for boundedly controlled CW complexes. The definition of $Wh^c(X,p)$ given in section 2 of this note (and in [2], or [5]) differs from the one given there. The reason is that we cannot prove the theory of [7] combinatorially invariant. Hence, even though [7] is logically correct, it does not provide a setting for an s-cobordism theorem.

The research reported on here goes back almost two years. The first named author wants to thank Odense Universitets Matematiske Institut for its hospitality during several visits. The second author extends similar thanks to the Mathematics Department of Syracuse University as well as to that of the University of Maryland, College Park, where he spent the year 1983-84 as a visiting professor.

Financial support has been received from the departments mentioned above as well as the NSF (grant number MCS-8201776), the Danish Natural

Science Research Council, and the Scientific Affairs Division of NATO (grant number 670/84). We thank all these sources.

The comparison of our $Wh^c(A \times \mathbb{R}^2, pr)$ and Siebenmann's $S(A \times \mathbb{R}^2)$ given in Theorem 7.4 arose as an answer to a question raised by Ib Madsen. On our way to the solution we had fruitful discussions with Ib Madsen, Erik K. Pedersen, Andrew Ranicki, and Mark Steinberger.

1. BOUNDEDLY CONTROLLED, FINITE CW COMPLEXES

In order to motivate the abstract definition of a boundedness control structure given later in this section, let Z be a metric space whose metric $\rho : Z \times Z \to R_+$ is proper in the sense that for every $z \in Z$, $\rho(z,-) : Z \to R_+$ is a proper map. Notice that this implies that for every $z \in Z$, $r \in R_+$, $B(z,r) = \{w \in Z | \rho(w,z) \leq r\}$ is compact. Hence, (Z,ρ) is locally compact and complete.

A boundedly controlled space (or, simply, a bc space) over (Z,ρ) is a pair (X,p) where X is a topological space and $p : X \to Z$ is a map. A bounded map $f : (X,p) \to (Y,q)$ is a map $f: X \to Y$ for which there exists an integer $d \geq 0$ such that for all $x \in X$, $\rho(qf(x),p(x)) \leq d$. The collection of bc spaces over Z and bounded maps forms a category called the category of bc spaces over Z and denoted $\underline{TOP}^c/Z$.

In terms of the family $P = \{B(z,r) | z \in Z, r \in R_+\}$ of closed balls in Z, $f : (X,p) \to (Y,q)$ is bounded if and only if there exists an integer $d' \geq 0$ such that for every $z \in Z$, $r \in R_+$, $fp^{-1}(B(z,r)) \subset q^{-1}(B(z,r+d'))$. By regarding the correspondence that sends $B(z,r)$ to $B(z,r+1)$ as an "enlargement operator" C on P, we are lead to the following definition:

A boundedness control structure on the space Z is a pair (P,C) where $P \subseteq 2^Z$ is a set of nonempty subsets of Z and $C : P \to P$ is a function such that the following axioms hold: for every K, L in P

1. $K \subseteq C(K)$, and if $K \subseteq L$, then $C(K) \subseteq C(L)$
2. $\cup_n C^n(K) = Z$
3. there is some minimal element of P, say K_0, with $K_0 \subseteq K$
4. there is some function $\Theta : \mathbb{Z}_+ \to \mathbb{Z}_+$ such that whenever $K_0 \in P$ is minimal and $K \cap C^d K_0 \neq \emptyset$, then $K_0 \subseteq C^{\Theta(d)}(K)$.

The triple (Z,P,C)--and by abuse of language the space Z--will be called a boundedness control space.

EXAMPLE 1.1. Let Z be a metric space with proper metric ρ and suppose

(A) If $B(z,r) \subseteq B(u,s)$, then $B(z,r+1) \subseteq B(u,s+1)$.

The metric boundedness control structure (P,C) on Z is defined by setting

$$P = \{B(z,r) \mid z \in Z,\ r \in R_+\}; \qquad C(B(z,r)) = B(z,r+1) .$$

We note that the technical condition (A) is imposed to insure that $C : P \to P$ is a well defined function satisfying condition 1 above.

Note that the box metric on $\mathbb{R}^k$, given by $\rho(\underline{x},\underline{y}) = \max_i |x_i - y_i|$, $(\underline{x},\underline{y} \in \mathbb{R}^k)$ certainly satisfies the above condition (A). In the sequel, when we talk about $\mathbb{R}^k$ as a boundedness control space, we assume it is endowed with the metric boundedness control structure arising from the above metric.

EXAMPLE 1.2. Let Z be any space. The indiscrete boundedness control structure on Z, (P_0,C_0), is given by

$$P_0 = \{Z\}, \qquad C_0(Z) = Z .$$

Other examples of boundedness control structures are given in [1, section 6].

By definition a boundedly controlled space (or, simply, a bc space) over Z is a pair (X,p) where X is a topological space and $p : X \to Z$ is a map. A boundedly controlled map (or, simply, a bc map) $f : (X,p) \to (Y,q)$ is a map $f : X \to Y$ for which there exists an integer $d \geq 0$ such that $f(p^{-1}K) \subseteq q^{-1}(C^dK)$ for any $K \in P$.

In a boundedness control space Z the radius of a subset A is defined to be the smallest $n \in \mathbb{Z}_+$ such that $A \subseteq C^nK_0$ for some minimal K_0 in P, (or ∞ if no such n exists). A collection $\{A_\lambda \mid \lambda \in \Lambda\}$ of subsets of Z is bounded if $\{\text{rad } A_\lambda \mid \lambda \in \Lambda\}$ is bounded. When (X,p) is a bc space over Z, a collection $\{X_\lambda \mid \lambda \in \Lambda\}$ of subsets of X is called bounded if $\{p(X_\lambda) \mid \lambda \in \Lambda\}$ is a bounded collection of subsets of Z.

If X is a CW complex and $\{\text{closure } (e) \mid e \text{ any cell of } X\}$ is bounded, then (X,p) is called a boundedly controlled CW complex (or, simply, a bc CW complex) over Z. If, in addition, X is finite dimensional and for each $K \in P$, $p^{-1}(K)$ is contained in a finite subcomplex of X, then (X,p) is a finite, boundedly controlled CW complex over Z. Finite, boundedly controlled CW complexes, together with bc maps, form the category $\underline{\underline{CW}}_f^c/Z$. Let I be the unit interval and put $(X,p) \times I = (X \times I, p\pi)$ where $\pi : X \times I \to X$ is the projection and $X \times I$ has the product cell structure. Using this cylinder construction one gets an obvious notion of homotopy in $\underline{\underline{CW}}_f^c/Z$.

2. THE BOUNDEDLY CONTROLLED WHITEHEAD GROUP

In this section we describe the construction of the boundedly controlled Whitehead group $Wh^c(X,p)$ of a finite, bc CW complex (X,p). A full accound of this construction is given in our companion paper [5] that appears elsewhere in these Proceedings.

Let $s : (X,p) \to (Y,q)$ be an inclusion in the category CW_f^c/Z; that is, s is a bc map which is a cellular isomorphism onto its image. We say that s is an *elementary expansion* if there exists a bounded collection of finite subcomplexes of Y, $\{Y_\lambda | \lambda \in \Lambda\}$, such that

1. $Y = X \cup \bigcup Y_\lambda$ and $Y_\lambda \cap Y_\mu \subseteq X$ if $\lambda \neq \mu$; and
2. For any λ, the inclusion $X_\lambda = X \cap Y_\lambda \to Y_\lambda$ is an (usual) expansion of finite CW complexes.

A finite composite of elementary expansion is called an *expansion*. An expansion is easily seen to be a bc homotopy equivalence. A homotopy inverse of an expansion is called a *contraction*.

Let (X,p) be a finite, bc CW comlex over Z and $f_i : (X,p) \to (Y_i,q_i)$, $(i = 1,2)$ be bc homotopy equivalences. We say that f_1 is *simply equivalent* to f_2 if there exists a sequence $s_k : (W_k,r_k) \to (W_{k+1},r_{k+1})$ $(1 \leq k \leq t-1)$ of expansions and/or contractions with $(W_1,r_1) = (Y_1,q_1)$ and $(W_t,r_t) = (Y_2,q_2)$ such that f_2 is bc homotopic to $s_{t-1} \cdots s_1 f_1$. Simple equivalence is clearly an equivalence relation.

DEFINITION 2.1. For any (X,p) in $\underline{\underline{CW}}_f^c/Z$, $Wh^c(X,p)$ is the set of simple equivalence classes of bc homotopy equivalences with domain (X,p).

In fact, the set $Wh^c(X,p)$ may be endowed with the structure of an abelian group. This is done by showing first, using mapping cylinders, that any element $x \in Wh^c(X,p)$ can be represented by an inclusion $j : (X,p) \to (Y,q)$ and then showing that if $j_i : (X,p) \to (Y_i,q_i)$ is an inclusion representing x_i $(i = 1,2)$, then there is a pushout diagram in $\underline{\underline{CW}}_f^c/Z$

$$\begin{array}{ccc} (X,p) & \xrightarrow{j_1} & (Y_1,q_1) \\ {\scriptstyle j_2}\downarrow & & \downarrow{\scriptstyle k_1} \\ (Y_2,q_2) & \xrightarrow[k_2]{} & (Y_1 \cup_X Y_2,q) \end{array}$$

The sum $x_1 + x_2$ is then represented by $k_1 j_1 = k_2 j_2$.

In a similar vein, a pushout argument is used to show that $f : (X,p) \to (Y,q)$ induces a homomorphism

$$f_* : Wh^c(X,p) \to Wh^c(Y,q) .$$

The main result in [5] is the following theorem:

THEOREM 2.2. $Wh^c : \underline{\underline{CW}}^c_f/Z \to \underline{\underline{Ab}}$ is a homotopy functor where $\underline{\underline{Ab}}$ is the category of abelian groups.

The following result is crucial in any application of simple homotopy theory:

THEOREM 2.3. Boundedly controlled simple homotopy type is a combinatorial invariant. More precisely, if X' is a subdivision of the bc CW complex (X,p), then the identity map $1 : (X,p) \to (X',p)$ represents zero in $Wh^c(X,p)$.

3. $R\underline{\underline{P}}(G)$-MODULES

Let R be a commutative ring with unit and let π be a group (usually the fundamental group of some space). The category of modules over the group ring $R[\pi]$ and various of its subcategories (e.g., finitely generated and/or projective or free or based modules) play important roles in classical (i.e., unbounded) algebraic and geometric topology. This is particularly so in the study of Whitehead torsion.

In this section we shall outline the definitions of the analogous categories which are used in doing algebraic topology for bc spaces (X,p), and in studying $Wh^c(X,p)$.

The boundedness control structure (P,C) on the space Z enters via the category $\underline{\underline{P}}$ whose objects are the sets in P and whose morphisms are the inclusions between such sets. The function $C : \underline{\underline{P}} \to \underline{\underline{P}}$ is a functor $C : P \to P$ and there is a natural transformation $\tau : Id_{\underline{\underline{P}}} \to C$ (because $K \subseteq CK$ for all $K \in P$). The triple $(\underline{\underline{C}}, P, \tau)$ is a category with endormphism in the sense of [1].

The "fundamental group data" of (X,p) enter into our algebra through the functor

$$G = G_1(X,p) : \underline{\underline{P}} \to \underline{\underline{Gpoid}}$$

into the category of small groupoids. It has G(K) = the fundamental groupoid of $p^{-1}(K)$ and is a functor in the obvious way.

From the category $\underline{\underline{P}}$ and any functor $G : \underline{\underline{P}} \to \underline{\underline{Gpoid}}$ we form a new category $\underline{\underline{P}}(G)$. An object is a pair (x,K) with $K \in P$ and $x \in |G(K)|$, the object set of G(K). A morphism $(\omega,i) : (x,K) \to (y,L)$ consists of an inclusion $i : K \subseteq L$ and a morphism $\omega \in G(L)(\bar{x},y)$ where $\bar{x}$ is the image of x under $G(i) : G(K) \to G(L)$. Composition is defined in the obvious way.

Note that when $G = G_1(X,p)$ one has $\bar{x} = x$ and ω is simply a path class (rel. endpoints) in $p^{-1}L$ from y to x.

A functor $C : \underline{\underline{P}}(G) \to \underline{\underline{P}}(G)$ and a natural transformation $\tau : Id_{\underline{\underline{P}}(G)} \to C$ are defined by

$$C(x,K) = (G(\tau_K)(x), CK) ,$$

$$\tau_{(x,K)} = (1_{G(\tau_K)(x)}, \tau_K) , \qquad (x,K) \in |\underline{\underline{P}}(G)| .$$

The triple $(\underline{\underline{P}}(G), C, \tau)$ is another category with endomorphism in the sense of [1].

We define an <u>$R\underline{\underline{P}}(G)$-module</u> to be a functor $M : \underline{\underline{P}}(G) \to R\text{-}\underline{\underline{mod}}$ into the category of R-modulus. A <u>morphism</u> $\varphi : M \to N$ of $R\underline{\underline{P}}(G)$-modules is defined to be an equivalence class of natural transformations $\varphi_d : M \to NC^d$ where the equivalence relation identifies φ_d with $\psi_e : M \to NC^e$ provided there is some $f \geq e,d$ such that

$$\begin{array}{ccc} M & \xrightarrow{\varphi_d} & NC^d \\ {\scriptstyle \psi_e}\downarrow & & \downarrow{\scriptstyle NC^d\tau^{(f-d)}} \\ NC^e & \xrightarrow[NC^e\tau^{(f-e)}]{} & NC^f \end{array}$$

commutes. With an obvious composition we have the category $R\underline{\underline{P}}(G)\text{-}\underline{\underline{mod}}$.

One may also define $R\underline{\underline{P}}(G)$-mod more formally as a category of fractions $R\text{-}\underline{\underline{mod}}^{\underline{\underline{P}}(G)}(\Sigma^{-1})$ where Σ is a set of morphisms in the functor category $R\text{-}\underline{\underline{mod}}^{\underline{\underline{P}}(G)}$, see [1]. This viewpoint is useful in proving

THEOREM 3.1. The category $R\underline{\underline{P}}(G)\text{-}\underline{\underline{mod}}$ is abelian.

EXAMPLE 3.2. Let Z be given the indiscrete boundedness control structure of Example 1.2. Since the corresponding category $\underline{\underline{P}}_0$ has only one morphism, a functor $G : \underline{\underline{P}}_0 \to$ Gpoid is simply a small groupoid. Thus as a special case of the above, we have defined the category of $R\underline{\underline{P}}_0(G)$-modules when G is a groupoid. If this groupoid G happens to be a group (i.e. if G has only one object), then it is easily seen that we recover the category of R[G]-modules.

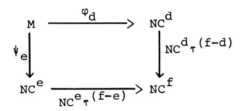

For a more general Z and any $G : \underline{\underline{P}} \to \underline{\text{Gpoid}}$, an $R\underline{\underline{P}}(G)$-module associates to each $(x,K) \in |\underline{\underline{P}}(G)|$ an R-module $M(x,K)$. The endomorphisms of (x,K) in $\underline{\underline{P}}(G)$ form a group (which we denote $\pi_1(G(K),x)$), and the functorial properties of M make M(x,K) into a module over the group ring $R[\pi_1(G(K),x)]$. Moreover, as (x,K) varies, these modules are related by maps $M(x,K) \to M(y,L)$ induced by inclusions $i : K \subseteq L$ and morphisms $\omega \in G(L)(\bar{x},y)$, $\bar{x}$ being $G(i)(x)$.

Classically, a basis for a free $R[\pi]$-module is simply a set. In our theory, a basis for a free $R\underline{\underline{P}}(G)$-module is a pair (S,σ) where S is a set and $\sigma : S \to |\underline{\underline{P}}(G)|$ is a function. We write $\sigma(s) = (x_s,K_s)$, $x_s \in |G(K_s)|$, $K_s \in P$. The idea is that "the generator corresponding to s sits in $p^{-1}K$ at the base point x_s" (cf. Theorem 3.9). Technically, one defines the free $R\underline{\underline{P}}(G)$-module $F(\sigma)$ with basis (S,σ) as follows: For any $b = (x,K) \in |\underline{\underline{P}}(G)|$, we let $F(\sigma)(b)$ be the free R-module generated by all (β,s) with $s \in S$ and $\beta \in \underline{\underline{P}}(G)(\sigma(s),b)$. If $c = (y,L)$ and $\gamma \in \underline{\underline{P}}(G)(b,c)$ then the induced R-homomorphism $\gamma_* : F(\sigma)(b) \to F(\sigma)(c)$ has $\gamma_*(\beta,s) = (\gamma\beta,s)$.

We call $F(\sigma)$ boundedly, finitely generated (bfg, for short) if

For each $K \in P, \{s \in S | K_s \subseteq K\}$ is finite, and (3.3)

$\{K_s | s \in S\}$ is bounded. (3.4)

The bfg free RP(G)-modules play the role classically awarded to the f.g. free $R[\pi]$-modules. Thus an arbitrary $R\underline{\underline{P}}(G)$-module M is called bfg if there is an epimorphism $\varphi : F(\sigma) \to M$. If M itself is free, the two definitions agree.

EXAMPLE 3.5. Let G be a small groupoid and consider it a functor $G : \underline{\underline{P}}_0 \to \underline{\text{Gpoid}}$ as in Example 3.2. Then a basis (S,σ) is simply a map $\sigma : S \to |G|$. It is bfg if and only if S is a finite set. If $x \in |G|$, then $F(\sigma)(x)$ is the free R-module on all $\beta_*(1,s)$ with $\sigma(s)$ in the same component of G as x and with β ranging over $G(\sigma(s),x)$. Thus, as an $R[G(x,x)]$-module, $F(\sigma)(x)$ is free with a basis consisting of all $\beta_{s*}(1,s)$ with s as above and β_s an arbitrarily chosen element of $G(\sigma(s),x)$.

We denote by bfg $R\underline{\underline{P}}(G)$-$\underline{\underline{\text{mod}}}$ the full subcategory of $R\underline{\underline{P}}(G)$-$\underline{\underline{\text{mod}}}$ consisting of all bfg modules. In [2] we prove

THEOREM 3.6. In bfg $R\underline{\underline{P}}(G)$-$\underline{\underline{\text{mod}}}$ an object M is projective if and only if it is a summand in some bfg $F(\sigma)$. There are enough projectives in the category, i.e., any bfg module is the surjective image of a bfg free module.

Two free $R[\pi]$-modules are, of course, isomorphic if and only if their bases are bijective correspondence. In our setting, an isomorphism of bases $(\alpha,\nu) : (S_1,\sigma_1) \to (S_2,\sigma_2)$ is defined to be a bijection $\alpha : S_1 \to S_2$ and, for some fixed $d \geq 0$ and for any $s \in S_1$, a morphism $\nu(s) \in \underline{\underline{P}}(G)(\sigma_1(s),C^d\sigma_2(\alpha(s)))$. If one restricts attention to bases satisfying (3.4), then such pairs are, indeed, the isomorphisms in a suitably defined category of bases, see [2]. In [2] we prove that isomorphism classes of bfg free $R\underline{\underline{P}}(G)$-modules correspond bijectively to isomorphism classes of bfg bases, i.e.,

THEOREM 3.7. If two based, bfg free $RP(G)$-modules are isomorphic, then so are their bases. Also, any isomorphism of bfg bases $(\alpha,\nu) : (S_1,\sigma_1) \to (S_2,\sigma_2)$ induces an isomorphism $F(\alpha,\nu) : F(\sigma_1) \to F(\sigma_2)$.

EXAMPLE 3.8. If $\underline{\underline{P}} = \underline{\underline{P}}_0$ consists of just one morphism and $G : \underline{\underline{P}}_0 \to \underline{\underline{\text{Gpoid}}}$ happens to be a group, then a basis (S,σ) is of course determined by the set S. However, a basis isomorphism $(\alpha,\nu) : (S_1,\sigma_1) \to (S_2,\sigma_2)$ is not just a bijection $\alpha : S_1 \to S_2$. One also has for each $s \in S_1$, the group element $\nu(s)$. The induced isomorphism $F(\sigma_1) \to F(\sigma_2)$ is the one that sends s to $\nu(s)\alpha(s)$, $s \in S_1$.

In [1] we defined, for each pair $((Y,q), (X,p))$ in $\underline{\underline{CW}}_f^c/Z$ the boundedly controlled relative homotopy module

$$\pi_n^c((Y,q),(X,p)) \in |\mathbb{Z}\underline{\underline{P}}(G_1(X,p))\text{-}\underline{\underline{\text{mod}}}| \ .$$

It is analogous to the classical relative homotopy group $\pi_n(Y,X)$ considered as a module over $\pi_1(X)$ (for $n \geq 3$). For example, one has

THEOREM 3.9. Suppose that the above pair has $Y = X \cup \bigcup_{s \in S} e_s^n$ with $n \geq 3$. Then

$$\pi_n^c((Y,q),(X,p)) \cong F(\sigma)$$

where $\sigma(s) = (x_s,K_s)$ with x_s any point on the boundary of e_s^n and $\{K_s | s \in S\}$ any bounded family in P with $q(\text{closure } (e_s^n)) \subseteq K_s$ for all s.

Furthermore, if $(X,p) \in |\underline{\underline{CW}}_f^c/Z|$ and $n \geq 3$ are given, then any bfg free $\mathbb{Z}\underline{\underline{P}}(G_1(X,p))$ module arises this way.

4. THE ALGEBRAIC WHITEHEAD GROUP

Let R, $\underline{\underline{P}}$, G be as in section 3. We define the abelian group $K_1(R\underline{\underline{P}}(G))$ to have one generator $[F,\alpha]$ for each automorphism of any bfg free $R\underline{\underline{P}}(G)$-module F and the usual relations; namely

For any isomorphism $\emptyset : F \to F'$, one has $[F,\alpha] = [F',\emptyset\alpha\emptyset^{-1}]$. (4.1)

For any pair of automorphisms $F \xrightarrow{\alpha} F \xrightarrow{\beta} F$, one has
$[F,\alpha] + [F,\beta] = [F,\beta\alpha]$. (4.2)

If (F,α) fits into a commutative diagram (4.3)

$$\begin{array}{ccccccccc} 0 & \longrightarrow & F' & \xrightarrow{i} & F & \xrightarrow{p} & F'' & \longrightarrow & 0 \\ & & \downarrow{\scriptstyle \alpha'} & & \downarrow{\scriptstyle \alpha} & & \downarrow{\scriptstyle \alpha''} & & \\ 0 & \longrightarrow & F' & \xrightarrow[i]{} & F & \xrightarrow[p]{} & F'' & \longrightarrow & 0 \end{array}$$

with exact rows, then $[F,\alpha] = [F',\alpha'] + [F'',\alpha'']$.

EXAMPLE 4.4. If $\underline{\underline{P}} = \underline{\underline{P}}_0$ and $G : \underline{\underline{P}}_0 \to \underline{\underline{\text{Gpoid}}}$ is a group, then $K_1(R\underline{\underline{P}}_0(G))$ coincides with the usual $K_1(R[G])$.

If $\underline{\underline{P}} = \underline{\underline{P}}_0$ and $G : \underline{\underline{P}}_0 \to \underline{\underline{\text{Gpoid}}}$ is any groupoid, then it is not difficult to show that

$$K_1(R\underline{\underline{P}}_0(G)) \cong \bigoplus_{(x)} K_1(R[\pi_1(G,x)])$$

where $\pi_1(G,x)$ is the endomorphism group of the object x in the groupoid G and x ranges over a set of representatives of the components of G.

We define $Wh(\mathbb{Z}\underline{\underline{P}}(G))$ to be the quotient of $K_1(\mathbb{Z}\underline{\underline{P}}(G))$ by certain sets of relations, the first two of which are

For any bfg free $\mathbb{Z}\underline{\underline{P}}(G)$-module F, $[F,-1_F] = 0$ in $Wh(\mathbb{Z}\underline{\underline{P}}(G))$. (4.5)

For any automorphism of bases $(\alpha,\nu) : (S,\sigma) \to (S,\sigma)$,
$[F(\sigma),F(\alpha,\nu)] = 0$ in $Wh(\mathbb{Z}\underline{\underline{P}}(G))$. (4.6)

In view of example 3.8, if $\underline{\underline{P}} = \underline{\underline{P}}_0$ and $G : \underline{\underline{P}}_0 \to \underline{\underline{\text{Gpoid}}}$ is a group, then the above relations correspond precisely to those which define Wh(G) in terms of $K_1(\mathbb{Z}G)$. Thus the following definition is natural.

If $\underline{\underline{P}} = \underline{\underline{P}}_0$ and $G : \underline{\underline{P}}_0 \to \underline{\underline{\text{Gpoid}}}$ is a groupoid, then we define $Wh(R\underline{\underline{P}}_0(G))$ to be $K_1(R\underline{\underline{P}}_0(G))$ modulo the relations (4.5) and (4.6). In view of Theorem 3.7, it follows that any isomorphism $\varphi : F(\sigma) \to F(\rho)$ between bfg modules over the groupoid G has a well defined invariant

$\tau(\varphi) \in Wh(\mathbb{Z}\underline{\underline{P}}_0(G))$. In fact $\tau(\varphi) = [F(\sigma), F(\alpha,\nu)^{-1}\varphi]$ for any isomorphism of bases $(\alpha,\nu) : (S,\sigma) \to (T,\rho)$.

For more general $\underline{\underline{P}}$, a third type of relations must be introduced. To do this, we assume given the following data:

1. a bounded, locally finite family $\{L_\lambda | \lambda \in \Lambda\} \subseteq P$; here locally finite means that $\{\lambda | L_\lambda \subseteq K\}$ is finite for each $K \in P$,

2. for each λ a bfg free $\mathbb{Z}\underline{\underline{P}}_0(G(L_\lambda))$-module with basis $\sigma'_\lambda : S_\lambda \to G(L_\lambda)$ (note that S_λ is then finite),

3. for each λ an automorphism $\varphi'_\lambda : F(\sigma'_\lambda) \to F(\sigma'_\lambda)$ of $\mathbb{Z}\underline{\underline{P}}_0(G(L_\lambda))$-modules, representing zero in $Wh(\mathbb{Z}\underline{\underline{P}}_0(G(L_\lambda)))$.

Define $\sigma_\lambda : S_\lambda \to \underline{\underline{P}}(G)$ by $\sigma_\lambda(s) = (\sigma'_\lambda(s), L_\lambda)$. There is a unique extension of φ'_λ to a natural isomorphism $\varphi_\lambda : F(\sigma_\lambda) \to F(\sigma_\lambda)$. Also let (S,σ) be the disjoint union of all the $(S_\lambda, \sigma_\lambda)$. Then we have the natural isomorphism

$$\varphi_0 = \oplus_\lambda \varphi_\lambda : F(\sigma) = \oplus_\lambda F(\sigma_\lambda) \to \oplus_\lambda F(\sigma_\lambda) = F(\sigma)$$

of functors $\underline{\underline{P}}(G) \to \mathbb{Z}$-$\underline{\underline{mod}}$. Let $\varphi : F(\sigma) \to F(\sigma)$ be the automorphism of the $\mathbb{Z}P(G)$-module $F(\sigma)$ represented by φ_0.

For any φ_0 constructed as above, $[F(\sigma), \varphi] = 0$ in $Wh(\mathbb{Z}\underline{\underline{P}}(G))$. (4.7)

One can easily check that when $\underline{\underline{P}} = \underline{\underline{P}}_0$ (or, more generally, when P has a maximal element, which must of course by Z) then (4.7) does not contribute any new relations.

Although the relations described in 4.7 are somewhat complicated, they are just the algebraic counterparts of the elementary expansions defined in section 2. Thus it is natural that they arise in $Wh(\mathbb{Z}\underline{\underline{P}}(G))$.

5. THE ISOMORPHISM $Wh^c(X,p) \cong Wh(\mathbb{Z}\underline{\underline{P}}(G_1(X,p)))$.

Let $(X,p) \in |\underline{\underline{Wh}}^c_f/Z|$. In [5] we define the boundedly controlled Whitehead group $Wh^c(X,p)$ and show that any element of it is represented by a pair $((Y,q), (X,p))$ in $\underline{\underline{CW}}^c_f/Z$ where (X,p) is a strong bc deformation retract of (Y,q). Let $G = G_1(X,p) : \underline{\underline{P}} \to \underline{\underline{Gpoid}}$. We associate to the pair above a chain complex $C_* = C_*((Y,q),(X,p))$ of $\mathbb{Z}\underline{\underline{P}}(G)$-modules in the following way:

For each $K \in P$ let Y_K be the smallest subcomplex of Y containing $q^{-1}(K)$. There is a functor

$$\widetilde{(Y,q)} : \underline{\underline{P}}(G) \to \underline{\underline{CW}}$$

which associates to any (x,K) the space $P(Y_K,x)/\simeq$ of path classes (rel. endpoints) in Y_K starting at x. Note that $P(Y_K,x)/\simeq$ is the standard construction of the universal cover of the component of Y_K containing x. To a morphism $(\omega,i) : (x,K) \to (u,L)$ in $\underline{\underline{P}}(G)$, $\widetilde{(Y,q)}$ associates the map

$$P(Y_K,x)/\simeq \xrightarrow{i_*} P(Y_L,x)/\simeq \xrightarrow{(-)\cdot\omega} P(Y_L,u)/\simeq .$$

There is a subfunctor $\overline{(X,p)} \leq \widetilde{(Y,q)}$ which has $\overline{(X,p)}(x,K)$ the restriction of the covering $P(Y_K,x)/\simeq \to Y_K$ to $X \cap Y_K$.

The $\mathbb{Z}\underline{\underline{P}}(G)$-module C_n is now defined to be the composite functor

$$\underline{\underline{P}}(G) \xrightarrow{\widetilde{(Y,q)},\overline{(X,p)}} \underline{\underline{\text{CW-pairs}}} \xrightarrow{C_n^{cell}} \underline{\underline{\mathbb{Z}\text{-mod}}}$$

where C_n^{cell} is the usual n^{th} cellular chain group of a CW pair.

The usual cellular boundary maps are natural transformations

$$\partial_{n,0} : C_n \to C_{n-1}$$

which represent the boundary morphisms

$$\partial_n : C_n \to C_{n-1}$$

in $\mathbb{Z}\underline{\underline{P}}(G)$-mod.

Let S_n be the set of n-cells of Y not in X and define $\sigma_n : S_n \to \underline{\underline{P}}(G)$ by $\sigma_n(e) = (x_e,K_e)$, where x_e is some point on the boundary of the n-cell e and $\{K_e | e \in S_n\}$ is a bounded collection in P with K_e containing q(closure e).

(For n = 0 one needs a small modification.)

It can be shown that there is an isomorphism $F(\sigma_n) \cong C_n$ which is unique modulo an automorphism of $F(\sigma_n)$ induced by an automorphism of the basis (S_n,σ_n). Thus C_* is an <u>acyclic</u>, <u>finite</u>, bfg, <u>based chain complex</u> of $\mathbb{Z}\underline{\underline{P}}(G)$-modules. Using standard "rolling up" technicues one can associate to any such chain complex C_* a torsion element $\tau(C_*) \in Wh(\mathbb{Z}\underline{\underline{P}}(G))$.

DEFINITION 5.1. If $x \in Wh^c(X,p)$ is represented by the pair ((Y,q),(X,p)) above, then we let $\tau(x) = \tau(C_*((Y,q),(X,p))) \in Wh(\mathbb{Z}\underline{\underline{P}}(G_1(X,p)))$.

In [2] we show

THEOREM 5.2. For any $(X,p) \in |\underline{\underline{CW}}_f^c/Z|$, the above construction defines an isomorphism $\tau : Wh^c(X,p) \to Wh(\mathbb{Z}P(G_1(X,p)))$.

6. THE BOUNDEDLY CONTROLLED s-COBORDISM THEOREM

In this section, we show how the simple homotopy theory outlined above can be applied to manifolds. In particular, we describe boundedly controlled versions of the s-cobordism theorem and a companion realization theorem. Throughout this section, we will assume the boundedness control structure (P,C) on Z is tame in the sense of the following definition.

DEFINITION 6.1. The boundedness control structure (P,C) on Z is <u>tame</u> if there exists an integer $d \geq 0$ such that

1. $\{\mathrm{Int}\, C^d(K_0) \mid K_0 \in P$ is a minimal element$\}$ is an open cover of Z;
2. for all $K \in P$, closure $(K) \subset \mathrm{Int}\, C^d K$; and
3. closure (K) is compact.

For example, the metric boundedness control structure of 1.1 is tame.

A <u>boundedly controlled manifold</u> (or, simply, a <u>bc manifold</u>) over Z is pair (M,p) where M is a manifold and $p : M \to Z$ is proper. Let (M,p) be a bc manifold over Z and suppose $\partial M = \partial_- M \amalg \partial_+ M$. If the inclusions $i_\pm : (\partial_\pm M, p_\pm) \to (M,p)$ $(p_\pm = p|\partial_\pm M)$ are bc homotopy equivalences, we call $(M,\partial_- M,\partial_+ M;p)$ a <u>bc</u> h-<u>cobordism</u>.

Let $(M,\partial_- M,\partial_+ M;p)$ a bc h-cobordism and suppose M supports a PL structure. Then it is possible to triangulate $(M,\partial_- M,\partial_+ M)$ by (L,L_-,L_+) so that the complexes (L,p), (L_-,p_-), and (L_+,p_+) are all finite, bc CW complexes. Since $i_- : (L_-,p_-) \to (L,p)$ is a bc homotopy equivalence, we may assign a torsion invariant $\tau(M,\partial_- M;p) \in Wh(\mathbb{Z}P(G_1(\partial_- M,p_-)))$ to $(M,\partial_- M;p)$ by setting $\tau(M,\partial_- M;p) = \tau(i_-)$. This invariant is well defined by the combinatorial invariance of torsion 2.3.

THEOREM 6.2. (Boundedly Controlled s-Cobordism Theorem). Let $n \geq 6$ and $(M^n,\partial_- M,\partial_+ M;p)$ be a bc PL h-cobordism. Then $(M,\partial_- M,\partial_+ M;p)$ is bc PL homeomorphic to $(\partial_- M \times I, \partial_- M \times 0, \partial_- M \times 1; p_- \pi)$ if and only if $\tau(M,\partial_- M;p) = 0$.

Here $\pi : \partial_- M \times I \to \partial_- M$ is the obvious projection.

THEOREM 6.3. (Realization Theorem). Let $n \geq 6$ and (N^{n-1},q) be a bc PL manifold. Then for any element $\tau_0 \in Wh(\mathbb{Z}\underline{\underline{P}}G_1(N,q))$ there exists a bc PL h-cobordism $(M,\partial_- M,\partial_+ M;p)$ with $(\partial_- M,p) = (N,q)$ such that $\tau(M,\partial_- M;p) = \tau_0$.

The proofs of these theorems are based on adaptations of the standard handlebody arguments to the boundedly controlled setting. They will be described in detail in [3].

7. TOWARDS COMPUTATIONS

In this section we take $\mathbb{Z} = \mathbb{R}^k$ with the metric boundedness control structure of example 1.1. We consider an $(X,p) \in \underline{\underline{CW}}^c_f/\mathbb{R}^k$ and we let $G = G_1(X,p)$.

DEFINITION 7.1. We say that (X,p) has <u>uniformly locally defined</u> fundamental group if for some $d \geq 0$ the following conditions hold

1. For any ball $B = B(z,r)$ and any $x,y \in p^{-1}(B)$, if x and y are joined by a path in X, then they are joined by a path in $B(z,r+d)$.

2. For any $x \in X$

$$\pi_1(p^{-1}(B(p(x),d)),x) \to \pi_1(X,x)$$

is onto.

3. For any $x \in X$ and any $r \geq 0$

$$\mathrm{Ker}[\pi_1(p^{-1}(B(p(x),r)),x) \to \pi_1(X,x)] \subseteq$$

$$\mathrm{Ker}[\pi_1(p^{-1}(B(p(x),r)),x) \to \pi_1(p^{-1}(B(p(x),r+d)),x)].$$

REMARK. In [10] E.K. Pedersen calls $\pi_1(X)$ d-bounded when the above holds.

THEOREM 7.2. Assume that

1. X is connected,
2. for some $d \geq 0$ $p^{-1}(B(z,d)) \neq \emptyset$ for each $z \in \mathbb{R}^k$.

Then there exists a homomorphism

$$\varphi : \mathrm{Wh}(\mathbb{Z}\underline{\underline{P}}(G_1(X,p))) \to \tilde{K}_{1-k}(\mathbb{Z}[\pi_1(X)])\ .$$

If furthermore

3. (X,p) has uniformly locally defined fundamental group, then φ is an isomorphism.

The proof of 7.2 utilizes E. K. Pedersen's description of $\tilde{K}_{1-k}(\mathbb{Z}[\pi_1(X)])$ [11] and the functoriality of $\mathrm{Wh}(\mathbb{Z}\underline{\underline{P}}(G))$ in the variable $G : \underline{\underline{P}} \to \underline{\underline{\mathrm{Gpoid}}}$, [2].

As an immediate consequence of Theorems 7.2 and 5.2, we have

COROLLARY 7.3. Let $(X,p) \in |\underline{\underline{CW}}_f^c/\mathbb{R}^k|$ satisfy the hypotheses of Theorem 7.2. Then $Wh^c(X,p) \cong \tilde{K}_{1-k}(\mathbb{Z}\pi_1(X))$.

This corollary was described by the first named author at the AMS summer meeting at Albany (Abstracts of the AMS 4 (1983), 373) and stimulated the work described here. Theorems 5.2 and 7.2 were described by the second author at the AMS winter meeting at Louisville 1984 (Abstracts of the AMS 5 (1984), 102).

If $(X,p) \in |\underline{\underline{CW}}_f^c/\mathbb{R}^k|$, X is a locally finite CW complex and one may form Siebenmann's group S(X) of proper simple homotopy types in the proper homotopy type of X, [14]. Furthermore, there is a "forgetful" homorphism

$$\square : Wh^c(X,p) \to S(X).$$

If A is a compact, connected CW complex, then $(A \times \mathbb{R}^k, pr : A \times \mathbb{R}^k \to \mathbb{R}^k)$ satisfies the conditions of Theorem 7.2. Also Siebenmann gives isomorphisms (see [14])

$$S(A \times \mathbb{R}^k) \cong \begin{cases} \tilde{K}_0(\mathbb{Z}\pi) & , \quad k = 1 \\ \mathrm{Ker}(\tilde{K}_0(\mathbb{Z}[\pi \times \mathbb{Z}]) \to \tilde{K}_0(\mathbb{Z}\pi)), & \quad k = 2 \\ 0 & , \quad k \geq 3 \end{cases}$$

where $\pi = \pi_1(A)$. In [6] we shall give the following description of $\square$, treating the isomorphisms above as identities.

THEOREM 7.4. For any compact, connected CW complex A the map $\square : Wh^c(A \times \mathbb{R}^k, pr) \to S(A \times \mathbb{R}^k)$ is the identity if k = 1 and the Bass-Heller-Swan injection if k = 2.

In view of the boundedly controlled s-cobordism theorem in section 6 and Siebenmann's proper s-cobordism theorem, we have the following surprising corollary:

COROLLARY 7.5. Let M be a compact PL manifold of dim $\geq$ 5-k and let (W,q) be a bc h-cobordism of $(M \times \mathbb{R}^k, pr)$, k = 1 or 2. If W admits a proper product structure, then (W,q) admits a bc product structure.

Lest the reader should start thinking that $\square : Wh^c(X,p) \to S(X)$ be injective in gnereal, we finish this section by mentioning that in [6] we shall also present an example (X,p) over $\mathbb{R}$ where s(X) = 0, but Wh(X,p) is uncountable.

REFERENCES

1. Anderson, D. R., and Munkholm, H.J., The Algebraic Topology of Controlled Spaces, Preprint, Odense Universitet, 1984.
2. Anderson, D. R., and Munkholm, H. J., The Simple Homotopy Theory of Controlled CW Complexes, Preprint (to appear), Odense Universitet.
3. Anderson, D. R., and Munkholm, H. J., The s-Cobordism Theorem for Boundedly Controlled Manifolds, in preparation.
4. Anderson, D. R., and Munkholm, H. J., The Algebraic and Geometric Topology of Boundedly Controlled Spaces, in preparation.
5. Anderson, D. R., and Munkholm, H. J., A Geometric Construction of the Boundedly Controlled Whitehead Group, these Proceedings.
6. Anderson, D. R., and Munkholm, H. J., Proper Simple Homotopy Theory Versus Simple Homotopy Theory Controlled over $\mathbb{R}^k$, in preparation.
7. Anderson, D. R., and Munkholm, H. J., The Simple Homotopy Theory of Controlled Spaces, an Announcement, Preprint, Odense Universitet, 1984.
8. Chapman, T. A., Controlled Simple Homotopy Theory and Applications, Springer Lecture Notes in Mathematics, vol. 1009, Springer-Verlag, Berlin, 1983.
9. Pedersen, E. K., K_{-i}-invariants of Chain Complexes, Springer Lecture Notes in Mathematics, vol. 1060, Springer Verlag, Berlin, 1984.
10. Pedersen, E. K., On the Bounded and Thin h-cobordism Theorem Parameterized by $\mathbb{R}^k$, Preprint, Gottingen, 1985.
11. Pedersen, E. K., On the K_{-i}-functors, J. of Alg., 90 (1984), 461-475.
12. Quinn, F., Ends of Maps, I, Ann. of Math., 110 (1979), 275-331.
13. Quinn, F., Ends of Maps, II, Inv. Math. 68 (1982), 353-424.
14. L. Siebenmann, Infinite Simple Homotopy Types, Indag. Math., 32 (1970), 479-495.

On Certain Branched Cyclic Covers of S^3

MARK D. BAKER / Department of Mathematics, Brown University, Providence, Rhode Island

1. Let $X = S^3 \setminus \mathring{N}(K)$ be the complement in S^3 of an open tubular neighborhood of the figure eight knot K. Denote by X_n the n-fold cyclic cover of X, and by Σ_n the n-fold branched cyclic cover of S^3 branched over K (cf., [4]). In this paper we consider the question of whether Σ_n has a finite cover with positive first Betti number. We prove:

THEOREM. For $n \geq 5$, Σ_n has a finite cover $\tilde{\Sigma}_n \to \Sigma_n$ with rank $H_1(\tilde{\Sigma}_n) > 0$.

NOTE. Since $\Sigma_1 = S^3$ and $\pi_1(\Sigma_2) \cong \mathbb{Z}/5\mathbb{Z}$, the problem is posed for $n \geq 3$. Hempel [2] has proved this result for n odd, using different methods.

The theorem is proved by constructing a cover $\tilde{X}_n \to X_n$ with the following two properties:

i) $\tilde{X}_n \to X_n$ extends to an (unbranched) cover $\tilde{\Sigma}_n \to \Sigma_n$ by Dehn filling on $\tilde{X}_n$ and X_n with respect to certain loops $\{\tilde{m}_i\}$ in $\partial\tilde{X}_n$ and m in ∂X_n.

ii) Rank $H_1(\tilde{X}_n) >$ number of components of $\partial\tilde{X}_n$.

By Dehn filling on a 3-manifold M with respect to a loop in a boundary torus we mean attaching a solid torus to ∂M so that this loop bounds a meridinal disk in the solid torus. Property (ii) implies that any manifold obtained by Dehn filling on $\tilde{X}_n$ (hence $\tilde{\Sigma}_n$) has positive first Betti number.

2. Recall that $\pi_1(X) = \langle x,y \mid (x^{-1}y\, x\, y^{-1})\, x\, (x^{-1}y\, x\, y^{-1})^{-1} = y\rangle$, and that a meridian and longitude of ∂X are given by $\mu = x$ and $\lambda = y^{-1}x\, y\, x^{-1}x^{-1}y\, x\, y^{-1}$. There is a discrete, faithful representation $\pi_1(X) \to PSL_2(\mathbb{Z}[\omega])$, where $\omega^3 = 1$, given by

$$x \mapsto \begin{bmatrix} 1 & 1 \\ 0 & 1 \end{bmatrix} \quad \text{and} \quad y \mapsto \begin{bmatrix} 1 & 0 \\ -\omega & 1 \end{bmatrix}.$$

A calculation shows

$$\lambda \mapsto \begin{bmatrix} 1 & 4\omega+2 \\ 0 & 1 \end{bmatrix}$$

(cf., [3]). It is clear that the matrices

$$\begin{bmatrix} 1 & n \\ 0 & 1 \end{bmatrix} \quad \text{and} \quad \begin{bmatrix} 1 & 4\omega+2 \\ 0 & 1 \end{bmatrix}$$

represent respectively a meridian, m, and a longitude, ℓ, for ∂X_n.

3. Denoting by $\Gamma_K(n)$ the image of $\pi_1(X_n) \subset \pi_1(X)$ in $PSL_2(\mathbb{Z}[\omega])$ under the representation above, consider the subgroup $\Gamma_K(n) \cap \Gamma(n) \subset \Gamma_K(n)$, where $\Gamma(n) \subset PSL_2(\mathbb{Z}[\omega])$ is the n-principal congruence subgroup, and let $\tilde{X}_n \to X_n$ be the cover corresponding to $\Gamma_K(n) \cap \Gamma(n)$.

PROPOSITION 1. a) $\tilde{X}_n \to X_n$ is regular.

b) The meridian loop m in ∂X_n lifts to loops on the boundary tori of $\tilde{X}_n$ (hence these lifted loops project homeomorphically to m).

Proof: $\Gamma(n)$ is normal in $PSL_2(\mathbb{Z}[\omega])$ thus $\Gamma_K(n) \cap \Gamma(n)$ is normal in $\Gamma_K(n)$, hence (a). The loop m in ∂X_n corresponds to a homotopy class represented by the matrix

$$\begin{bmatrix} 1 & n \\ 0 & 1 \end{bmatrix} \in \Gamma_K(n) \cap \Gamma(n)$$

hence (b).

On each boundary torus of $\tilde{X}_n$ choose one of the lifts of m and denote these loops by $\{\tilde{m}_i\}$. Then it is easy to show using proposition 1 that:

PROPOSITION 2. $\tilde{X}_n \to X_n$ extends to a regular (unbranched) cover $\tilde{\Sigma}_n \to \Sigma_n$ by Dehn filling on $\tilde{X}_n$ and X_n with respect to loops $\{\tilde{m}_i\}$ in $\partial\tilde{X}_n$ and m in ∂X_n.

4. We now show that if $n \geq 5$, the cover $\tilde{X}_n \to X_n$ constructed above satisfies property (ii), i.e., that rank $H_1(\tilde{X}_n)$ > number of components of $\partial\tilde{X}_n$. Let $i : \partial\tilde{X}_n \to \tilde{X}_n$ denote the inclusion map. Then (ii) is equivalent to $\text{rank}[H_1(\tilde{X}_n)/i_*(H_1(\partial\tilde{X}_n))] > 0$.

Given a group $\Gamma \subset PSL_2(\mathbb{Z}[\omega])$ of finite index, denote by U_Γ the (normal) subgroup generated by the parabolic matrices (trace = ± 2).

DEFINITION. $d(\Gamma) = \dim_{\mathbb{Q}}((\Gamma/U_\Gamma)^{ab} \otimes_{\mathbb{Z}} \mathbb{Q})$.

PROPOSITION 3. a) If $\Gamma' \subset \Gamma$ is of finite index, then $d(\Gamma') \geq d(\Gamma)$.

b) $d(\Gamma_K(n) \cap \Gamma(n)) = \text{rank}[H_1(\tilde{X}_n)/i_*(H_1(\partial\tilde{X}_n))]$.

Proof: The homomorphism $(\Gamma'/U_{\Gamma'})^{ab} \otimes_{\mathbb{Z}} \mathbb{Q} \to (\Gamma/U_\Gamma)^{ab} \otimes_{\mathbb{Z}} \mathbb{Q}$ is surjective since $\Gamma' \subset \Gamma$ is of finite index, hence (a). Part (b) follows from the isomorphism $\pi_1(\tilde{X}_n) \cong \Gamma_K(n) \cap \Gamma(n)$ and the correspondence between homotopy classes of loops in $\partial\tilde{X}_n$ and parabolic matrices of $\Gamma_K(n) \cap \Gamma(n)$.

Since $\Gamma_K(n) \cap \Gamma(n) \subset \Gamma(n)$, property (ii) follows from proposition 3 and:

PROPOSITION 4. For $n \geq 5$, $d(\Gamma(n)) > 0$.

Proof: Let $\mathbb{Z}_n[\omega] \subset \mathbb{Z}[\omega]$ denote the order of index n. Since $\Gamma(n) \subset PSL_2(\mathbb{Z}_n[\omega])$ is of finite index, it suffices to prove the proposition for $PSL_2(Z_n[\omega])$, $n \geq 5$. Here we can apply results of Grunewald and Schwermer [1] who show that $d(PSL_2(\mathbb{Z}_n[\omega])) \geq \text{card}(W) - 1$, where W is the set of natural numbers m satisfying the following conditions:

a) $m > 0$, $m \neq 2$, $(m,n) = 1$.

b) $4m^2 \leq 3n^2 - 3$.

c) Every prime divisor of m is inert in $\mathbb{Z}[\omega]$.

If $n \geq 6$ and $(5,n) = 1$, then $\{1,5\} \subset W$ and $d(\Gamma(n)) \geq d(PSL_2(\mathbb{Z}_n[\omega])) \geq 1$. If $5|n$, then $PSL_2(\mathbb{Z}_n[\omega]) \subset PSL_2(\mathbb{Z}_5[\omega])$ and a computation shows that $d(PSL_2(\mathbb{Z}_5[\omega])) \geq 1$. (One shows that in this case the prime 2 can be included in W so $\{1,2\} \subset W$).

ACKNOWLEDGEMENTS

The author was supported in part by NSF Grant DMS-85-03861.

REFERENCES

1. Grunewald, F., and Schwermer, J., A nonvanishing theorem for the cuspidal cohomology of SL_2 over imaginary quadratic integers, Math. Ann. 258 (1981), 183-200.

2. Hempel, J., Coverings of Dehn fillings of surface bundles, II, preprint.

3. Riley, R., A quadratic parabolic group, Math. Proc. Camb. Phil. Soc. 77 (1975), 281-288.

4. Rolfsen, D., Knots and Links, Publish or Perish, Inc., Berkeley, CA, 1976.

A Geometric Interpretation of Siebenmann's Periodicity Phenomenon

SYLVAIN CAPPELL / Courant Institute of Mathematical Sciences, New York University, New York, New York

SHMUEL WEINBERGER / Department of Mathematics, The University of Chicago, Chicago, Illinois

1. STATEMENT OF MAIN THEOREM

If M is a manifold, $\mathcal{S}(M)$, the structure set of M is defined, following Sullivan and Wall [11], as the set of pairs (N,f) consisting of a manifold N and a simple homotopy equivalence $f : N \to M$ that restricts to a homeomorphism on the boundary. Two pairs, (N_1,f_1) and (N_2,f_2), represent the same element if there is a homeomorphism $H : N_1 \to N_2$ such that $f_1 \sim f_2 H$ rel ∂. One of the most beautiful results in the theory is Siebenmann's periodicity theorem that $\mathcal{S}(M) \cong \mathcal{S}(M \times D^4)$ for most M, e.g. all M with nonempty boundary (see [7] and the paper of Nicas [4] for a correction). Siebenmann's proof was rather indirect; it proceeded by constructing a simplicial set whose π_0 was $\mathcal{S}(M)$ and which was the fiber of a fibration which had periodicity properties. While this is enough for Siebenmann's (and others') applications, (such as providing a group structure on $\mathcal{S}(M)$ via the obvious one on $\mathcal{S}(M \times D^4)$ analogous to the definition of π_4), its indirectness is mysterious (see [7]). (This map is <u>not</u> just crossing with D^4; one does not then get a homeomorphism on the boundary.) In this paper we shall give a geometrically defined map $\mathcal{S}(M) \to \mathcal{S}(M \times D^4)$ (or actually to $\mathcal{S}(E)$ for a class of total spaces of four-plane bundles) that has all the properties of Siebenmann periodicity, by means of embedding theory. In addition to whatever aesthetic advantages there may be in the gained geometricity, there is also at least one practical pay-off. As an application we shall show that many homotopy $\mathbb{C}P^n$'s have locally smooth S^1-actions.

Before proceeding it is important to note that periodicity is very much a topological phenomenon that reflects both the periodicity of surgery groups $L_n(\pi) = L_{n+4}(\pi)$ and of the classifying space $\mathbb{Z} \times G/Top \approx \Omega^4 G/Top$. (The $\mathbb{Z}$ factor accounts for the exceptions to periodicity among closed manifolds.) In particular, it fails even in the PL category since G/PL has one twisted k-invariant at the prime 2. On the other hand, that is the whole (small) difference. Moreover, the tools that we will use include embedding theory, block bundles, transversality to subpolyhedra which are only available in the PL category. Nonetheless, the experts should be able to use mapping cylinders, approximate fibrations, and torus tricks to make these ideas (although not the details) work topologically.

Our construction is an analogue of the classical notion of the branched cyclic cover of a manifold W branched over (or along) a codimension two (locally flat) submanifold M. Here one starts with a k-fold cyclic regular cover on W-M. The well known fact is that if the restriction of the cover to a circle meridionally linking M is the usual $z \to z^k$ cover of S^1 to itself, then there is a canonical manifold compactification of the cover, obtained by appropriately "filling in" M. The covering translates extend to a $\mathbb{Z}_k$ action on the whole branched cover with fixed set M, and the quotient is W. (One way to do this is to remove the interior of any regular neighborhood of M; the boundary has the structure of an S^1-block bundle. The k-fold cover restricts to a k-fold cover on each block resulting in a new S^1-block bundle over M. One can now (inductively over a triangulation, as usual) cone down to obtain a manifold. The key to doing this is that the total space of the fibration restricted to a (linking meridinal) sphere is again a sphere.

There are, of course, other fibrations over spheres with total spaces spheres - namely the Hopf fibrations $S^1 \to S^3 \to S^2$ and $S^3 \to S^7 \to S^4$ and one can use these to get notions of branched fibrations. Only the first is relevant to this paper (although everything we say applies equally to the second). To repeat, in other words, whenever $M \subset W$ is codimension three and one has a principal S^1-bundle on W-M which restricts to the Hopf bundle on a linking S^2 (i.e., the first Chern class evaluated on $[S^2]$ is ± 1) then the total space can be canonically compactified by "filling in" M in such a way that M is now a codimension four submanifold, and the S^1 action (fiberwise, by rotation) on the (open) principal bundle piece extends to one on the compactification with M as fixed set. We call this space the branched S^1-fibration of W along M.

One way to analyze branched S^1-fibrations is to make use of the S^1 action and the relations that exist between (Mischenko-Ranicki symmetric or higher) signatures of a manifold and the fixed set of any S^1 action on it. (see [11]). There are unfortunately some thorny transversality problems which make the discussion at the prime 2 complicated. We shall be slightly more indirect, motivated by [2].

We are now ready to give the construction of the periodicity map. The fundamental theorems of embedding theory (see eg. [10, §11]) reduce PL embedding problems in codimension at least three to problems in the homotopy category; in particular, if $(N,f) \in \mathscr{S}(M)$ then one can homotop the composite $i \circ f : N \xrightarrow{f} M \xrightarrow{i} M \times D^3$ to an embedding. Since $i \circ f$ and i are homotopic the S^1-branched fibrations of $M \times D^3$ along them are homotopy equivalent, and since the only difference between the branch fibrations is what goes on in the neighborhood of $M \times \{0\}$ in which N embeds, there is no change on the boundary. It is an enlightening exercise in Whitehead torsions and this construction to calculate the torsion of the homotopy equivalence between branched S^1-fibrations in terms of $\tau(f)$. Since f is simple, the branched fibrations are in fact simple homotopy equivalent and one obtains an element of $\mathscr{S}(M \times D^4)$.

THEOREM. Branched S^1-fibrations define a homomorphism so that one has an exact sequence if $\dim M \geq 5$ or if $\pi_1 M$ is "small" ([3]) and $\dim M = 4$

$$0 \to \mathscr{S}(M) \to \mathscr{S}(M \times D^4) \to L_0(0) = \mathbb{Z} .$$

The last arrow is trivial if $\partial M \neq \emptyset$.

This interprets the corrected formulation of [7] given in [4].

2. PROOF OF THEOREM

An embedding, as in §1, $e : M' \to M \times D^3$, which is a simple homotopy equivalence, gives rise to, by consideration of the complement of the interior of a regular neighborhood of $e(M')$, an s-cobordism between $\partial(M \times D^3) = M \times S^2$ and the boundary of that regular neighborhood. (This equation should be reinterpreted in the obvious way if M itself has boundary.) S-cobordisms have product structures and boundaries of regular neighborhoods have block (sphere) bundle structures. As a result $M \times S^2$ has been canonically given the structure of an S^2-block bundle over M'. One can now take fiberwise Hopf bundles to produce $\mathring{\mathbb{C}P}^2$ block bundle structure over M' on $M \times \mathring{\mathbb{C}P}^2$ (where $\mathring{\mathbb{C}P}^2$ is the complex projective plane with an open disk removed). Let us summarize this in the homotopy commutative diagram

$$\begin{array}{ccc} E(\mathring{\mathbb{C}P}^2) & \xrightarrow{\text{homeo}} & M \times \mathring{\mathbb{C}P}^2 \\ \downarrow & & \downarrow \\ M' & \xrightarrow{h} & M \end{array}$$

(A symbol E(F) is the total space of an F-block bundle.) Now glue in the D^4 bundle over M' constructed in §1 to $E(\mathring{\mathbb{C}P}^2)$ and $M \times D^4$ to $M \times \mathring{\mathbb{C}P}^2$ (recall that these bundles are simple homotopy equivalent rel a homeomorphism on the boundary). This produces a diagram

$$\begin{array}{ccc} E(\mathbb{C}P^2) & \longrightarrow & M \times \mathbb{C}P^2 \\ \downarrow & & \downarrow \pi \\ M' & \longrightarrow & M \end{array}$$

Using this we shall verify injectivity of the branched cover map $\mathcal{S}(M) \to \mathcal{S}(M \times D^4)$. Very similar considerations yield the whole result.

Therefore, suppose now that $[h] \in \mathcal{S}(M)$ vanishes in $\mathcal{S}(M \times D^4)$, then the D^4 bundle over M' is homotopic rel ∂ to a homeomorphism to $M \times D^4$ so that $E(\mathbb{C}P^2) \to M \times \mathbb{C}P^2$ is homotopic (even rel $E(\mathring{\mathbb{C}P}^2)$) to a homeomorphism. We now essentially consider the block fibering obstructions [6] [1] for $M \times \mathbb{C}P^2 \xrightarrow{\pi} M \xrightarrow{h^{-1}} M'$ to deduce first that M' is normally cobordant to M.

Notice that $M \times \mathbb{C}P^2$ over M' is fiber homotopy trivial. Let $K \subset M'$ be part of a characteristic variety [8] for M'. Assume $h^{-1} \pitchfork K$. Now consider $(h^{-1}\pi)^{-1}K \to K \times \mathbb{C}P^2$. Since $h^{-1}\pi$ is homotopic to a block fibration the surgery obstruction is zero. On the other hand $(h^{-1}\pi)^{-1}K = \pi^{-1}(h^{-1})^{-1}K$ $= ((h^{-1})^{-1}K) \times \mathbb{C}P^2$ so the surgery obstruction is that of $(h^{-1})^{-1}K \to K$, which must therefore vanish.

Since the surgery obstructions vanish for all pieces of the characteristic variety [8], h^{-1} has trivial normal invariant. Hence h is normally cobordant to the identity as well.

Now we must show that h is, in fact, homotopic to a homeomorphism. Consider any normal cobordism N from M' to M. Cross it with D^3. The π-π

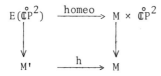

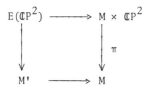

theorem applies to show that $M' \times D^3$ is homeomorphic to $M \times D^3$. Therefore all the block bundles considered throughout are trivial. In particular, $M \times \mathbb{C}P^2 = E(\mathbb{C}P^2) = M' \times \mathbb{C}P^2$. Consider now the surgery exact sequences

$$\ldots \to [\Sigma M : G/Top] \to L_{n+1}(\pi) \to \mathscr{S}(M)$$

$$\ldots \to [\Sigma(M \times \mathbb{C}P^2) : G/Top] \to L_{n+5}(\pi) \to \mathscr{S}(M \times \mathbb{C}P^2)$$

We know that the obstruction for $N \times \mathbb{C}P^2$ lies in the image of $[\Sigma(M \times \mathbb{C}P^2) : G/Top]$ and want to deduce that it actually lies in the image of $[\Sigma M : G/Top]$. This is true because they have the same image : indeed, according to [9] the image only depends on the images of $\mathbb{H}(M \times \mathbb{C}P^2;\mathbb{L}_0)$ and $\mathbb{H}(M;\mathbb{L}_0)$ in $\mathbb{H}(B\pi;\mathbb{L}_0)$ which are visibly identical.

3. AMPLIFICATION AND APPLICATIONS

3.1. Periodicity holds equally for other D^4 bundles over M. Our methods interpret this if the bundle has a semifree S^1-action fixing the zero-section.

3.2. Branched S^3-fibrations interpret the periodicity $\mathscr{S}(M) \to \mathscr{S}(M \times D^8)$.

3.3. One can use the theorem to construct group actions. Consider the S^1 action on $\mathbb{C}P^n$ given by $\theta(z_1 \cdots z_{n+1}) = (\theta z_1, \theta z_2, z_3, z_4, \cdots, z_{n+1})$. It has fixed set $\mathbb{C}P^1 \sqcup \mathbb{C}P^{n-2}$. Taking the quotient, embedding a homotopy $\mathbb{C}P^{n-2}$ in it and taking the branched S^1-cover produces a locally linear S^1 action on a homotopy $\mathbb{C}P^n$. Its splitting invariants [8] can be computed using the theorem (and remark 1) from those of the homotopy $\mathbb{C}P^{n-2}$. In particular if the bottom two splitting invariants of a homotopy $\mathbb{C}P^n$ vanish then it has a locally linear S^1 action equivariantly homotopy equivalent to this linear one. Compare [5] for smooth results and conjectures.

3.4. However, if the fixed set is nullhomotopic in the ambient manifold, the group actions constructed will be on that manifold. For example $\mathscr{S}(S^4 \times S^4) = \mathbb{Z} \oplus \mathbb{Z}$ and each element is the fixed set of a locally linear S^1 action on $S^6 \times S^6$.

3.5. These results can be extended, however not from the embedding theory point of view, to other codimensions. This will be the subject of another paper [2].

ACKNOWLEDGEMENTS

The first author was partially supported by an NSF Grant. The second author was partially supported by an NSF Postdoctoral Fellowship.

REFERENCES

1. Burghelea, D., Lashof, R., and Rothenberg, M., Groups of Automorphisms of Manifolds. LNM 473 (1975).
2. Cappell, S., and Weinberger, S., Replacement theorems for S^1-actions, (in preparation).
3. Freedman, M., The Disk Theorem for 4-dimensional Manifolds, Proc. I.C.M. (1983) Warsaw, 647-663.
4. Nicas, A., Induction Theorems for Groups of Homotopy Manifold Structure Sets, Memoirs AMS 267 (1982).
5. Petrie, T., Smooth S^1 Actions on Homotopy Complex Projective Spaces and Related Topics, BAMS 78 (1972), 105-153.
6. Quinn, F., A Geometric Formulation of Surgery in Topology of Manifolds, ed. J. C. Cantrell and C. H. Edwards 1970, 500-511, Markham.
7. Siebenmann, L., Periodicity in Topological Surgery, Appendix C to Essay V in Foundational Essays on Topological Manifolds, Smoothings, and Triangulations by R. Kirby and L. Siebenmann, 1977, Princeton University Press.
8. Sullivan, D., Geometric Topology Seminar Note, Princeton, 1965.
9. Taylor, L., and Williams, B., Surgery Spaces : Formulae and Structure, LNM 741 (1979), 170-195.
10. Wall, C.T.C., Surgery on Compact Manifolds, 1970, Academic Press.
11. Weinberger, S., Group Actions and Higher Signatures II, (Preprint).

Regular Convex Cell Complexes

MICHAEL W. DAVIS / Department of Mathematics, Ohio State University, Columbus, Ohio

INTRODUCTION

As $\varepsilon = +1$, 0, or -1, let Y^n_ε stand for the n-sphere, Euclidean n-space, or hyperbolic n-space. The study of regular tessellations of Y^n_ε by convex cells is a classical topic. Such tessellations have been completely classified (e.g., see [2] and [3]). The theory of regular tessellations of the n-sphere is essentially identical with the theory of regular convex polyhedra of dimension n + 1. In the case of hyperbolic space, regular tessellations exist only in dimensions 2, 3, and 4 (cf. [2]).

There is a close connection between the theory of regular tessellations of Y^n_ε and the theory of Coxeter groups: the group of isometric symmetries of such a tessellation is a group generated by the reflections across the faces of an n-simplex in the barycentric subdivision of the tessellation; such reflection groups are Coxeter groups. To a large extent this relationship is of a purely combinatorial nature. This paper is a systematic exposition of the combinatorial aspects of this relationship. Most of this material is classical; however, some new results do emerge.

Suppose that the geometric realization of a convex cell complex K is a PL-manifold of dimension n. We shall say that K is <u>symmetrically regular</u> if its group of combinatorial symmetries acts transitively on the set of n-simplices in its derived complex K'. More generally, K is said to be <u>regular</u> if there is an n-tuple $(m_1, \cdots, m_n)$ of integers ≥ 3 such that (a) the boundary of each 2-cell in K is an m_1-gon, (b) the link of each (n-2)-cell in K is an m_n-gon, and (c) for $2 \leq i \leq n-1$, for each (i+1)-cell F_{i+1} in K, and for each (i-2)-face F_{i-2} of F_{i+1}, the link of F_{i-2} in

∂F_{i+1} is an m_i-gon. The n-tuple $(m_1, \cdots, m_n)$ is called the Schläfli symbol of K. (It is easy to see that symmetric regularity implies regularity. We give a proof in (2.7).) We note that any covering space of a regular convex cell complex naturally has the structure of a regular convex cell complex.

We shall prove the following result in section 3.

THEOREM. Suppose that K is a regular convex cell complex and that K is a connected PL n-manifold. Then the universal cover of K is combinatorially equivalent to a classical regular tessellation of Y^n_ε by convex cells, for some $\varepsilon \in \{+1, 0, -1\}$. The fundamental group π of K is then identified with a subgroup of the group of isometric symmetries of this tessellation of Y^n_ε and K is combinatorially equivalent to the induced tessellation of Y^n_ε/π.

In particular, this result implies that if a manifold admits the structure of a regular convex cell complex, then it must be PL-homeomorphic to a complete Riemannian manifold of constant sectional curvature.

In dimension 2, the above theorem was proved by Edmonds, Ewing, and Kulkarni in [6]. In the special case where K is the boundary complex of a convex (n+1)-cell and where K is symmetrically regular, it is due to McMullen [9]. The theorem was proved in full generality by Kato in [7].

Actually, we shall carry out the whole theory in the broader context where K is a connected n-dimensional pseudo-manifold and where the link of each i-cell, $i \leq n-2$, in K is connected. Regular convex cell complexes are classified in this generality; there are some further possibilities besides the classical tessellations.

Here is a sketch of the main argument. The derived complex of K is a simplicial complex with a natural projection p to the standard n-simplex Δ^n. One associates to $(m_1, \cdots, m_n)$ a Coxeter group W, the diagram of which is a connected line segment. To an n-simplex Σ in K', one associates a subgroup $\pi(K', \Sigma)$ of W. There is a close analogy with the theory of covering spaces: the projection $p : K' \to \Delta^n$ plays the role of a covering projection, the Coxeter group W plays the role of the fundamental group of the base, and $\pi(K', \Sigma)$ plays the role of the fundamental group of K' (when K is a PL-manifold it actually is the fundamental group). It turns out that the role of the universal cover of the base is played by the so-called "Coxeter complex" of W. The theorem is proved by showing that K' is a PL-manifold only in the cases where the Coxeter complex is naturally identified with Y_ε^n.

The preceding paragraph suggests that we generalize the situation by studying simplical complexes over Δ^n to which an arbitrary Coxeter group can be associated (rather than restricting ourselves to Coxeter groups with diagrams connected line segments). This is done in section 2.

As we have already mentioned, the theorem stated above was proved in [7] (by somewhat different methods than those of this paper). Tits' paper [12] is concerned with a generalization of the material discussed here to the theory of buildings; the methods of [12] are very similar to those of this paper.

1. CONVEX CELL COMPLEXES

1.1. Suppose that E is a convex cell in some finite-dimensional real vector space V. Let V_E denote the linear subspace of V consisting of all vectors of the form $t(x-y)$, where $x, y \in E$ and $t \in \mathbb{R}$. In other words, V_E is the linear subspace parallel to the affine subspace supported by E. For $x \in E$, denote by $C_{E,x}$ the set of v in V_E such that $x + tv$ lies in E for some t in $[0,\varepsilon)$ and $\varepsilon > 0$. Suppose that F is a proper face of E (written as $F < E$). Let $\mathring{F}$ denote the relative interior of F. If $x \in \mathring{F}$ and $y \in F$, then $C_{E,y} \subset C_{E,x}$, with equality if and only if $y \in \mathring{F}$. If $x, y \in \mathring{F}$, then $C_{E,x}$ and $C_{E,y}$ have the same image in V_E/V_F. This common image is denoted by Cone(F,E); it is a convex polyhedral cone in V_E/V_F.

The <u>unit sphere</u> S(V) in a real vector space V is the quotient space $(V - \{0\})/\mathbb{R}_+$.

1.2. Suppose that E is a convex cell and that $F < E$. The <u>link of F in E</u>, denoted Link(F,E), is the image of (Cone(F,E) - {0}) in $S(V_E/V_F)$; it is a convex cell in $S(V_E/V_F)$. (The link of a simplex in a simplicial complex is usually defined in another way; our definition is especially for use in convex cell complexes.) If $F_1 < F_2 < E$, then the inclusions $V_{F_2} \subset V_E$, $V_{F_2}/V_{F_1} \subset V_E/V_{F_1}$, and $C_{F_2,y} \subset C_{E,y}$ induce a natural identification of $\mathrm{Link}(F_1,F_2)$ with a face of $\mathrm{Link}(F_1,E)$. Thus, the set of faces of Link(F,E) is in bijective correspondence with the set of faces of E which properly contain F.

1.3. Classically a convex cell complex K is a set of convex cells in some finite dimensional vector space satisfying the following two conditions:

(i) If $E \in K$ and F is a face of E, then $F \in K$.

(ii) If E, $F \in K$, then either $E \cap F = \phi$ or $E \cap F$ is a common face of E and of F.

A convex cell complex, in the above sense, has the structure of a poset: the partial ordering is given by inclusion of faces.

1.3.1. Let K be any poset. If $E \in K$, then let $K_{\leq E}$ denote the subposet $\{F \in K \mid F \leq E\}$. Generalizing the definition in (1.3), we shall say that a poset K is a convex cell complex if the following two conditions are satisfied:

(i') If $E \in K$, then $K_{\leq E}$ is isomorphic to the set of faces of some convex cell.

(ii') If E and F are elements of K, then either $K_{\leq E} \cap K_{\leq F} = \phi$ or else there exists an element F' in this intersection such that $K_{\leq E} \cap K_{\leq F} = K_{\leq F'}$.

For example, an abstract simplicial complex is a convex cell complex in this sense. An element of K is called a cell. A convex cell complex K is n-dimensional if it contains cells of dimension n but none of dimension n+1.

1.4. Associated to a convex cell complex K, there is a topological space called its geometric realization: this is the polyhedron formed by pasting together convex cells, one for each element of K, in the obvious fashion. As is common practice, we shall use the same symbol K to stand for a convex cell complex and its geometric realization.

1.5. The derived complex of K, denoted by K', is the poset of all finite chains in K. (A chain in a poset is a totally ordered nonempty subset.) An element of K' is a simplex; it is a k-simplex if it consists of k + 1 elements of K. The poset K' is an abstract simplicial complex; its vertex set can be identified with K. The geometric realization of K' is naturally identified with the barcentric subdivision of K.

1.6. If K is a convex cell complex and if F is a cell in K, then the link of F in K, denoted Link(F,K), is the convex cell complex consisting of all cells of the form Link(F,E), where $F < E \in K$, and where, of course, whenever $F_1 < F_2 < F_3 \in K$, we identify $\text{Link}(F_1,F_2)$ with the corresponding face of $\text{Link}(F_1,F_3)$.

The poset Link(F,K) is in bijective correspondence with the poset of cells in K which properly contain F. It follows that the simplicial complex Link(F,K)' can be identified with the subcomplex of K' consisting of all simplices $\sigma = \{F_0, F_1, \cdots, F_k\}$, such that $F < F_0 < F_1 \cdots < F_k \in K$.

1.7. A combinatorial equivalence f from a convex cell complex K to another one L is an isomorphism of posets f : K --> L. Such an equivalence f induces a simplicial isomorphism f' : K' --> L'. Hence, a combinatorial equivalence induces a PL-homeomorphism of geometric realizations. A combinatorial self-equivalence of K is called a combinatorial symmetry of K (or sometimes simply a "symmetry"). The group of combinatorial symmetries of K will be denoted by Aut(K).

1.8. Suppose that K is connected and that p : $\tilde{K}$ --> K is a covering projection. Since cells are simply connected, each cell of K is evenly covered by p. Thus, K inherits the structure of a convex cell complex. Let Γ denote the group of covering transformations. Then Γ is a subgroup of Aut($\tilde{K}$) and Γ freely permutes the cells of $\tilde{K}$. If p : $\tilde{K}$ -->K is a regular covering (i.e., if $K \cong \tilde{K}/\Gamma$), then Γ is a normal subgroup of Aut($\tilde{K}$) and Aut($\tilde{K}$)/Γ can be identified with the group of combinatorial symmetries of K.

1.9. In a similar vein, suppose that K is a convex cell complex and that Γ is a subgroup of Aut($\tilde{K}$) which freely permutes the cells of $\tilde{K}$. By an abuse of language, we shall also call the quotient space $\tilde{K}/\Gamma$ a "convex cell complex." (Strictly speaking, $\tilde{K}/\Gamma$ might not be a convex cell complex as defined in (1.3), since distinct faces of a cell in $\tilde{K}$ might be identified by an element of Γ.) However, by passing back to $\tilde{K}$, the notion of the link of a cell in $\tilde{K}/\Gamma$ still makes sense.

1.10. In the remaining sections of this paper we shall often impose the following conditions on an n-dimensional convex cell complex K (or an appropriate cover).

1.10.1. Each cell in K is a face of an n-cell. Each (n-1)-cell in K is a face of precisely two n-cells.

1.10.2. K is connected and for each cell F in K of dimension $\leq$ n-2, Link(F,K) is connected.

1.11. Condition (1.10.1) means that K is an n-dimensional pseudo-manifold. It follows from (1.10.1) that for each k-cell F in K, $0 \leq k \leq n-1$, Link(F,K) is also a pseudo-manifold of dimension n - k - 1. The complex K is a PL-manifold if for each k-cell F, $0 \leq k \leq n-1$, Link(F,K) is PL-homeomorphic to S^{n-k-1} with its standard PL structure. Conditions (1.10.1) and (1.10.2) imply that the link of any i-cell in K has the same number of path components as the link of an i-cell in a PL n-manifold. We shall sometimes also want to impose the following condition.

1.11.1. The link of each (n-2)-cell in K is a circle and the link of each cell of dimension $\leq$ n-3 is simply connected.

This condition means that the link of any i-cell in K has the same fundamental group as the link of an i-cell in a PL n-manifold.

Next, suppose that L is an n-dimensional simplicial complex.

1.12. An n-simplex in L is called a chamber. Let Chamb(L) denote the set of chambers in L. Also, for each simplex σ in L, let $\text{Chamb}_\sigma(L)$ denote the set of chambers in L which have σ as a face.

Two distinct chambers are adjacent if their intersection is an (n-1)-simplex. A gallery in L is a sequence of adjacent chambers. The gallery $(\Sigma_0, \cdots, \Sigma_m)$ is said to begin at Σ_0, to end at Σ_m, and to connect Σ_0 to Σ_m.

1.12.1. If L satisfies (1.10.1), then given any $\Sigma \in \text{Chamb}(L)$ and any (n-1)-dimensional face σ of Σ, there is a unique chamber Σ' adjacent to Σ with $\Sigma \cap \Sigma' = \sigma$.

1.12.2. If, in addition, L satisfies the connectivity conditions in (1.10.2), then any two chambers in L can be connected by a gallery; moreover, if $\sigma \in L$, then any two chambers in $\text{Chamb}_\sigma(L)$ can be connected by a gallery of chambers in $\text{Chamb}_\sigma(L)$.

1.13. The standard n-simplex Δ^n is the poset of all nonempty subsets of $\{0, 1, \cdots, n\}$. It is an abstract simplicial complex. If $\sigma \in \Delta^n$, then put $\text{Type}(\sigma) = \{0,1, \cdots, n\} - \sigma$. Thus, if σ is a k-simplex, Type(σ) is a proper subset of $\{0,1, \cdots, n\}$ of cardinality n - k.

1.14. A projection from L onto Δ^n is a simplicial map $q : L \dashrightarrow \Delta^n$ such that the restriction of q to each simplex is injective. The pair (L,q) is called a simplicial complex over Δ^n. If (L,q) is a simplicial complex over Δ^n, then for any simplex σ in L, put $\text{Type}(\sigma) = \text{Type}(q(\sigma))$.

Note that if Σ is a chamber in L and I is a proper subset of $\{0,1, \cdots, n\}$, then, since $q|\Sigma : \Sigma \dashrightarrow \Delta^n$ is an isomorphism, Σ has a unique face of type I.

Suppose that (L_1,q_1) and (L_2,q_2) are simplicial complexes over Δ^n. A simplicial map $f : L_1 \dashrightarrow L_2$ is called a map over Δ^n if the following diagram commutes.

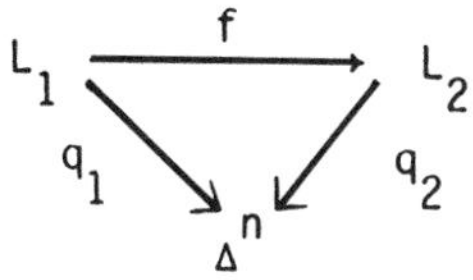

1.15. If $f : L_1 \dashrightarrow L_2$ is a map over Δ^n, then it follows easily from the definitions, that f preserves the dimension of simplices. In particular, there is an induced map $\bar{f} : \mathrm{Chamb}(L_1) \dashrightarrow \mathrm{Chamb}(L_2)$.

LEMMA 1.16. Suppose that (L_1,q_1) and (L_2,q_2) are simplicial complexes over Δ^n, that L_1 satisfies (1.10), and that f and g are two simplicial maps over Δ^n from L_1 to L_2. If there exists a chamber Σ in $\mathrm{Chamb}(L_1)$ such that $\bar{f}(\Sigma) = \bar{g}(\Sigma)$, then $f = g$.

Proof. Suppose that $\bar{f}(\Sigma) = \bar{g}(\Sigma)$. Let v_i denote the vertex of Σ which projects to the vertex $\{i\} \in \Delta^n$, so that $\Sigma = \{v_0, \cdots, v_n\}$. Since f and g preserve type, we must have $f(v_i) = g(v_i)$, $0 \le i \le n$. Thus, f and g agree on the geometric realization of Σ. In particular, they agree on every (n-1)-dimensional face of Σ. It follows from (1.12.1) that f and g agree on every chamber adjacent to Σ. By (1.12.2), this implies that f and g agree on every chamber of L_1 and hence, that $f = g$.

COROLLARY 1.17. Suppose that (L,q) is a simplicial complex over Δ^n and that L satisfies (1.10). Then the group of automorphisms over Δ^n of L, denoted by Aut(L,q), acts freely on Chamb(L).

1.18. Suppose that (L,q) is a simplicial complex over Δ^n and that L satisfies (1.10). Let $\Sigma \in \mathrm{Chamb}(L)$. For each pair of integers i and j, with $0 \le i, j \le n$ and $i \ne j$, let $\Sigma_{\{i,j\}}$ be the (n-2)-face of Σ of type $\{i,j\}$ and put

1.18.1. $L_{ij}(\Sigma) = \mathrm{Link}(\Sigma_{\{i,j\}}, L)$.

It follows from (1.10) that $L_{ij}(\Sigma)$ is a connected 1-manifold; hence, it is either a polygon with a finite number of sides or the tessellation of $\mathbb{R}$ by intervals. The map q induces a simplicial map

$$q_{ij} : L_{ij}(\Sigma) \dashrightarrow \mathrm{Link}(\Delta^n_{\{i,j\}}, \Delta^n) \cong \Delta^1.$$

Thus, $L_{ij}(\Sigma)$ is a complex over Δ^1. The vertices of $L_{ij}(\Sigma)$ correspond to (n-1)-simplices which contain $\Sigma_{\{i,j\}}$. The existence of the map q_{ij} shows that the vertices of $L_{ij}(\Sigma)$ alternate between type $\{i\}$ and type $\{j\}$ (i.e., the corresponding (n-1)-simplices alternate between these types). Hence, if $L_{ij}(\Sigma)$ is a finite polygon, it must have an even number of sides. If $L_{ij}(\Sigma)$ is finite, then let $m_{ij}(\Sigma)$ denote 1/2 the number of its sides; if $L_{ij}(\Sigma)$ is infinite, then put $m_{ij}(\Sigma) = \infty$. Also, for $0 \leq i \leq n$, put $m_{ii}(\Sigma) = 1$. In this way, we obtain an (n+1) by (n+1) matrix $(m_{ij}(\Sigma))_{0 \leq i,j \leq n}$. Its diagonal entries are all equal to 1; its off-diagonal entries are integers ≥ 2 or ∞.

1.19. Let $X = \{x_0, \cdots, x_n\}$ be a set of n + 1 symbols and let G(X) (resp. $G_+(X)$) denote the free group (resp. free monoid) on X. Let (L,q) be as above. For each chamber Σ in L and for $0 \leq i \leq n$, let Σx_i denote the unique chamber in L such that Σ and Σx_i are adjacent and such that the (n-1)-simplex $\Sigma \cap \Sigma x_i$ is of type $\{i\}$. This defines an injective function from X into the group of permutations of Chamb(L). By the universal property of G(X), this function extends to a homomorphism defined on G(X). Hence, we get an action of G(X) (from the right) on Chamb(L). (N.B. This action is <u>not</u> induced from a simplicial action on L.)

1.19.1. Suppose $(\Sigma_0, \cdots, \Sigma_m)$ is a gallery in L. Then $\Sigma_i = \Sigma_{i-1} y_i$ for some $y_i \in X$. Hence, the gallery can be rewritten as

$$(\Sigma_0, \Sigma_0 y_1, \Sigma_0 y_1 y_2, \cdots, \Sigma_0 y_1 \cdots y_m)$$

where the element $g = y_1 \cdots y_m$ is in $G_+(X)$. Conversely, any element $g \in G_+(X)$ yields a gallery from Σ_0 to $\Sigma_0 g$. Since any two chambers can be connected by a gallery, the group G(X) acts transitively on Chamb(L).

1.19.2. Suppose that σ is a simplex in L of type I for some proper subset I of $\{0,1, \cdots, n\}$. Put $X_I = \{x_i \in X \mid i \in I\}$. For any $\Sigma \in \mathrm{Chamb}_\sigma(L)$ and any $x \in X_I$, we have $\Sigma x \in \mathrm{Chamb}_\sigma(L)$. It follows that for $g \in G_+(X_I)$ the corresponding gallery from Σ to Σg is a gallery of chambers in $\mathrm{Chamb}_\sigma(L)$. Thus, $G(X_I)$ stabilizes $\mathrm{Chamb}_\sigma(L)$. It follows from (1.12.2) that $G(X_I)$ acts transitively on $\mathrm{Chamb}_\sigma(L)$.

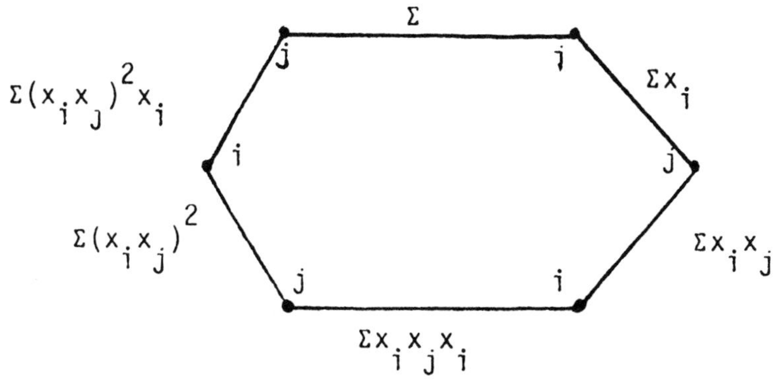

FIGURE 1

Moreover, with σ and Σ as above, we clearly have that

$$\sigma = \Sigma \cap \bigcap_{i \in I} \Sigma x_i; \tag{1.19.3}$$

hence,

$$\sigma = \bigcap_{g \in G_+(X_I)} \Sigma g. \tag{1.19.4}$$

1.20. With notation as above the following formula holds:

$$\Sigma(x_ix_j)^{m_{ij}(\Sigma)} = \Sigma, \tag{1.20.1}$$

for $0 \leq i, j \leq n$ and $m_{ij}(\Sigma) < \infty$. Moreover, if $m_{ij}(\Sigma) < \infty$, then it is the smallest positive integer m such that $\Sigma(x_ix_j)^m = \Sigma$; while, if $m_{ij}(\Sigma) = \infty$ then there is no integer m with this property. Since $\text{Chamb}_{\Sigma_{\{i,j\}}}(\Sigma)$ is isomorphic to $\text{Chamb}(L_{ij}(\Sigma))$, to prove these assertions it suffices to consider the case of a polygon with 2m sides. But these assertions are obvious in this case. (See Figure 1 for the picture when m = 3.)

LEMMA 1.21. Suppose that (L_1,q_1) and (L_2,q_2) are simplicial complexes over Δ^n satisfying (1.10) and that $f : L_1 \dashrightarrow L_2$ is a simplicial map over Δ^n. Let $\bar{f} : \text{Chamb}(L_1) \dashrightarrow \text{Chamb}(L_2)$ be the induced map. Then $\bar{f}$ is G(X)-equivariant.

Proof. Let $\Sigma \in \text{Chamb}(L_1)$. It suffices to prove that for each x_i in X, we have $f(\Sigma x_i) = (f(\Sigma))x_i$. The chambers $f(\Sigma)$ and $f(\Sigma x_i)$ are adjacent and their intersection has type $\{i\}$ (since f preserves adjacency and type). Therefore, $f(\Sigma x_i) = (f(\Sigma))x_i$.

Conversely, we have the following result.

THEOREM 1.22. Suppose that (L_1,q_1) and (L_2,q_2) are simplicial complexes over Δ^n and that ϕ : Chamb(L_1) --> Chamb(L_2) is a map of G(X)-sets. Then there is a unique simplicial map over Δ^n denoted $f_\phi : L_1 \to L_2$, such that $\bar{f}_\phi = \phi$.

In other words, there is a natural bijection between the set of simplicial maps over Δ^n from (L_1,q_1) to (L_2,q_2) and the set of G(X)-equivariant maps from Chamb(L_1) to Chamb(L_2).

Proof. We first remark that since ϕ is G(X)-equivariant, $\phi(\Sigma x_i) = (\phi(\Sigma))x_i$. Thus, the chambers $\phi(\Sigma)$ and $\phi(\Sigma x_i)$ are adjacent and their intersection is an (n-1)-simplex of type $\{i\}$. Let σ be a k-simplex in L_1. Put I = Type(σ). Choose a chamber Σ such that $\sigma < \Sigma$. Let

$$\alpha = \phi(\Sigma) \cap \bigcap_{i\in I} \phi(\Sigma x_i)$$

It follows from the above remarks that α is k-simplex of type I. Furthermore, by (1.19.4),

$$\alpha = \bigcap_{g\in G_+(X_I)} (\phi(\Sigma))g.$$

From this last equation we see that the definition of α is independent of the choice of $\Sigma \in \text{Chamb}_\sigma(L_1)$. Define $f_\phi : L_1 \to L_2$ by $f_\phi(\sigma) = \alpha$. It is clear that f_ϕ is a simplicial map over Δ^n and that $\bar{f}_\phi = \phi$. Uniqueness follows from Lemma (1.16).

Now suppose that K is an n-dimensional convex cell complex. We shall apply paragraphs (1.12) through (1.22) to the special case L = K'.

1.23. If $F \in K$, then $\{F\}$ is a vertex of K'. If $i \in \{0,1,\cdots,n\}$, then $\{i\}$ is a vertex of Δ^n. The map of vertex sets Vert(K') --> Vert(Δ^n) defined by $\{F\}$ --> $\{\dim F\}$ yields a simplicial projection p' : K' --> Δ^n. Thus, _K' has a canonical structure of a simplicial complex over_ Δ^n.

1.24. Let K_1, K_2 be n-dimensional convex cell complexes. If $f : K_1 \to K_2$ is a combinatorial equivalence, then $f' : K_1' \to K_2'$ is a simplicial isomorphism over Δ^n. (In particular, f' preserves type; consequently, f preserves the dimension of cells. Thus, a combinatorial equivalence preserves the dimension of cells.)

Conversely, suppose that $g : K_1' \to K_2'$ is a simplicial map over Δ^n. Let $f : K_1 \to K_2$ be the induced map of vertex sets. Since g is compatible with the projections to Δ^n, it follows that f is a map of posets and that $f' = g$. If g is an isomorphism, then so is f. Hence, _the set of combinatorial equivalences from K_1 to K_2 is naturally bijective with the set of simplicial isomorphisms over Δ^n from K_1' to K_2'._

1.25. Applying this in the special case $K = K_1 = K_2$, we see that Aut(K), the group of combinatorial symmetries of K, is canonically isomorphic to Aut(K',p), the group of simplicial isomorphisms of K' over Δ^n.

From now on, we suppose that K satisfies (1.10). First we note the following.

1.26. K' also satisfies (1.10).

Using (1.25), corollary (1.17) can be translated as follows.

LEMMA 1.27. If K satisfies (1.10), then the group Aut(K) acts freely on Chamb(K').

1.28. Suppose that $\Sigma = \{F_0 < \cdots < F_n\}$ is a chamber in K'. For $1 \leq i \leq n$, let $L_i(\Sigma)$ be the 1-dimensional convex cell complex defined as follows:

$$L_1(\Sigma) = \partial F_2$$

$$L_i(\Sigma) = \mathrm{Link}(F_{i-2}, \partial F_{i+1}),\ 2 \leq i \leq n-1,$$

$$L_n(\Sigma) = \mathrm{Link}(F_{n-2}, K),\ n \geq 2.$$

For $1 \leq i \leq n-1$, $L_i(\Sigma)$ is a polygon; let $m_i(\Sigma)$ denote the number of its sides. The complex $L_n(\Sigma)$ is either a polygon, in which case $m_n(\Sigma)$ denotes the number of its sides, or the tessellation of $\mathbb{R}$ by intervals, in which case put $m_n(\Sigma) = \infty$.

Recall that for $0 \leq i, j \leq n$, $i \neq j$, in (1.18.1) we define complexes $L_{ij}(\Sigma) = \mathrm{Link}(\Sigma_{\{i,j\}}, K')$, where $\Sigma_{\{i,j\}}$ denotes the (n-2)-simplex $\Sigma - \{F_i, F_j\}$. For $\{i,j\} = \{k-1,k\}$, $L_{ij}(\Sigma)$ can be identified with the derived complex of $L_k(\Sigma)$. Thus,

$$m_{ij}(\Sigma) = m_k(\Sigma),\ \text{whenever}\ \{i,j\} = \{k-1,k\}\ .$$

In the notation of (1.19) we have that, for $0 \leq i \leq n$, $\Sigma x_i = (\Sigma - \{F_i\}) \cup \hat{F}_i$ for some i-cell F_i in K. Suppose $|i-j| \geq 2$. Put $\Sigma(i,j) = (\Sigma - \{F_i, F_j\}) \cup \{\hat{F}_i, \hat{F}_j\}$. It follows from the hypothesis $|i-j| \geq 2$, that $\Sigma(i,j)$ is a chamber of K'. Moreover, $\Sigma(i,j)$ is the adjacent chamber to Σx_i (resp. to Σx_j) across the face of type $\{j\}$ (resp. type $\{i\}$). Thus, $\Sigma x_i x_j = \Sigma(i,j) = \Sigma x_j x_i$, from which it follows that $\Sigma(x_i x_j)^2 = \Sigma$.

In other words, $m_{ij}(\Sigma) = 2$ and $L_{ij}(\Sigma)$ is a quadrilateral, whenever $|i-j| \geq 2$. In summary,

$$m_{ij}(\Sigma) = \begin{cases} 1 & , \text{ if } i = j \\ 2 & , \text{ if } |i-j| \geq 2 \\ m_k(\Sigma) & , \text{ if } \{i,j\} = \{k-1,k\} . \end{cases}$$

2. REGULAR SIMPLICIAL COMPLEXES OVER Δ^n

2.1. A Coxeter matrix of degree n + 1 is a symmetric (n+1) by (n+1) matrix $M = (m_{ij})_{0 \leq i,j \leq n}$, with each diagonal entry equal to 1 and with each off-diagonal entry either an integer ≥ 2 or ∞.

2.2. To a Coxeter matrix $M = (m_{ij})_{0 \leq i,j \leq n}$, one can associate a graph as follows. The graph has one vertex z_i for each integer i, $0 \leq i \leq n$. Distinct vertices z_i and z_j are connected by an edge if and only if $m_{ij} \neq 2$. The edge corresponding to $\{z_i, z_j\}$ is labelled m_{ij} if $m_{ij} > 3$ and it is left unlabelled if $m_{ij} = 3$. The resulting graph with edge labels is called the Coxeter diagram associated to M. The matrix M is clearly determined by its diagram, up to permutations of the indices.

2.3. Suppose that $X = \{x_0, \cdots, x_n\}$ is a set of n + 1 symbols and that $M = (m_{ij})_{0 \leq i,j \leq n}$ is a Coxeter matrix. The Coxeter group W associated to M is the quotient of the free group on X (denoted G(X)) by the normal subgroup generated by $\{(x_i x_j)^{m_{ij}}\}$, where $0 \leq i, j \leq n$, and $m_{ij} \neq \infty$.

2.4. Associated to M there is another matrix $C_M = (c_{ij})_{0 \leq i,j \leq n}$, called its cosine matrix, defined by the formula

$$c_{ij} = -\cos(\pi/m_{ij}).$$

2.5. Regard $\mathbb{R}^{n+1}$ as the set of all functions from $\{0,1, \cdots, n\}$ to $\mathbb{R}$ and let $\{e_0, e_1, \cdots, e_n\}$ be the standard basis. Let M, C_M, X, W be as above. Put $B_M(e_i, e_j) = c_{ij}$. This extends to a bilinear form

$$B_M : \mathbb{R}^{n+1} \times \mathbb{R}^{n+1} \dashrightarrow \mathbb{R}.$$

For $0 \leq i \leq n$, define a linear reflection s_i on $\mathbb{R}^{n+1}$ by $s_i y = y - 2B_M(y, e_i)e_i$. (This is not a standard orthogonal reflection with respect to a hyperplane, rather it is orthogonal with respect to the symmetric bilinear form $B_M(\ ,\ )$. This bilinear form could be degenerate and/or indefinite.) It can be checked that the order of $(s_i s_j)$ in $GL(n+1, \mathbb{R})$ is m_{ij}. Hence, the map $X \dashrightarrow GL(n+1, \mathbb{R})$ defined by $x_i \dashrightarrow s_i$ extends to a representation $\phi : W \dashrightarrow GL(n+1, \mathbb{R})$ called the canonical representation of W. (For more details on the canonical representation, see Ch. V of [1].)

It follows from the existence of the representation ϕ that the natural map X --> W is injective and that it takes each element of X to a non-trivial element of W. Henceforth, we shall identify X with its image in W. It also follows from the existence of ϕ that the element $(x_i x_j)$ has order m_{ij} in W (rather than just dividing m_{ij}). The pair (W,X) is called the Coxeter system associated to M. Any pair isomorphic to (W,X) is also called a Coxeter system.

2.6. Suppose that (L,q) is a simplicial complex over Δ^n (cf. (1.14)) and that L satisfies (1.10). From (1.18), we get a function M : Chamb(L) --> {Coxeter matrices} defined by $M(\Sigma) = (m_{ij}(\Sigma))_{0 \leq i,j \leq n}$. We say that (L,q) is regular if M is a constant function. We denote its value by $M(L) = (m_{ij}(L))_{0 \leq i,j \leq n}$. We say that L is of type M(L).

2.7. Again, suppose that (L,q) is a simplicial complex over Δ^n and that L satisfies (1.10). Then (L,q) is symmetrically regular if its automorphism group Aut(L,q) acts transitively on Chamb(L).

Suppose (L,q) is symmetrically regular. Obviously, for any $a \in \mathrm{Aut}(L,q)$ and $\Sigma \in \mathrm{Chamb}(L)$, we have $M(a\Sigma) = M(\Sigma)$. Hence, symmetric regularity implies regularity.

We suppose for the remainder of this section that (L,q) is a regular simplicial complex over Δ^n.

2.8. Let (W,X) be the Coxeter system associated to the Coxeter matrix M(L). In (1.19) we defined a transitive nonsimplicial action of G(X) on Chamb(L). It follows from formula (1.20.1) that this action factors through W. Hence, W acts transitively (from the right) on Chamb(L). For any $\Sigma \in \mathrm{Chamb}(L)$, let $\pi(L,\Sigma)$ denote the isotropy subgroup at Σ. We call $\pi(L,\Sigma)$ the group of L at Σ. As a right W-set, Chamb(L) is isomorphic to $\pi(L,\Sigma)\backslash W$. Hence, for each $w \in W$, $\pi(L,\Sigma w) = w^{-1}\pi(L,\Sigma)w$. In other words, as Σ ranges over Chamb(L), $\pi(L,\Sigma)$ ranges over the conjugacy class of a subgroup in W.

Next we state a few easy consequences of theorem 1.22.

PROPOSITION 2.9. For i = 1, 2, suppose that (L_i,q_i) is a regular simplical complex over Δ^n of type M and that $\Sigma_i \in \text{Chamb}(L_i)$. Let (W,X) be the Coxeter system associated to M. Then there is a simplicial map $f : L_1 \dashrightarrow L_2$ over Δ^n such that $f(\Sigma_1) = \Sigma_2$ if and only if $\pi(L_1,\Sigma_1)$ is a subgroup of $\pi(L_2,\Sigma_2)$. Moreover, f is an isomorphism if and only if $\pi(L_1,\Sigma_1) = \pi(L_2,\Sigma_2)$.

COROLLARY 2.10. For i = 1, 2, suppose that (L_i,q_i) is a regular simplicial complex over Δ^n and that $\Sigma_i \in \text{Chamb}(L_i)$. Then (L_1,q_1) and (L_2,q_2) are isomorphic over Δ^n if and only if $M(L_1) = M(L_2)$ and $\pi(L_1,\Sigma_1)$ and $\pi(L_2,\Sigma_2)$ are conjugate subgroups of the associated Coxeter group W.

THEOREM 2.11. Suppose that (L,q) is a regular simplicial complex over Δ^n and that $\Sigma \in \text{Chamb}(L)$.

(i) The automorphism group, Aut(L,q), is isomorphic to $N(\pi)/\pi$, where $\pi = \pi(L,\Sigma)$, and $N(\pi)$ denotes the normalizer of π in W.

(ii) (L,q) is symmetrically regular if and only if π is normal in W (and hence, $\text{Aut}(L,q) \cong W/\pi$).

2.12. We shall say that (L,q) is universal of type M if it has the following property: given any other regular simplicial complex (L_1,q_1) over Δ^n of type M and chambers $\Sigma \in \text{Chamb}(L)$ and $\Sigma_1 \in \text{Chamb}(L_1)$, there exists a unique $f : L \dashrightarrow L_1$ which is a simplicial map over Δ^n and which takes Σ to Σ_1. Clearly, a universal regular simplicial complex over Δ^n of type M is unique up to isomorphism over Δ^n.

Consider the following two conditions on a regular simplicial complex (L,q) of type M.

(i) $\pi(L,\Sigma)$ is the trivial group.

(ii) $\text{Aut}(L,q) \cong W$.

From (2.9) and (2.11) we see, just as in the theory of covering spaces, that these two conditions are equivalent and that either implies that (L,q) is universal.

2.13. We turn now to the question of existence of such universal complexes. Let $M = (m_{ij})_{0 \le i,j \le n}$ be a Coxeter matrix and (W,X) the associated Coxeter system. It turns out that the universal complex of type M coincides with the well-known "Coxeter complex" of (W,X). We shall now recall its construction. For any proper subset I of $\{0,1,\cdots,n\}$, let W_I denote the subgroup of W generated by X_I, where $X_I = \{x_i \in X \mid i \in I\}$. Consider the simplicial complex $W \times \Delta^n$. Define an equivalence relation $\sim$ on $W \times \Delta^n$ by $(w,\sigma) \sim (w',\sigma') \iff$ the simplices σ and σ' are equal and $w^{-1}w' \in W_I$, $I = \text{Type}(\sigma)$. ($\text{Type}(\sigma)$ is defined in (1.13).) The quotient of $W \times \Delta^n$ by $\sim$ is a simplicial complex which we shall denote by U_M. (U_M is the Coxeter complex.) (The geometric realization of U_M is the space formed by pasting together n-simplices, one for each element of W, in the obvious manner.) For the moment, write $[w,\sigma]$ for the simplex in U_M which is the equivalence class of (w,σ). We can identify Δ^n with the subcomplex in U_M consisting of all simplices of the form $[1,\sigma]$, $\sigma \in \Delta^n$. The natural projection $W \times \Delta^n \dashrightarrow \Delta^n$ induces a projection $p : U_M \dashrightarrow \Delta^n$. Thus, U_M is a simplicial complex over Δ^n. Define an action (from the left) of W on U_M by $w[w',\sigma] = [ww',\sigma]$. It is clear that W acts as a group of automorphisms over Δ^n and that W acts transitively on $\text{Chamb}(U_M)$ ($= \{w\Delta^n\}_{w \in W}$); hence, $W = \text{Aut}(U_M,p)$ and U_M is symmetrically regular. Thus, we have proved the following result.

2.13.1. The Coxeter complex U_M equipped with its canonical projection $p : U_M \dashrightarrow \Delta^n$ is the universal regular simplicial complex over Δ^n of type M.

Here is another property of U_M.

2.13.2. If W is finite, then U_M is homeomorphic to S^n. If W is infinite, then U_M is contractible. (For a proof, see [11], p. 108, or [5].)

2.14. If π is any subgroup of W, then U_M/π is naturally a simplicial complex over Δ^n; however, the quotient may fail to satisfy condition (1.10). (For example, if $\pi = W$, then $U_M/\pi \cong \Delta^n$, which does not satisfy (1.10).) Even if the quotient does satisfy (1.10), it might still fail to be regular of type M. We shall now determine the conditions on the subgroup π so that the quotient is regular of type M.

For $0 \le i, j \le n$, $i \neq j$, let W_{ij} denote the subgroup of W generated by $\{x_i,x_j\}$. Then W_{ij} is a dihedral group of order $2m_{ij}$.

2.14.1. If (L,q) is regular of type M, then for each Σ in Chamb(L), the intersection of $\pi(L,\Sigma)$ and W_{ij} is the trivial group. (To see this, note that the subgroup W_{ij} acts freely and transitively on $\text{Chamb}(L_{ij}(\Sigma))$, where $L_{ij}(\Sigma)$ is defined in (1.18.1); hence, the intersection of W_{ij} with the isotropy subgroup $\pi(L,\Sigma)$ must be trivial.)

Consider the following condition on a subgroup π of W.

2.14.2. For $0 \leq i, j \leq n$, $i \neq j$, and for each $w \in W$, the intersection of $w^{-1}\pi w$ and W_{ij} is trivial.

It is now easy to see that we have the following result.

LEMMA 2.14.3. Let M be a Coxeter matrix and π a subgroup of W. Then U_M/π is a regular simplicial complex over Δ^n of type M if and only if π satisfies (2.14.2).

COROLLARY 2.15. Let M be a Coxeter matrix and (W,X) the associated Coxeter system. Then the set of isomorphism classes of regular simplicial complexes over Δ^n of type M is naturally bijective with the set of conjugacy classes of subgroups π of W such that π satisfies (2.14.2).

PROPOSITION 2.16. Let M be a Coxeter matrix. Then there exists a finite regular simplicial complex over Δ^n of type M if and only if each entry of M is $< \infty$.

Proof. If (L,q) is of type M and L is a finite complex, then obviously each m_{ij} is $< \infty$. Conversely, suppose each m_{ij} is $< \infty$. Since the associated Coxeter group W has a faithful representation in $GL(n+1,\mathbb{R})$, it is virtually torsion-free (cf. [10]). Hence, there is a torsion-free subgroup π of finite index in W. Since W_{ij} is finite and π is torsion-free the subgroup π satisfies (2.14.2). The proposition follows.

2.17. Next let us consider the action of the automorphism group W on U_M. For any proper subset I of $\{0,1, \cdots ,n\}$ let (W_I, X_i) be as in (2.13). The maximal proper subsets of $\{0,1, \cdots ,n\}$ are of the form $I(i) = \{0,1, \cdots ,n\} - \{i\}$ for some i, with $0 \leq i \leq n$. To simplify notation, put $X_{(i)} = X_{I(i)}$ and $W_{(i)} = W_{I(i)}$. The isotropy subgroup at a simplex σ in Δ^n $(\Delta^n \subset U_M)$ is $W_{Type(\sigma)}$. Similarly, the isotropy subgroup at $w\sigma$, $w \in W$, is $wW_{Type(\sigma)}w^{-1}$. It is easy to see that W acts properly on the geometric realization of U_M if and only if each isotropy subgroup is finite. In other words,

2.17.1. W acts properly on the geometric realization of U_M if and only if for each proper subset I of $\{0,1, \cdots ,n\}$, the group W_I is finite.

Since each maximal subgroup of the form W_I is of the form $W_{(i)}$, we can rephrase (2.17.1) as follows: W acts properly on the geometric realization of U_M if and only if it satisfies the finiteness condition (FC) below.

(FC) For $0 \leq i \leq n$, the group $W_{(i)}$ is finite.

Next, suppose that π is any subgroup of W. The isotropy subgroup of π at a simplex $w\sigma$ in U_M is $\pi \cap wW_{Type(\sigma)}w^{-1}$. Hence, the π-action on U_M is free and if and only if each of these intersections is the trivial group. Moreover, if this condition holds then it is also easy to see that the π-action is proper. Thus, π acts freely and properly on U_M if and only if it satisfies the following condition.

2.17.2. For $0 \leq i \leq n$ and for each $w \in W$, $w^{-1}\pi w \cap W_{(i)}$ is the trivial group.

LEMMA 2.18. Suppose that M is a Coxeter matrix, W is the associated Coxeter group, and that π is a subgroup of W.

(i) The natural map $q : U_M \to U_M/\pi$ is a topological covering if and only if π satisfies (2.17.2).

(ii) If π satisfies (2.17.2), then the fundamental group of U_M/π is isomorphic to π, (provided $n \geq 2$).

Proof. Statement (i) follows from the remarks in (2.17). Statement (ii) follows from the fact that U_M is simply connected (cf. (2.13.2)), if $n \geq 2$.

2.19. Let M be a Coxeter matrix and C_M $(= (c_{ij})_{0\leq i,j\leq n})$ its associated cosine matrix (cf. (2.2)). Let $C_{(i)}$ be the n by n matrix obtained from C_M by deleting the ith row and i^{th} column. We suppose that, for $0 \leq i \leq n$, $C_{(i)}$ is positive definite. Then C must be one of the following three types: positive definite (type (+1)), positive semi-definite with 1-dimensional null-space (type (0)), or nondegenerate and indefinite of signature (n,1) (type (-1)). We shall say that M is of <u>type ε</u>, $\varepsilon \in \{+1,0,-1\}$, if C_M is of type ε.

2.20. The complete simply connected Riemannian-manifolds, $n \geq 2$, of constant sectional curvature are the n-sphere, Euclidean n-space, and hyperbolic n-space (denoted H^n). Let Y^n_ε stand for S^n, $\mathbb{R}^n$, or H^n as $\varepsilon = +1$, 0, or -1.

Let $(y_0, \cdots, y_n)$ be linear coordinates on $\mathbb{R}^{n+1}$. For $\varepsilon \in \{+1,0,-1\}$, let q_ε be the quadratic form defined by $q_\varepsilon(y) = (y_0)^2 + \cdots + (y_{n-1})^2 + \varepsilon(y_n)^2$, and let $B_\varepsilon(\ ,\ )$ be the symmetric bilinear form associated to q_ε. For $\varepsilon = \pm 1$, we can identify Y^n with the hypersurface $q^n_\varepsilon(y) = \varepsilon$ in $\mathbb{R}^{n+1}$. (When $\varepsilon = -1$, we also require $y_n > 0$.) We identify Y^n_0 with the affine hyperplane $y_n = 1$. For $y \in Y^n_\varepsilon$, the tangent space of Y^n_ε can be identified with a linear hyperplane in $\mathbb{R}^{n+1}$; for $\varepsilon = \pm 1$, this is the hyperplane which is

B_ε-orthogonal to y. The Riemannian inner product on the tangent space, $T_y(Y^n_\varepsilon)$, is obtained by restricting B to the corresponding hyperplane. It follows that we can identify the group of isometries of Y^n_ε, denoted $\mathrm{Isom}(Y^n_\varepsilon)$, with the subgroup of $GL(n+1,\mathbb{R})$ which preserves q_ε.

By a <u>hyperplane</u> in Y^n_ε we shall mean the intersection of the linear hyperplane in $\mathbb{R}^{n+1}$ with Y^n_ε. (Such a hyperplane in Y^n_ε is a complete, totally geodesic submanifold of codimension one.) It follows that an isometric reflection on Y^n_ε across a hyperplane is the restriction to Y^n_ε of a linear reflection on $\mathbb{R}^{n+1}$.

2.21. Suppose that (W,X) is a Coxeter system of rank n+1. We are interested in finding a representation $\theta : W \to \mathrm{Isom}(Y^n_\varepsilon)$ satisfying the following condition.

2.21.1. The group $\theta(W)$ acts as a (discrete) reflection group on Y^n_ε (cf. [5]). Moreover, there is a convex n-simplex Σ in Y^n_ε with codimension-one faces denoted by $\Sigma_{(i)}$, $0 \le i \le n$, such that Σ is a fundamental chamber for $\theta(W)$ on Y^n_ε and such that for each $x_i \in X$, $\theta(x_i)$ is the reflection across the hyperplane supported by $\Sigma_{(i)}$.

We shall say that $\Sigma_{(i)}$ has type $\{i\}$. More generally, for any proper subset I of $\{0,1,\cdots,n\}$, the simplex $\Sigma_I = \cap_{i\in I}\Sigma_{(i)}$ is said to be of type I. This gives a isomorphism from Σ to the standard simplex Δ^n.

If W acts on Y^n_ε as in (2.21.1), then the translates of Σ by W give a triangulation of Y^n_ε by convex simplices. It is well-known that the underlying simplicial complex of this triangulation can be identified with the Coxeter complex U_M, where M is the Coxeter matrix associated to (W,X). (See, for example, Prop. 15.1, p. 318 in [5].)

The following result is classical (e.g., see [1] or [13]).

THEOREM 2.22. Let (W,X) be a Coxeter system of rank n+1 and C its associated cosine matrix. A necessary and sufficient condition for there to be a representation $\theta : W \to \mathrm{Isom}\,(Y^n_\varepsilon)$, $\varepsilon \in \{+1,0,-1\}$, as in (2.21.1) is that C be of type ε. Moreover, if θ exists and we regard it as a representation into $GL(n+1,\mathbb{R})$, then it is equivalent to the dual of the canonical representation (cf. (2.5)).

Proof. First we show necessity. Let Σ be as in (2.21.1). For $0 \le i \le n$, let u_i be the unit normal to $\Sigma_{(i)}$ which points inward. Since $\Sigma_{(i)}$ and $\Sigma_{(j)}$ must make a dihedral angle of π/m_{ij}, we have that

$$B_\varepsilon(u_i,u_j) = \langle u_i,u_j\rangle = -\cos(\pi/m_{ij}) = c_{ij}.$$

In other words, the cosine matrix C can be identified with the matrix of inner products $(B_\varepsilon(u_i,u_j))_{0\leq i,j\leq n}$. Let v_i denote the vertex of Σ of type $\{0,1,\cdots,n\} - \{i\}$. The set $\{u_j\}_{j\neq i}$ is a basis for $T_{v_i}(Y^n_\varepsilon)$. Hence, the matrix of dot products of this basis is positive definite; i.e., $C_{(i)}$ is positive definite. If $\varepsilon = \pm 1$, then $\{u_0,\cdots,u_n\}$ is a basis for $\mathbb{R}^{n+1}$ Since C is the matrix representation for B_ε with respect to this basis, it must be of Type(ε). The case $\varepsilon = 0$ also follows easily.

To prove the converse, suppose C is of Type(ε). To simplify the discussion, suppose $\varepsilon \neq 0$. Then we can find a basis $\{u_0,\cdots,u_n\}$ for $\mathbb{R}^{n+1}$ such that $B_\varepsilon(u_i,u_j) = c_{ij}$. The intersection of the half-spaces $B_\varepsilon(u_i,y) \geq 0$ with Y^n_ε is a simplex Σ. For each $x_i \in X$ let $\theta(x_i) \in \mathrm{Isom}(Y^n_\varepsilon)$ be the reflection across the hyperplane $B_\varepsilon(u_i,y) = 0$ given by $z \dashrightarrow z - 2B_\varepsilon(u_i,z)$. The map $\theta : X \to \mathrm{Isom}(Y^n_\varepsilon)$ extends to a representation $\theta : W \to \mathrm{Isom}(Y^n_\varepsilon)$ as in (2.21.1). Moreover, θ is obviously equivalent to the canonical representation (which is self-dual since B is nondegenerate). The argument must be modified somewhat when $\varepsilon = 0$. (See Ch. V, 4.9 of [1].)

COROLLARY 2.23. Let (W,X) be a Coxeter system and C its cosine matrix.

(i) The group W is finite if and only if C is positive define.

(ii) (W,X) satisfies condition (FC) (cf. (2.17)) if and only if C is of type ε for some $\varepsilon \in \{+1,0,-1\}$.

2.24. Suppose that M is a Coxeter matrix of degree (n+1) and of type ε. We shall say that the corresponding Coxeter complex U_M is a classical regular triangulation of Y^n_ε over Δ^n. More generally, if π is a subgroup of W such that each conjugate of π intersects $W_{(i)}$ trivially (i.e., if π satisfies (2.17.2)), then U_M/π is called a classical regular triangulation of Y^n_ε/π over Δ^n.

LEMMA 2.25. Suppose that (L,q) is a regular simplicial complex over Δ^n of type M, $n \geq 2$, and that (W,X) is the associated Coxeter system. Also, suppose that L is a finite complex. The following statements are equivalent.

(i) L is simply connected and satisfies (1.11.1).

(ii) L is PL-homeomorphic to S^n.

(iii) W is finite and the natural projection $p : U_M \to L$ is an isomorphism.

The implication (ii) $\Rightarrow$ (iii) means that any regular triangulation of S^n over Δ^n is combinatorially equivalent to a classical one.

Proof. It is fairly clear that (iii) $\Rightarrow$ (ii) $\Rightarrow$ (i). (The implication (iii) $\Rightarrow$ (ii) follows from theorem (2.22) and from (2.13.1).)

Suppose, by induction, that the implication (i) $\Rightarrow$ (iii) holds in all dimensions ≥ 2 and $< n$. Let $\Sigma \in$ Chamb(L) and put $\pi = \pi(L,\Sigma)$. For $0 \leq i \leq n$, let v_i be the vertex of Σ which projects to $\{i\}$ in Δ^n. The complex Link(v_i,L) is a regular simplicial complex over Δ^{n-1} ($\Delta^{n-1} \cong$ Link$(\{i\},\Delta^n)$); its associated Coxeter system is $(W_{(i)}, X_{(i)})$. Since L satisfies (1.11.1) so does Link(v_i,L). If $n > 2$, it follows from the inductive hypothesis that $W_{(i)} \cap \pi = \{1\}$. If $n = 2$, the same conclusion follows from (2.14.3). Thus, π satisfies (2.17.2). By Lemma (2.18) (i), $p : U_M \to L$ is a covering projection. Since L is simply connected, p must be an isomorphism (i.e., π is trivial). Since L (and hence, U_M) is a finite complex and since W is bijective with Chamb(U_M), the group W is finite. Thus, (iii) holds.

THEOREM 2.26. Suppose that (L,q) is a regular simplicial complex over Δ^n of type M and that (W,X) is the associated Coxeter system. The following statements are equivalent.

(i) L satisfies (1.11.1).

(ii) L is a PL-manifold.

(iii) (W,X) satisfies condition (FC) and the natural projection $p : U_M \to L$ is a topological covering.

(iv) M is of type ε, for some $\varepsilon \in \{+1,0,-1\}$, and L is equivalent to a classical regular triangulation of Y^n_ε/π (where $\pi \cong \pi(L,\Sigma)$).

In other words, if L is a PL-manifold, then it is isomorphic to a classical triangulation of a complete manifold of constant sectional curvature.

Proof. It follows from theorem 2.22 that (iii) $\Leftrightarrow$ (iv). Obviously, (iv) $\Rightarrow$ (ii) $\Rightarrow$ (i). We shall show that (i) $\Rightarrow$ (iii). Suppose that L satisfies (1.11.1). Let $\Sigma \in$ Chamb(L), $\pi = \pi(L,\Sigma)$, and for $0 \leq i \leq n$, let v_i be the vertex of Σ which projects to $\{i\} \in \Delta^n$. Since Link(v_i,L) is simply connected, it follows from the previous lemma, that $W_{(i)}$ is finite and that $W_{(i)} \cap \pi = \{1\}$. In other words, (W,X) satisfies (FC) and π satisfies 2.17.2. By lemma (2.18), $p : U_M \to L$ is a covering. Thus, (iii) holds.

3. REGULAR CONVEX CELL COMPLEXES

3.1. Suppose that K is an n-dimensional convex cell complex satisfying (1.10). Then K' is a simplicial complex over Δ^n (cf. (1.23)) and K' satisfies (1.10)(cf. (1.26)). We shall say that K is a regular convex cell complex if (K',p) is a regular simplicial complex over Δ^n.

3.2. Here is an equivalent definition. In (1.27), we associated to each Σ in Chamb(K') an n-tuple $(m_1(\Sigma), \cdots, m_n(\Sigma))$, where $m_i(\Sigma)$ is an integer ≥ 3 for $1 \leq i \leq n-1$, and where $m_n(\Sigma)$ is either an integer ≥ 3 or ∞. The Coxeter matrix $(m_{ij}(\Sigma))_{0 \leq i,j \leq n}$ is given by (1.28.1). It follows that K is regular if and only if the function from Chamb(K') to n-tuples is constant; it is then denoted by $(m_1(K), \cdots, m_n(K))$ (or simply by $(m_1, \cdots, m_n)$) and called the Schläfli symbol of K.

3.3. An n-dimensional convex cell complex K satisfying (1.10) is symmetrically regular if Aut(K) acts transitively on Chamb(K'). Since Aut(K) = Aut(K',p) (cf. (1.25)), we see that K is symmetrically regular if and only if (K',p) is symmetrically regular over Δ^n. It follows from (2.7) that the symmetric regularity of K implies its regularity.

From now on we suppose that K is a regular convex cell complex of dimension n.

3.4. The Coxeter matrix of K is the Coxeter matrix $M = (m_{ij})_{0 \leq i,j \leq n}$ associated to (K',p). It is related to the Schläfli symbol by formula (1.28.1) which we restate here:

$$m_{ij} = \begin{cases} 1 & , \quad \text{if } i = j \\ 2 & , \quad \text{if } |i - j| \geq 2 \\ m_k & , \quad \text{if } \{i,j\} = \{k-1,k\} \end{cases} \qquad (3.4.1)$$

It follows that the Coxeter diagram of M is a connected line segment. Similarly, the Coxeter system of K is the Coxeter system (W,X) associated to M.

Using (1.23), (2.10) and (2.11) can be translated as follows.

PROPOSITION 3.5. For i = 1,2, suppose that K_i is a regular convex cell complex and that $\Sigma_i \in \mathrm{Chamb}(K_i')$. Then K_1 and K_2 are combinatorially equivalent if and only if their Schläfli symbols are equal and $\pi(K_1',\Sigma_1)$ and $\pi(K_2',\Sigma_2)$ are conjugate subgroups of the associated Coxeter group.

PROPOSITION 3.6. (i) Aut(K) is isomorphic to $N(\pi)/\pi$, where $\pi = \pi(K',\Sigma)$ and $N(\pi)$ is the normalizer of π in W.

(ii) K is symmetrically regular if and only if π is normal in W (and hence, $\mathrm{Aut}(K) \cong W/\pi$).

3.7. A convex (n+1)-cell E is combinatorially regular if its boundary complex, denoted by ∂E, is regular in the sense of (3.1). A convex (n+1)-cell E in $\mathbb{R}^{n+1}$ is a classical regular convex polyhedron, if its group of isometric symmetries, denoted by Isom(E), acts transitively on $\mathrm{Chamb}((\partial E)')$. If E is a classical regular polyhedron, then radial projection from the center gives a tessellation of S^n by convex cells such that Isom(E) acts transitively on the chambers in its barycentric subdivision. We call such a tessellation a classical regular tessellation of S^n, by convex cells. Conversely, given a classical regular tessellation of S^n, by taking the convex hull of its vertex set, we obtain a classical regular polyhedron in $\mathbb{R}^{n+1}$.

3.8. Suppose that K is a classical regular tessellation of S^n by convex cells. Let (W,X) be the associated Coxeter system, then $W \cong \mathrm{Aut}(K)$ and we can identify W with a finite reflection group in $\mathrm{Isom}(S^n)$ $(= O(n+1))$. The Coxeter diagram of (W,X) is a line segment. Conversely, there is the following classical result (cf. [3]).

THEOREM 3.9. Let (W,X) be a Coxeter system with Coxeter matrix $M = (m_{ij})_{0\le i,j\le n}$. Suppose that W is finite and that the Coxeter diagram of M is a line segment. For $1 \le k \le n$, put $m_k = m_{ij}$, where $\{i,j\} = \{k-1,k\}$. Then there is a classical regular tessellation K of S^n by convex cells with Schläfli symbol $(m_1, \cdots, m_n)$. Moreover, K is unique up to isometries of S^n.

Proof. By theorem (2.22), there is an orthogonal action of W on S^n such that the corresponding triangulation can be identified with U_M. We want to show that the spherical simplices in S^n can be assembled into convex cells in S^n giving a convex cell complex K, with $K' = U_M$. Since the diagram of W is a line segment, the diagram of $W_{(i)}$ is either a line segment (if i = 0 or n) or two line segments (if $1 \le i \le n-1$). It follows that $W_{(i)} = G_i \times H_i$ where G_i is the subgroup generated by $x_0, \cdots, x_{i-1}$ and H_i is the subgroup generated by $x_{i+1}, \cdots, x_n$. Let Σ be a chamber in S^n as in (2.21.1) (so that Σ is a spherical n-simplex) and for $0 \le i \le n$, let v_i be the vertex of Σ which projects to $\{i\} \in \Delta^n$. Let σ_i be the face of σ spanned by $\{v_0, \cdots, v_i\}$. Put

$$E_i = \bigcup_{\omega \in G_i} \omega\sigma_i .$$

We claim that, for $0 \leq i \leq n$,

(a) E_i is a convex i-cell in the i-sphere fixed by H_i.

(b) The triangulation of E_i by the i-simplices $\omega\sigma_i$, $\omega \in G_i$, is the barycentric subdivision.

Once we have established (a) and (b) it then follows easily that the translates of these cells under W give a classical regular tessellation K of S^n and that $K' \cong U_M$. Let S^i be the i-sphere fixed by H_i. Obviously, $\sigma_i \subset S^i$. Since the G_i- and H_i-actions commute, we also have $E_i \subset S^i$. The convexity of E_i is then clear, i.e., (a) holds. Suppose, by induction, that the triangulation of E_i by translates of σ_i is the barycentric subdivision of E_i for $i < n$. (The case, $i = 0$, is trivial.) In particular, this holds for E_{n-1} and all of its translates under G_n. But the union of these translates in ∂E_n. Also, v_n is the central point of E_n. Therefore, (b) holds. The uniqueness of K follows from the last sentence of theorem (2.22).

3.10. The theorem above shows that the classification of classical regular convex polyhedra follows from the classification of finite Coxeter groups. Given a finite Coxeter group with diagram a line segment, we obtain a Schläfli symbol $(m_1, \cdots, m_n)$ by the formula $m_i = m_{ij}$, where $j = i-1$. Since we get the same diagram if we reverse the order of the indices, this Schläfli symbol is only well-defined up to reversing the order of the m_i's. Thus, a given diagram corresponds to classical regular polyhedron and its dual. The regular polyhedron is self-dual if and only if $(m_1, \cdots, m_n) = (m_n, \cdots, m_1)$

The diagram $\circ\overset{p}{—}\circ$, $p \geq 3$, has Schläfli symbol (p) corresponding to the regular p-gon which is self-dual. The diagram $\circ—\circ \cdots \circ—\circ$ (A_n) has Schläfli symbol $(3,3,\cdots,3)$ corresponding to the regular n-simplex (self-dual). The diagram $\circ\overset{4}{—}\circ \cdots \circ—\circ$ (B_n) has Schläfli symbol $(4,3,\cdots,3)$ or $(3,3,\cdots,4)$ corresponding to the regular n-cube or regular n-octahedron, respectively. The diagram $\circ\overset{5}{—}\circ—\circ$ (H_3) has symbol (5,3) or (3,5) corresponding to the dodecahedron or icosahedron. The diagram $\circ\overset{5}{—}\circ—\circ—\circ$ (H_4) has symbols (5,3,3) or (3,3,5); the corresponding regular 4-dimensional polyhedra are called, by Coxeter, the "120-cell" and the "600-cell" respectively. The diagram $\circ—\circ\overset{4}{—}\circ—\circ$ (F_4) has symbol (3,4,3) and corresponds to the self-dual regular 4-dimensional polyhedron called the "24-cell" by Coxeter. Since these are the only finite Coxeter groups with diagram a line segment (see Table 1), we have listed above, all of the classical regular polyhedra.

Combining theorem (3.9) with lemma (2.25) we get the following result.

PROPOSITION 3.11. Suppose that K is a regular convex cell complex and that K is PL-homeomorphic to S^n. Then K is equivalent to a classical regular tessellation of S^n by convex cells.

COROLLARY 3.12. Suppose that E is a combinatorially regular convex cell. Then E is combinatorially equivalent to a classical regular convex polyhedron.

This corollary was proved by McMullen [9], under the slightly stronger hypothesis of combinatorial symmetric regularity. (See Chapter V of [8] for a somewhat different argument.)

3.13. Let $(m_1, \cdots, m_n)$ be an n-tuple of numbers. The <u>initial part</u> (resp. <u>final part</u>) of this n-tuple is the (n-1)-tuple $(m_1, \cdots, m_{n-1})$ (resp. $(m_2, \cdots, m_n)$). The initial part (resp. final part) is <u>spherical</u> if it is the Schläfli symbol of a classical regular convex polyhedron of dimension n. We shall say that $(m_1, \cdots, m_n)$ is an <u>admissible Schläfli symbol</u> if its initial part is spherical.

PROPOSITION 3.14. Let K be a regular convex cell complex with Schläfli symbol $(m_1, \cdots, m_n)$.

(i) Each k-cell in K, $0 \leq k \leq n$, is combinatorially regular and has Schläfli symbol $(m_1, \cdots, m_{k-1})$.

(i)' The link of each k-cell in K, $0 \leq k \leq n-2$, is a regular convex cell complex with Schläfli symbol $(m_{k+2}, \cdots, m_n)$.

(ii) Any two k-cells in K are combinatorially equivalent.

(ii)' Put $\pi = \pi(K', \Sigma)$. If each conjugate of π has trivial intersection with H_k (defined in the proof of Theorem (3.9)), $0 \leq k \leq n-2$, then the links of any two k-cells in K are combinatorially equivalent.

(iii) The Schläfli symbol of K is admissible.

Proof. Statements (i) and (i)' are immediate from the definitions. It follows from corollary (3.12) that any k-cell in K is classically regular with automorphism group G_k (defined in the proof of (3.9)). It follows that the group of such a k-cell is trivial; thus, statement (ii) follows from (i) and proposition (3.5). Similarly, (ii)' follows from (i)' and proposition (3.5). Statement (iii) follows from (i) (when $k = n$) and Corollary (3.12).

LEMMA 3.15. Suppose that (W,X) is the Coxeter system of K. Let $\Sigma \in \mathrm{Chamb}(K')$ and put $\pi = \pi(K',\Sigma)$. (π is a subgroup of W.) Then π satisfies the following two conditions:

3.15.1. Each conjugate of π has trivial intersection with $W_{(n)}$. (Recall that $W_{(n)}$ is the subgroup of W generated by $\{x_0, \cdots, x_{n-1}\}$.)

3.15.2. Each conjugate of π has trivial intersection with $W_{n-1,n}$, the subgroup of W generated by $\{x_{n-1}, x_n\}$.

Proof. The group $W_{(n)}$ is the Coxeter group associated to an n-cell in K. Since such an n-cell is classically regular (cf. (3.12)) we have condition (3.15.1). Condition (3.15.2) is a special case of (2.14.1) which holds for (K',p).

3.16. Let $(m_1, \cdots, m_n)$ be an admissible Schläfli symbol, M the associated Coxeter matrix, (W,X) the associated Coxeter system, and U_M the Coxeter complex. Let v_n be a vertex in U_M which projects to $\{n\} \in \Delta^n$. Then $\mathrm{Link}(v_n,U_M)$ is the Coxeter complex of the finite Coxeter group $W_{(n)}$. It follows that $\mathrm{Star}(v_n,U_M)$ can be identified with the barycentric subdivision of a regular n-cell, the Schläfli symbol of which is $(m_1, \cdots, m_{n-1})$. By assemblying the chambers of U_M which meet at each vertex of type $\{0, \cdots, n-1\}$ in this fashion, we obtain a convex cell complex $U(m_1, \cdots, m_n)$, the derived complex of which is U_M.

THEOREM 3.17. (i) Let K be a regular convex cell complex with Schläfli symbol $(m_1, \cdots, m_n)$ and let $\pi(= \pi(L,\Sigma))$ be its group. Then K is combinatorially equivalent to $U(m_1, \cdots, m_n)/\pi$.

(ii) Let $(m_1, \cdots, m_n)$ be an admissible Schläfli symbol, (W,X) the associated Coxeter system, and π a subgroup of W satisfying (3.15.1) and (3.15.2). Then $U(m_1, \cdots, m_n)/\pi$ is a regular convex cell complex with symbol $(m_1, \cdots, m_n)$.

Proof. Statement (i) follows from (2.13.1). Consider (ii). The fact that π satisfies (3.15.1) means that each cell in $U(m_1, \cdots, m_n)$ has trivial stabilizer in π. It follows that $U(m_1, \cdots, m_n)/\pi$ is a convex cell complex and that for each chamber Σ in its derived complex $m_i(\Sigma) = m_i$, $0 \le i \le n-1$. Condition (3.15.2) implies that $m_n(\Sigma) = m_n$ for each Σ. This proves (ii).

COROLLARY 3.18 (compare (2.15)). Let $(m_1, \cdots, m_n)$ be an admissible Schläfli symbol and (W,X) the associated Coxeter system. The set of combinatorial equivalence classes of regular convex cell complexes with symbol $(m_1, \cdots, m_n)$ is naturally bijective with the set of conjugacy classes of subgroups π of W satisfying (3.15.1) and (3.15.2).

Proposition (2.16) can be translated as follows:

PROPOSITION 3.19. Let $(m_1, \cdots, m_n)$ be an admissible Schläfli symbol and suppose that $m_n \neq \infty$. Then there exists a finite symmetrically regular convex cell complex with the given Schläfli symbol.

Proof. If W is finite, then choose π to be the trivial group. If W is infinite, then let π be any torsion-free normal subgroup of finite index in W (such π exist). Then $U(m_1, \cdots, m_n)/\pi$ is the desired finite convex cell complex.

3.20. Let $(m_1, \cdots, m_n)$ be an admissible Schläfli symbol, M the associated Coxeter matrix and (W,X) the associated Coxeter system. Suppose that the final part of $(m_1, \cdots, m_n)$ is also spherical (cf. (3.13)). This means that the subgroups $W_{(n)}$ and $W_{(0)}$ of W are finite. Since the diagram of W is a line segment, this implies that $W_{(i)}$ is finite for $0 \leq i \leq n$. In other words, (W,X) satisfies condition (FC) of (2.17). Therefore, M is of type ε for some $\varepsilon \in \{+1,0,-1\}$, (cf. (2.19)). We shall say that $(m_1, \cdots, m_n)$ is of type ε.

3.21. Suppose that $(m_1, \cdots, m_n)$ is of type ε. In theorem (2.22) we showed that U_M can be identified with a classical regular triangulation of Y^n_ε (cf. (2.24)). The proof of Theorem (3.9) shows that the simplices in Y^n_ε can be assembled into convex cells in Y^n_ε corresponding to the cells of $U(m_1, \cdots, m_n)$. In this way we identify W with a subgroup of $\mathrm{Isom}(Y^n_\varepsilon)$ (unique up to conjugation). We shall say that $U(m_1, \cdots, m_n)$ is a <u>classical regular tessellation of Y^n_ε by convex cells</u>. The Schläfli symbols of type ε are listed in Table 4. (We note from Table 4 that if $\varepsilon = -1$, then $n = 2$, 3, or 4. Thus, hyperbolic n-space admits a regular tessellation by convex cells only when $n = 2$, 3, or 4.) A subgroup π of W satisfies (2.17.2) if and only if it satisfies condition (3.15.1) and the following condition.

3.21.1. Each conjugate of π intersects $W_{(0)}$ trivially.

A subgroup π satisfying (3.15.1) and (3.21.1) acts freely on Y^n_ε. Conversely, any subgroup of W which acts freely on Y^n_ε satisfies (3.15.1) and (3.21.1). We note that if $\varepsilon = 0$ or -1, this means that a subgroup π satisfies (3.15.1) and (3.21.1) if and only if it is torsion-free.

If π satisfies (3.15.1) and (3.21.1), then we shall say that $U(m_1, \cdots, m_n)/\pi$ is a <u>classical regular tessellation of</u> Y^n_ε/π <u>by convex cells</u>.

From theorem (2.26) we immediately get the following result.

THEOREM 3.22. Suppose that K is a regular convex cell complex with Schläfli symbol $(m_1, \cdots, m_n)$ and associated Coxeter system (W,X). The following statements are equivalent.

(i) K satisfies (1.11.1).

(ii) K is a PL-manifold.

(iii) The final part of $(m_1, \cdots, m_n)$ is spherical and K is equivalent to $U(m_1, \cdots, m_n)/\pi$, where $\pi = \pi(K', \Sigma)$.

(iv) The symbol $(m_1, \cdots, m_n)$ is of type ε for some $\varepsilon \in \{+1, 0, -1\}$ and K is equivalent to a classical regular tessellation of Y^n_ε/π by convex cells.

REMARK. This result implies the theorem in the Introduction. It was first proved by Kato [7] by a different method.

EXAMPLES 3.23. (i) Suppose that K is a classical tessellation of S^n by convex cells and that W contains the element -1. ($W \cong \mathrm{Isom}(K) \subset O(n+1)$.) (This always happens except in the case where K is the boundary of a regular $(n+1)$-simplex or in the case where $n = 1$ and K is a polygon with an odd number of sides, cf. [3].) Then $K/\{\pm 1\}$ is a symmetrically regular tessellation of $\mathbb{RP}^n$.

(ii) A $(p,q,2)$-triangle group W gives a classical tesselation of Y^2_ε with Schläfli symbol (p,q), where $\varepsilon = +1$, 0, or -1 as $(p^{-1} + q^{-1} + 2^{-1})$ is greater than, equal to, or less than 1, respectively. For $\varepsilon = 0$ or -1 and π any torsion-free subgroup of finite index in W, the complex K/π is a tessellation of the closed surface Y^n_ε/π. (See [6].)

(iii) Suppose that K is a classical tessellation of $\mathbb{R}^n$ and that π ($\cong \mathbb{Z}^n$) is the subgroup of translations in W. Then K/π is a symmetrically regular tessellation of a torus.

(iv) Let K be the classical tesselation of S^3 with Schläfli symbol $(5,3,3)$ and let K^* be the dual tessellation with symbol $(3,3,5)$. The automorphism group W contains a subgroup π which is isomorphic to the binary icosahedral group and which acts freely on S^3. The quotient S^3/π is Poincaré's homology 3-sphere. It follows that K/π and K^*/π are regular convex cell complexes (at least, in the sense of (1.9)). The complex K/π has only one 3-cell, a dodecahedron; the complex K^*/π is tessellated by 5 icosahedra. These complexes are not symmetrically regular (by theorem (4.7) in the next section).

If K is not required to be a PL-manifold or to satisfy (1.11.1), then nothing prevents one of the above examples from occuring as the link of a cell.

4. SYMMETRIC REGULARITY

The conclusions of theorems (2.26) and (3.22) can be substantially improved if we add the hypothesis of symmetric regularity. (see theorem (4.7) below.)

4.1. A <u>pre-Coxeter system</u> (G,S) of rank n+1 consists of a group G and a set $S = \{s_0, \cdots, s_n\}$ of involutions in G such that S generates G. Associated to a pre-Coxeter system (G,S) there is a Coxeter matrix $M = (m_{ij})_{0 \le i,j \le n}$ given by the formula, $m_{ij} = \text{order}(s_i s_j)$. If (W,X) is the Coxeter system associated to M, then the map $X \to S$ given by $x_i \to s_i$ extends to an epimorphism $\Lambda : W \to G$. Let π denote the kernel of Λ. Let U(G,S) denote the quotient of the Coxeter complex U_M by π. The group G acts on U(G,S) and there is a natural projection $p : U(G,S) \to \Delta^n$. Thus, U(G,S) is a symmetrically regular simplicial complex over Δ^n (cf. (2.7)).

Let R denote the set of conjugates of S in G. Put U = U(G,S). For each r in R, the fixed point set of r on U is denoted by U_r and is called a <u>wall</u> of U.

LEMMA 4.1.1. Suppose that r and r' are elements of R such that $U_r = U_{r'}$. Then r = r'.

Proof. By definition r is conjugate to some element s in S. Choose g in G so that $g^{-1}rg = s$. Also, choose a point x in the relative interior of the (n-1)-simplex $\Delta^n \cap s\Delta^n$. The isotropy subgroup at x is the cyclic group of order two generated by s. Hence, the isotropy subgroup at gx is the cyclic group of order two generated by r. But r' fixes gx (since $gx \in U_{r'}$) and this forces r' = r.

The next result gives a characterization of Coxeter systems among pre-Coxeter systems.

LEMMA 4.2. Suppose that (G,S) is a pre-Coxeter system. The following statements are equivalent.

(i) For each s in S, $U - U_s$ is not connected.

(ii) (G,S) is a Coxeter system.

Moreover, if one of these conditions holds, then $U - U_r$ has exactly two components, for each $r \in R$.

Proof. (i) $\Rightarrow$ (ii). On p. 18 of [1] we find the following result.

LEMMA 4.2.1 (Bourbaki). Let (G,S) be a pre-Coxeter system and let $(P_s)_{s \in S}$ be a family of subsets of G satisfying the following conditions.

(A) For each $s \in S$, $1 \in P_s$.

(B) For each $s \in S$, $P_s \cap sP_s = \phi$.

(C) For elements s,s' in S and g in G, if $g \in P_s$ and $gs' \notin P_s$, then $sg = gs'$.

Then (G,S) is a Coxeter system. Moreover, $P_s = \{g \in G | \ell(sg) > \ell(g)\}$, where $\ell(g)$ denotes the word length of g with respect to the generating set S.

Supposing that (i) holds, we apply this lemma as follows. For each $s \in S$, let P_s denote the set of g in G such that the open chambers $\mathring{\Delta}^n$ and $g\mathring{\Delta}^n$ belong to the same component of $U - U_s$. Condition (A) holds trivially. Since U is connected, s must permute the components of $U - U_s$; hence, $g\mathring{\Delta}$ and $sg\mathring{\Delta}$ are contained in different components of $U - U_s$. This implies (B). To verify (C), suppose $g \in P_s$ and $gs' \notin P_s$. Then $g\mathring{\Delta}^n$ and $gs'\mathring{\Delta}^n$ lie in different components of $U - U_s$. Hence, $\mathring{\Delta}^n$ lie in different components of $U - g^{-1}U_s$. Putting $r = g^{-1}sg$, we have $g^{-1}U_s = U_r$; hence, U_r separates the adjacent chambers Δ^n and $s'\Delta^n$. But $U_{s'}$ is the unique wall with this property. Therefore, $U_r = U_{s'}$. Using Lemma (4.1.1), this implies that $r = s'$, i.e., $g^{-1}sg = s'$. This verifies (C). Consequently, Lemma (4.2.1) shows that (i) $\Rightarrow$ (ii).

(ii) $\Rightarrow$ (i). We suppose that (G,S) is a Coxeter system. Here are some basic facts about Coxeter systems, the proofs of which can be found in [1].

4.2.2. For each $g \in G$ and $r \in R$, the parity of the number of times a gallery in U from Δ^n to $g\Delta^n$ crosses U_r depends only on g and r (and not on the gallery). This gives a mapping $n : G \times R \to \{\pm 1\}$ defined by $n(g,r) = -1$ (resp. +1) if a gallery from Δ^n to $g\Delta^n$ crosses U_r an odd (resp. even) number of times.

4.2.3. If $n(g,r) = -1$ (resp. $+1$), then a minimal gallery from Δ^n to $g\Delta^n$ crosses U_r exactly once (resp. does not cross U_r).

4.2.4. We have $n(g,r) = -1$ if and only if $\ell(rg) < \ell(g)$. From (4.2.2) and (4.2.3), we see that $\mathring{\Delta}^n$ and $g\mathring{\Delta}^n$ (resp. $sg\mathring{\Delta}^n$) belong to the same component of $U - U_s$ if and only if $n(g,s) = +1$ (resp. $n(g,s) = -1$). It follows that $U - U_s$ has exactly two components.

This shows that (ii) $\Rightarrow$ (i) and it also proves the last sentence of lemma (4.2).

4.3. Suppose that (L,q) is a symmetrically regular simplicial complex over Δ^n, that M is its Coxeter matrix, and that (W,X) is its Coxeter system. Put $G = \mathrm{Aut}(L,q)$. Then $G \cong W/\pi$, where π is the normal subgroup $\pi(L,\Sigma)$. Let S denote the image of X in G. Then (G,S) is a pre-Coxeter system, and (W,X) is its associated Coxeter system. Moreover, L and $U(G,S)$ are isomorphic over Δ^n (cf. (2.13.1)).

PROPOSITION 4.4. Let (L,q) be a symmetrically regular simplicial complex over Δ^n of type M. Suppose that L is a PL-manifold and that one of the following two conditions holds:

(a) $\dim L = 1$, or

(b) $H_1(L; \mathbb{Z}/2) = 0$.

Then (G,S) is a Coxeter system and $L \cong U_M$ (and consequently, L is a classical triangulation of Y^n_ε (cf. (2.24)).

Proof. If (G,S) is a Coxeter system, then $W \cong G$ and consequently, π is trivial and $L \cong U_M$. By lemma (4.2), it suffices to prove that for each $s \in S$, $L - L_s$ is not connected. (L_s denotes the fixed point set of s.) If (a) holds, this is obvious. In general, if follows from Smith theory that L_s is a $\mathbb{Z}/2$-homology manifold. Let $\hat{L}_s$ be the component of L_s containing Δ^n_s, the $(n-1)$-face of Δ^n corresponding to s. Then $\hat{L}_s$ is a $\mathbb{Z}/2$-homology manifold of codimension one in L. If $L - \hat{L}_s$ is connected, then the fundamental class in $H^n_c(L; \mathbb{Z}/2)$ must be in the image of $H^{n-1}_c(L_s; \mathbb{Z}/2)$, and hence, the Poincaré dual of $\hat{L}_s$ must represent a nonzero element of $H_1(L; \mathbb{Z}/2)$. It follows that condition (b) implies that $L - \hat{L}_s$ is not connected. This completes the proof.

4.5. The following condition on a n-dimensional simplicial complex L is a weak version of (1.11.1).

4.5.1. For each k-simplex σ in L, with $k \leq n-3$, $H_1(\mathrm{Link}(\sigma,L); \mathbb{Z}/2) = 0$.

LEMMA 4.6 (Compare lemma (2.25)). Suppose that (L,q) is a symmetrically regular simplicial complex over Δ^n, $n \geq 2$, that L is a finite complex, that L satisfies (4.5.1), and that $H_1(L; \mathbb{Z}/2) = 0$. Then L is isomorphic to a classical triangulation of S^n. (In particular, L is PL-homeomorphic to S^n.)

Proof. It is easy to see that since L is symmetrically regular, then so is the link of each simplex in L. Suppose, by induction, that the lemma holds in dimensions $< n$. Since $\mathrm{Link}(\sigma,L)$ satisfies the inductive hypothesis for each k-simplex σ, $k \leq n-3$, $\mathrm{Link}(\sigma,L) = S^{n-k-1}$. Thus, L is a PL-manifold. Since $H_1(L; \mathbb{Z}/2) = 0$, it follows from the previous proposition, that L is isomorphic to the Coxeter complex U_M. Since L is finite, the associated Coxeter group W (= Aut(L,q)) must also be finite. Thus, U_M is a classical triangulation of S^n (cf. (2.22)).

THEOREM 4.7. Suppose that (L,q) is a symmetrically regular simplicial complex over Δ^n, that L is locally finite and that L satisfies (4.5.1). Then L is a PL-manifold. Consequently, L is equivalent to a classical triangulation of Y^n_ε/π for some $\varepsilon \in \{+1,0,-1\}$.

Proof. This follows from the previous lemma and (2.26).

REFERENCES

1. Bourbaki, N., Groupes et Algèbres de Lie, Ch. [IV-VI], Herman, Paris 1968.
2. Coxeter, H.S.M., Regular honeycombs in hyperbolic space, Twelve Geometric Essays, Southern Illinois Univ. Press, Carbondale, Ill., 1968, pp. 200-214.
3. Coxeter, H.S.M., Regular Polytopes, 3rd ed, Dover, New York, 1978.
4. Coxeter, H.S.M., Regular Complex Polytopes, Cambridge Univ. Press, Cambridge, 1973.
5. Davis, M., Groups generated by reflections and aspherical manifolds not covered by Euclidean space, Annals of Math. 117 (1983), 293-324.

6. Edmonds, A. L., Ewing, J. H., and Kulkarni, R. S., Regular tessellations of surfaces and (p,q,2)-triangle groups, Annals of Math. 116 (1982), 113-132.

7. Kato, M., On Combinatorial Space Forms, Scientific Papers of the College of General Education, University of Tokyo 30 (1980), 107-146.

8. Lyndon, R.C., Groups and Geometry, London Math. Society Lecture Note Series, no. 101, Cambridge Univ. Press, Cambridge, 1985.

9. McMullen, P., Combinatorially Regular Polytopes, Mathematika 14 (1967), 142-150.

10. Selberg, A., On Discontinuous Groups in Higher Dimensional Symmetric Spaces, International Colloquim on Function Theory, Tata Institute, Bombay, 1960.

11. Serre, J-P, Cohomologie des groupes discrets, in Prospects in Mathematics, Ann. Math. Studies Vol. 70 (1971), 77-169, Princeton Univ. Press, Princeton.

12. Tits, J., A local approach to buildings, in The Geometric Vein: The Coxeter Festschrift (Edited by C. Davis, B. Grunbaum, and F. A. Sherk), pp. 519-547, Springer-Verlag, New York, Heidelberg, Berlin, 1982.

13. Vinberg, Discrete Linear Groups Generated by Reflectons, Math. USSR Izvestija 5, No. 5 (1971), 1083-1119.

APPENDIX

TABLE 1. Elliptic Coxeter Systems ($\varepsilon = +1$)

The Irreducible Diagrams

A_n

B_n

D_n

$I_2(p)$

H_3

F_4

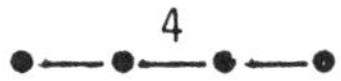

E_6

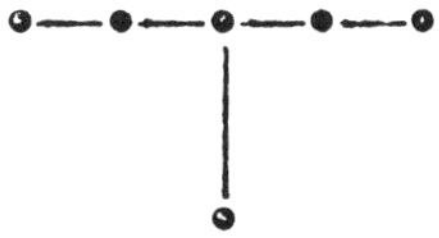

E_7

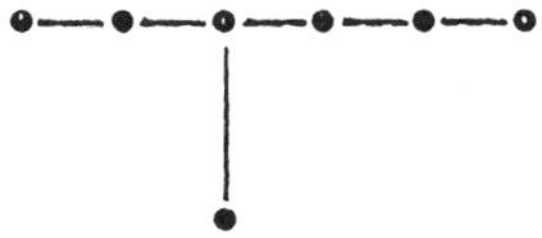

E_8

TABLE 2. Flat Coxeter Systems ($\varepsilon = 0$) with Fundamental Chamber an n-simplex

$\widehat{A}_n (n \geq 2)$	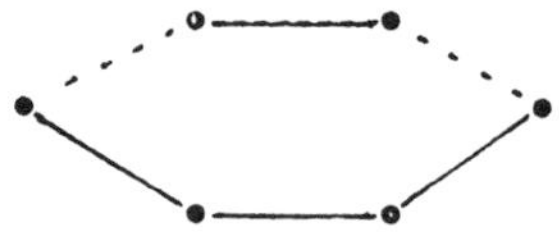
$\widehat{B}_n (n \geq 3)$	
$\widehat{C}_n (n \geq 3)$	
$\widehat{D}_n (n \geq 4)$	
$\widehat{A}_1$	
$\widehat{B}_2$	
$\widehat{G}_2$	
$\widehat{F}_4$	
$\widehat{E}_6$	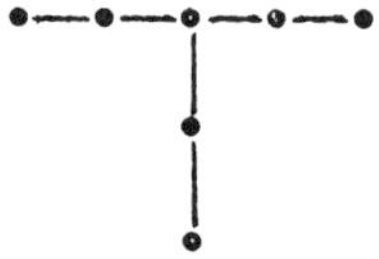
$\widehat{E}_7$	
$\widehat{E}_8$	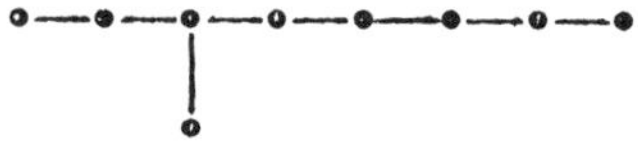

TABLE 3. Hyperbolic Coxeter Systems ($\varepsilon = -1$) with Fundamental Chamber an n-simplex

n = 2

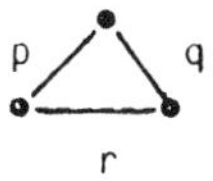

with $(p^{-1} + q^{-1} + r^{-1}) < 1$

n = 3

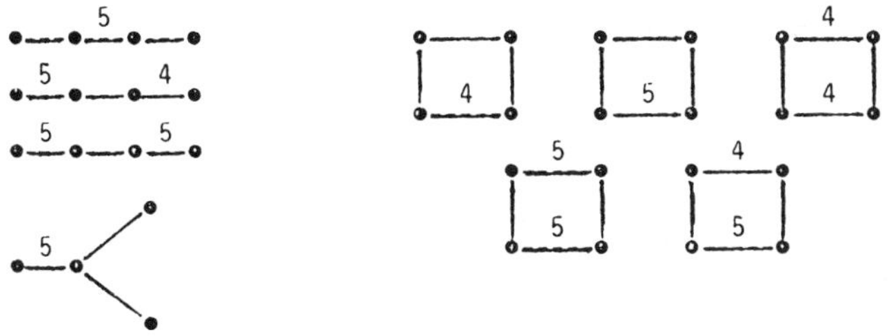

n = 4

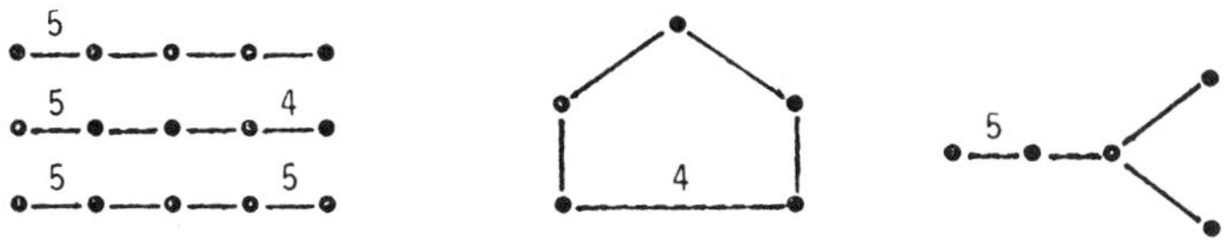

TABLE 4. The Classical Schläfli Symbols (only one is listed for each pair of dual tessellations).

n	ε = +1	ε = 0	ε = -1
1	(p)	(∞)	
2	(3,3) (4,3) (5,3)	(4,4) (6,3)	(p,q) , with $(p^{-1}+q^{-1}+2^{-1}) < 1$
3	(3,3,3) (4,3,3) (5,3,3) (3,4,3)	(4,3,4)	(3,5,3) (5,3,4) (5,3,5)
4	(3,3,3,3) (4,3,3,3)	(4,3,3,4) (3,4,3,3)	(5,3,3,3) (5,3,3,4) (5,3,3,5)
$n \geq 5$	(3,3,...,3) (4,3,...,3)	(4,3,...,3,4)	

ACKNOWLEDGEMENT

My lecture at the conference was on a different topic, namely, on the material in the two preprints, "The homology of a space on which a reflection group acts" and "Some aspherical manifolds".

The author was partially supported by NSF grant DMS8412891.

Gauge Theory and Smooth Structures on 4-Manifolds

S. K. DONALDSON / Mathematical Institute, University of Oxford, Oxford, England

The classical solutions to the first order Yang-Mills equations seems set to become precision tools in problems of 4-manifold topology. It appears quite likely that any general theory of smooth 4-manifolds developed in the future will build in the geometry of gauge fields in a fundamental way. I will review here some of the techniques which have been used and the main results they yield.

It is useful to contrast these techniques, relating non-linear P.D.E.'s and topology, with four aspects of the well understood theory of linear differential operators.

1. If D is a linear elliptic operator over a compact manifold the two equations $Df = 0$, $D^*f = 0$ have finite dimensional solution spaces. In typical examples from geometry these would be the harmonic forms, spinors, etc.

2. The dimensions of Ker D, Ker D* may change if D is varied slightly but the index:

$$\text{ind } D = \dim \text{Ker } D - \dim \text{Ker } D^*$$

is a deformation invariant.

3. If $\{D_t\}_{t \in T}$ is a family of such operators depending upon the parameter t one can define an "index of the family" in K(T). Again, this is unchanged by continuous deformations of the family.

4. There are explicit topological formulae expressing the invariants of (2), (3) in terms of homotopy properties of a bundle constructed from the 'symbol', or highest order part, of D.

Thus, in this linear theory, retaining a small part of the global geometry of the operator yields a rigid invariant, but in principle these can be calculated without using analysis.

Now let $E \to X$ be a bundle with structure group G over a closed, oriented Riemannian 4-manifold. Gauge theory studies connections on E (locally, in trivialisations, these are represented by 1-forms on X with values in the Lie algebra of $\underset{\sim}{G}$). A connection A is anti-self-dual (ASD) if its curvature F_A satisfies:

$$F_A + * F_A = 0 ,$$

where the Hodge *-operator is extended to act on bundle valued 2-forms. We define the moduli space M to be the space of ASD connections (solutions of a P.D.E.) divided by the action of the symmetry group Aut E. If E is a complex line bundle, and $G = S^1$, the classification of the solutions can be read off from the Hodge Theory of harmonic 2-forms. For brevity we will stick to the simplest interesting case when G is the group SU(2). Then the bundle E is determined topologically by an integer $k = c_2(E)[X]$ which is necessarily positive if E admits ASD connections. So we have a family of moduli spaces $M_k(X)$ $(k \geq 0)$ depending on the Riemannian metric used to define the *-operator on X.

For special metrics there are ways to "solve" the ASD equations explicitly and describe the moduli spaces in detail, analogous to finding harmonic forms explicitly in linear theory. Three such examples will illustrate the general pattern:

1. ([2]) If X is S^4 with the standard metric M_0 is a point, corresponding to the unique flat connection, and M_1 is an open 5-ball. In standard conformal co-ordinates, $S^4 \setminus \{\infty\} \cong \mathbb{R}^4$ a 1-parameter family of connections $I_\lambda (\lambda > o)$ in M_1 is given by:

$$I_\lambda = \frac{1}{\lambda^2 + |x|^2} (\theta_1 \underline{i} + \theta_2 \underline{j} + \theta_3 \underline{k})$$

where $\theta_1 = x_1\, dx_2 - x_2 dx_1 - x_3 dx_4 + x_4 dx_3$, etc. The ASD condition is conformally invariant and the extra 4 parameters come from conformal transformations corresponding to translations in $\mathbb{R}^4$.

2. ([4]) If $X = \overline{\mathbb{CP}^2}$ (with standard metric but orientation reversed) then M_1 is an open cone over $\mathbb{CP}^2$. The vertex of the cone corresponds to the standard connection on the Hopf line bundle, viewed as an SU(2) bundle via the inclusion $S^1 \to SU(2)$. The isometries of $\mathbb{CP}^2$ act transitively on points of a fixed distance $t < 1$ from the vertex, and in standard affine co-ordinates the corresponding connection may be written:

$$\frac{1}{(1-t^2) + |x|^2} (\theta_1 \underline{i} + t(\theta_2 \underline{j} + \theta_3 \underline{k}))$$

For $|x| > 0$ the connection-form $\frac{1}{|x|^2} (\theta_1 \underline{i} + \theta_2 \underline{j} + \theta_3 \underline{k})$ represents the flat connection, with zero curvature. Comparing the formulae of (1) and (2) we see that when λ and $(1 - t)$ are small the connections are approximately localized at the origins of the co-ordinate systems, where their curvature is large. Moreover they have approximately the same form in this region if we match $1 - t^2$ with λ^2.

3. $X = \mathbb{CP}^2$ [3], [6] (with any Kähler metric and the usual complex of orientation), M_0 is a point, M_1 is empty and M_2 is the open subset of $\mathbb{CP}^2$ obtained by deleting the image of the map : $\mathbb{CP}^2 \times \mathbb{CP}^2 \to \mathbb{CP}^5 = \mathbb{P}(S^2(\mathbb{C}^3))$ induced by $(x,y) \to xy$.

In this case we do not have explicit formulae for the solutions, only for the moduli space. More generally, if X is any complex algebraic surface with a Hodge metric then the ASD connections can be identified with certain holomorphic vector bundles [6] and the moduli spaces approached through algebraic geometry.

Although the number of such explicit examples known is small, many general properties of the moduli spaces $M_k(X)$ are understood. They have a "virtual" dimension given via the Atiyah-Singer Index theorem:

$$\dim M_k = 8k - 3(1 - b_1 + b_2^+) \quad .$$

Here b_1 is the first Betti number of X and b_2^+ is the dimension of a maximal subspace on which the intersection form Q on $H^2(X)$ is positive. Freed and Uhlenbeck show in [11] that for $k > 0$ and generic metrics on X the moduli spaces will be smooth manifolds of the proper dimension except perhaps for the existence of singularities associated to S^1 reductions of the bundle, as in example (2). These singularities depend on the position of the b_2^+ dimensional subspace of H_+^2 of cohomology classes realized by self-dual harmonic forms, relative to the integer lattice in H^2.

The ASD moduli spaces are typically not compact, but using Uhlenbeck's Theorem [15] one can define a natural compactification $\overline{M}_{k,x} \supset M_{k,x}$ by adjoining points representing pairs $((x_1, \cdots, x_\ell), A)$ when $(x_1, \cdots, x_\ell)$ is in the symmetric product $S^\ell(X)$ and A is a connection in a "lower" moduli space $M_{k-\ell}(X)$. In the three examples above the compactified spaces are respectively, the closed 5-ball, the closed cone over $\mathbb{CP}^2$, and $\mathbb{CP}^5 = M_2(\mathbb{CP}^2) \cup S^2(\mathbb{CP}^2)$. Combining analysis of Taubes ([13], [14]) with the Kuranishi deformation theory it is possible to understand the attaching maps between the different "strata" in $\overline{M}_k$ quite well. The basic idea of Taubes is that, as in Example (2), sequences in M_k converging to a point in $M_{k-\ell} \times S^\ell(X)$ represent connections converging off a finite set in X and localized in an approximately standard form near these points.

We get in this way a large supply of inter-related spaces $M_{k,x}$ associated to a Riemannian 4-manifold X. The first topological application of these took the point of view of cobordism but one gets more precise results by combining information about the moduli spaces with certain "universal" cohomology classes on them, that is, classes extending over any family of connections.

Consider, for a simple example, connections on a S^1 bundle over a Riemann surface Y. To any loop γ in Y and connection A we associate the holonomy $h_\gamma(A)$ of the connection round the loop. If C is any family of (isomorphism classes of) connections we get a map $h_\gamma : C \to S^1$, and the correspondence $[\gamma] \to [h_\gamma]$ defines a linear map $\mu : H_1(Y) \to H^1(C)$.

To each SU(2) bundle E over the 4-manifold X we can associate the infinite dimensional space $\mathbb{B}$ of all gauge equivalence classes of connections and an open subset $\hat{\mathbb{B}} \subset \mathbb{B}$ of irreducible connections. Suppose for simplicity that X is 1-connected. Then the homotopy type of $\hat{\mathbb{B}}$ can be analyzed with the methods of Atiyah and Bott [1]. There is a monomorphism:

$$\mu : H_2(X; \mathbb{Z}) \to H^2(\hat{\mathbb{B}}; \mathbb{Z}) ,$$

analogous to that in the Riemann surface example above, and the rational cohomology of $\hat{\mathbb{B}}$ is the polynomial algebra generated by $\mu(H_2(X))$ and a single further class in $H_4(\hat{\mathbb{B}})$.

To illustrate these cohomology classes take the case when X is a simply connected 4-manifold with negative definite intersection form. Then, as in examples (1), (2), $M_1(X)$ is a 5 dimensional space with cone singularities labelled by classes $\pm$ e where $e \in H^2(X; \mathbb{Z})$ has $e^2 = -1$ ([5], [11]). By definition M_1 is contained in $\mathbb{B}$ and the complement of the singular points lies in $\hat{\mathbb{B}}$. The compactification $\overline{M}_1(X)$ is obtained by adjoining the 4-manifolds X as a boundary using a collar $X \times (0,1) \subset M_1$. If γ is in $H_2(X)$ one can show that $\mu(\gamma)$ restricts to a level set $X \times (1/2)$ in the collar as the Poincaré dual of γ and restricts to a link P_e (a copy of $\mathbb{CP}^2$) of a singularity as $e(\gamma)$ times the generator of $H^2(\mathbb{CP}^2)$. [7]. The moduli space is orientable and, viewing it as a homology between the boundary X and the links P_e, leads immediately to the formula:

$$\gamma_1 \cdot \gamma_2 = \mu(\gamma_1) \cup \mu(\gamma_2)\ [X \times (1/2)] = -\frac{1}{2} \sum_{e^2=-1} e(\gamma_1)\ e(\gamma_2)$$

saying that the intersection form is diagonal, as proved in [5] and [11].

One can extend the same scheme of proof to study 4-manifolds whose intersection form has a positive part of rank 1 or 2 [7], obtaining restrictions on the existence of smooth structures. Very roughly, we assert "$[\partial M] = 0$ in $H_*(\mathbb{B})$" and investigate the consequences. Questions of uniqueness of smooth structure can be tackled in a complimentary fashion - roughly we assert that the homology class [M] in $H_*(\mathbb{B})$ is an invariant and investigate the information this contains.

If, for a given Riemannian metric g_0 a Yang-Mills moduli space $M = M(X, g_0)$ is a compact smooth manifold of even dimension 2d (given by the index formula), and if it avoids the reducible connections, $\mathbb{B} \setminus \hat{\mathbb{B}}$ then its fundamental class in $H_{2d}(\hat{\mathbb{B}})$ has a component which defines a polynomial q of degree d in $H^2(X; \mathbb{Z})/\text{Torsion}$. That is, for α in $H_2(X)$, $q(\alpha) = \langle [\mu(\alpha)]^d, M\rangle$. If g_t is a path of metrics defining moduli spaces $M(X,g_t)$ which are compact and avoid reducibles, and if $M(X,g_1)$ is smooth then the polynomials obtained by calculating from $M(X,g_0)$, $M(X,g_1)$ will be equal. This is true even though the topological type of the moduli spaces may change, and comes from a standard transversality argument. Even when M is not compact one can make sense of the pairing $\langle \mu(\alpha)^d, M\rangle$ provided the codimension of the other strata in the compactification $\overline{M}$ is at least 2. This is done by choosing representatives for the $\mu(\alpha)$ so that $\text{supp}\langle[\mu(\alpha)]^d, M\rangle$ is compact. To fix the sign of the polynomials a rule for orienting the moduli spaces is required. It can be shown [9] that a canonical orientation of all the SU(2) moduli spaces $M_k(X)$ is fixed by a non-zero element α_X in:

$$(\Lambda^{max} H^1(X) \otimes \Lambda^{max} H^2_+(X))/R^+ .$$

We call such a choice a "homology orientation," it does not depend on the maximal positive subspace chosen. One then gets:

THEOREM 1. Let X be a 1-connected closed oriented 4-manifold with $b_2^+(X) = 2p + 1$ odd and $p > 0$ and with homology orientation α_X. Then for $k > 1/4\ (6(1+p) + 1)$ the homology of the moduli spaces $M_k(X)$ define polynomials q_k of degree $4k-3(1+p)$ in $H^2(X)$ which depend only on the differentiable structure of X.

These invariants may be seen as analogous to indices of elliptic operators in that they reflect that part of the structure of the solution set of a P.D.E. which is unchanged by perturbation. It is not at all obvious that they contain useful information going beyond the homotopy or homeomorphism type of X. A simple example showing that they do comes from a problem on the diffeomorphisms of a K3 surface K, suggested to the author by A. Beauville.

Any orientation preserving diffeomorphism $f : X \to X$ of a 4-manifold determines an isometry $f^* \in Aut(H^2(X),Q)$. If the form is indefinite we associate to f^* a sign $\alpha_f \in \{\pm 1\}$ by the positive orientation homomorphism $\varepsilon_+ : Aut(H^2,Q) \to \{\pm 1\}$. α_f is the sign by which f acts on the homology orientations α_X. If X satisfies the hypotheses of Theorem 1 we have:

THEOREM 2. The invariants q_k transform by

$$f^*(q_k) = \alpha_f \cdot q_k .$$

The proof of this amounts to considering the fibration $X \to \mathcal{X}_f \to S^1$ defined by f, with a family of metrics on the fibres and a corresponding family of ASD moduli spaces, analogous to a family of linear operators. More generally one could define cohomology classes in $B[Diff(X)]_1$ by considering multi-parameter families of metrics.

Now take X to be a K3 surface K (with $b_2^+ = 3$). If some polynomial q_k is non-zero it follows from Theorem 2 that α_f is 1 for any diffeomorphism f of K, and so the image of $Diff(K) \to Aut(H_2(K),Q)$ lies in a subgroup of index 2 (conversely it follows from the Torrelli Theorem for K3 surfaces that Diff(K) is all of Ker ε_+).

Using the complex geometry of a K3 surface one can find moduli spaces with the desired non-zero homology invariants q_k. In fact it is easier to use SO(3) connections (as in the work of Fintushel and Stern [10]) and a 0-dimensional moduli space consisting of a single point. This economical argument leads easily to a proof that no $S^2 \times S^2$ factor can be split off from K (taking the direct sum of diffeomorphisms to get a contradiction). On the other hand it disproves the h cobordism conjecture for 4-manifolds since there is a 5-dimensional h-cobordism W with ends copies of K:

$$K \xrightarrow[i_0]{} W \xleftarrow[i_1]{} K$$

such that $i_0^*(i_1^*)^{-1} : H^2(K) \to H^2(K)$ is not realized by a diffeomorphism, so W is not a product.

More dramatic examples of the failure of the h-cobordism conjecture are provided by the Dolgachev surfaces $D_{p,q}$. These form an infinite family of complex algebraic surfaces, each one <u>homeomorphic</u> to a connected sum $D_{1,1} = \mathbb{CP}^2 \# 9\, \overline{\mathbb{CP}^2}$ Since $b_2^+(D_{p,q}) = 1$ they do not fit into the framework of Theorem 1, but one can still get useful information from the 2-dimensional spaces $M_1(D_{p,q})$. Two complications appear: first, for typical metrics, the natural boundary or lower stratum has codimension 1 - corresponding to a 1-dimensional sub-manifold of X. The moduli spaces themselves do not carry a fundamental class relative to the cohomology $\mu(H_2(X))$ but it is possible to identify a suitable correction term which, together with the moduli space, associates a class in $H^2(X)$ to generic metrics on X. The second complication is that this class varies with the metric, since there is now a codimension 1 set of metrics for which reducible connections appear in the moduli space. Thus the resulting invariant, described in [8] is not very easy to use but is still good enough to distinguish the smooth structures of the manifolds $D_{1,1}$ and $D_{2,3}$, say. Again the key point is the possibility of finding the moduli spaces explicitly using algebraic geometry. R. Friedman and J. Morgan have used this approach to show that among the Dolgachev surfaces there is an infinite set of distinct smooth manifolds. (A. Van de Ven and C. Okonek have obtained similar results independently.)

There are two general properties of the invariants q_k of Theorem 1 which stand in stark contrast and emphasize the sensitivity of these ASD moduli spaces to the differentiable structure of the base 4-manifold. First, for decomposable manifolds we have:

THEOREM 3. Suppose X satisfies the hypotheses of Theorem 1 and can be decomposed as a connected sum $X = X_1 \# X_2$ with $b_2^+(X_1), b_2^+(X_2) > 0$. Then all the invariants q_k of Theorem 1 vanish.

This is proved using the analysis in [7] of moduli spaces over connected sums. When the "neck" of the connecting region is made very small the intersection $M_k(X_1 \# X_2) \cap \bigcap_{i=1}^{d} \operatorname{supp} \mu(\alpha_i)$ may be described in terms of $M_p(X_1)$, $M_q(X_2)$ $(p+q=k)$ and the homology classes calculated from this description.

Second, if X is an algebraic surface the moduli spaces M, interpreted via the holomorphic bundles, are quasi projective varieties [12]. M may have a singular set $\Sigma(M)$ but one can show that for some constant c(X):

$$\dim_{\mathbb{C}}\Sigma(M) \leq 3k + c(X)$$

Hence when k is large M contains an open subset of the "proper" dimension 2d. If $[H] \in H_2(X)$ is a hyperplane class then $\mu([H])|_M$ is the first Chern class of an ample line bundle, represented by a positive Kahler form, and:

$$\langle \mu(H)^d, M \rangle = \operatorname{vol}(M) > 0$$

This familiar link between homology and complex structures gives:

THEOREM 4. If X is an algebraic surface satisfying the hypotheses of Theorem 1 and $[H] \in H_2(X)$ a hyperplane class then for large enough k $q_k([H]) > 0$.

These two facts fit in neatly with a Theorem of Wall [16] which asserts that 1-connected, homotopy equivalent, 4-manifolds become diffeomorphic after taking connected sums with sufficiently many copies of $S^2 \times S^2$. Taking such a sum kills the information from the moduli spaces. Conversely, algebraic surfaces with Pg > 0 cannot split off $S^2 \times S^2$ summands smoothly, although they do topologically, and this gives many more examples where the gauge field invariants detect different smooth structures on 4-manifolds.

At present our knowledge of these invariants is very small. In the analogy with linear elliptic equations we are missing the fourth step by which the rigid invariants coming from the analysis are given entirely topological interpretation in terms only of the local manifestation of the operator in its symbol. It is not even clear if all the invariants q_k are independent. For minimal models of algebraic surfaces it seems likely that the q_k can always be written as polynomials in the intersection form Q (regarded as an element of $S^2(H^2(X))$ and the canonical class $c_1(K_X)$. But this still leaves infinitely many numerical co-efficients to be determined as k increases. Perhaps there are some universal relations between these which will carry over to general 4-manifolds.

Since these invariants are generated by the 2-dimensional homology of X they do not seem to give much help with the smooth Poincaré conjecture. Indeed it is a strange fact that these methods prefer simply connected manifolds with b_2^+ odd (ensuring that the moduli spaces are even dimensional). Of course this condition always holds for algebraic surfaces.

REFERENCES

1. Atiyah, M. F. and Bott, R., The Yang-Mills Equation over Riemann Surfaces. Phil. Trans. Roy. Soc. London A 308 (1982), 523-615.
2. Atiyah, M. F., Hitchin, N. J., and Singer, I. M., Self-duality in Four Dimensional Riemannian Geometry. Proc. Roy. Soc. London A 362 (1978), 425-61.
3. Barth, W., Moduli of Vector Bundles on the Projective Plane. Invent. Math. 42 (1977), 63-91.
4. Buchdahl, N., Instantons on $\mathbb{C}\mathbb{P}^2$. To appear in Journal. Diff. Geom.
5. Donaldson, S. K., An Application of Gauge Theory to Four Dimensional Topology. Journal Diff. Geom. 18 (1983), 279-315.
6. Donaldson, S. K., Anti Self Dual Yang Mills Connections over Complex Algebraic Surfaces and Stable Vector Bundles. Proc. Lond. Math. Soc. 50 (1985), 1-26.
7. Donaldson, S. K., Connections, Cohomology, and the Intersection Forms of 4-manifolds. Preprint, submitted to Journal Diff. Geom.
8. Donaldson, S. K., La Topologic Differentielle des Surfaces Complexes. C. R. Acad. Sciences. Paris t. 301 1 no. 6 (1985), 317-320.
9. Donaldson, S. K., The Orientation of Yang-Mills Moduli Spaces and Fundamental Groups of 4-manifolds. In preparation.
10. Fintushel, R., and Stern, R. J., SO(3) Connections and the Topology of 4-manifolds. Journal Diff. Geom. 20 (1984), 523-539.
11. Freed, D. S., and Uhlenbeck, K. K., Instantons and Four Manifolds. M.S.R.I. publications, Springer (New York) 1984.

12. Maruyama, M., Stable Vector Bundles on Algebraic Surfaces. Nagoya Math. J. 58 (1975), 25-68.

13. Taubes, C. H., Self-dual Yang-Mills Connections on Non Self-dual Four Manifolds. Journal Diff. Geom. 17 (1982), 139-170.

14. Taubes, C. H., Self-dual Connections on 4-manifolds with Indefinite Intersection Form. Journal Diff. Geom. 19 (1984), 517-560.

15. Uhlenbeck, K. K., Connections with L^p bounds on Curvature. Commun. Math. Phys. 83 (1982), 31-42.

16. Wall, C. T. C., On Simply Connected 4-manifolds. Journal Lond. Math. Soc. 39 (1964), 141-149.

Algebraic Varieties Which Are a Disjoint Union of Subvarieties

ALAN H. DURFEE / Department of Mathematics, Mount Holyoke College, South Hadley, Massachusetts

The purpose of this note is to prove the following result:

THEOREM 1. Let X and Y be smooth projective complex varieties, and suppose that $X = X_1 \cup \cdots \cup X_n$ and $Y = Y_1 \cup \cdots \cup Y_n$ are a disjoint union of quasiprojective subvarieties. Suppose that X_i is algebraically isomorphic to Y_i, for all i. Then the Betti numbers of X and Y are equal, and in fact their Hodge numbers are equal.

The result for the Betti numbers is part of the mathematical folklore; it has been used, for instance, in [1] and implicitly in [4]. Apparently it was originally noticed by Serre, who proved it by reducing the variety modulo p, counting points and using the solution to the Weil conjectures. The proof given in this paper uses mixed Hodge theory rather than the Weil conjectures; the general yoga of Hodge theory and the Weil conjectures suggest that such a proof should exist.

The theorem is in fact a simple consequence of a sum formula for the Hodge numbers of the mixed Hodge structure of an arbitrary variety in terms of those of the pieces of a decomposition. (Throughout this paper, the term "variety" will mean "quasiprojective complex variety.") First, some definitions: According to [2], the rational cohomology groups of a complex algebraic variety have a weight filtration W and a Hodge filtration F, and so do its rational cohomology groups with compact support. The associated graded objects of these filtrations are denoted Gr^W and Gr_F, respectively. We refine the concept of Euler characteristic by using these filtrations: Let

$$\chi(X) = \Sigma(-1)^k \dim H^k(X)$$

$$\chi_m(X) = \Sigma(-1)^k \dim Gr_m^W H^k(X)$$

$$\chi^{pq}(X) = \Sigma(-1)^k \dim Gr_F^p Gr_{p+q}^W H^k(X)$$

Similarly, define χ^c, χ_m^c, and χ_c^{pq} using $H_c^k(X)$ (cohomology with compact supports) in place of $H^k(X)$. These graded Euler characteristics satisfy

$$\chi(X) = \Sigma_m \chi_m(X)$$

$$\chi_m(X) = \Sigma_{p+q=m} \chi^{pq}(X)$$

and so forth. If X is smooth of dimension n, the Poincaré duality implies that

$$\chi_c^{pq}(X) = \chi^{n-p,n-q}(X)$$

If X is smooth and projective, then

$$\chi_m(X) = (-1)^m \dim H^m(X)$$

The next theorem immediately implies Theorem 1:

THEOREM 2. Let X be a complex quasiprojective variety. Suppose that X is a finite disjoint union $X_1 \cup \cdots \cup X_n$, where the X_i are quasiprojective subvarieties. Then

$$\chi_c^{pq}(X) = \Sigma_i \chi_c^{pq}(X_i)$$

EXAMPLE. Let X be a smooth projective variety, and suppose that the pieces X_i are smooth and contractible. The formula then reduces to

$$\dim H_m(X) = \#\{X_i : 2 \dim X_i = m\}$$

which is familar from cellular homology theory.

REMARK. The varieties X and Y in Theorem 1 may not have the same homotopy type. For example, let $X = P^1 \times P^1$ and let Y be a P^1 bundle over P^1 with Chern class -1. Since Y is homeomorphic to $P^2\#(-P^2)$, the spaces X and Y have different homotopy types. On the other hand, we can write both X and Y as the union

(cross-section at infinity) $\cup$ (affine fiber) $\cup$ (trivial C-bundle over C) i.e., as

$$P^1 \cup C \cup C^2$$

This decomposition can easily be seen by the usual blowing up and down to go from $P^1 \times P^1$ to P^2. Thus Theorem 1 applies.

Now let us prove Theorem 2. First let us recall the construction of a mixed Hodge structure on the cohomology with compact supports: Let X be a variety. Let $X \subset \bar{X}$ be an inclusion into a compact variety $\bar{X}$, and let $X' = \bar{X} - X$. Then $H_c^{\bullet}(X) \cong H^{\bullet}(\bar{X}, X')$. Since relative cohomology groups have a mixed Hodge structure, the mixed Hodge structure of $H_c^k(X)$ is defined to be the one induced by the above isomorphism. It is independent of the choice of compactification by the usual argument.

LEMMA. Let X be a quasiprojective variety and let A be a closed quasiprojective subvariety. Let $U = X - A$. Then there is a long exact sequence of mixed Hodge structures:

$$\to H_c^k(U) \to H_c^k(X) \to H_c^k(A) \to H_c^{k+1}(U) \to$$

Proof: Choose a compactification $X \subset \bar{X}$ and let $X' = \bar{X} - X$. Let $\bar{A}$ be the closure of A in X and let $A' = \bar{A} - A$. The above exact sequence becomes

$$\to H^k(\bar{X}, X' \cup \bar{A}) \to H^k(\bar{X}, X') \to H^k(\bar{A}, A') \to$$

After replacing (by excision) the last group by the isomorphic group $H^k(X' \cup \bar{A}, X')$, the sequence becomes the exact sequence of the triple $(\bar{X}, X' \cup \bar{A}, X')$. However, the exact sequence of a triple is an exact sequence of mixed Hodge structures: One only needs to check that the connecting homomorphism preserves mixed Hodge structure, but this follows since it factors through the connecting homomorphism of the exact sequence of a pair.

To prove Theorem 2, it clearly suffices to do the case $n = 2$. The result follows directly from the lemma when X_1 or X_2 is closed. In general, let $\bar{X}_1$ be the closure of X_1 in X and let $C = \bar{X}_1 \cap X_2$; both these are quasiprojective varieties. Also, X_1 is open in $\bar{X}_1$ (since X_1 is quasiprojective), $\bar{X}_1$ is closed in X, and C is closed in X_2. Thus applying the lemma three times gives

$$\chi_c^{pq}(X) = \chi_c^{pq}(\bar{X}_1) + \chi_c^{pq}(X_2-C)$$

$$= \chi_c^{pq}(X_1) + \chi_c^{pq}(C) + \chi_c^{pq}(X_2-C)$$

$$= \chi_c^{pq}(X_1) + \chi_c^{pq}(X_2)$$

This completes the proof.

REMARK. There is a simple relation between the graded Euler characteristic of a complex variety as defined above and the number of points on a reduction of this variety over a finite field: Let X be such a variety. There is a ring R, finitely generated over Z, such that X is defined over R. For a maximal ideal $\underline{m} \subset R$, $R/\underline{m}$ is isomorphic to a finite field F_q, with q elements. Let $\bar{X}$ be the variety obtained by reducing X modulo m and extending scalars to the algebraic closure of F_q. The following statements are true for almost all $\underline{m}$. If X is smooth and projective, then $\dim H^m(X)$ = # poles [resp. #zeros] of the zeta function of $\bar{X}$ with absolute value $q^{-m/2}$ if m even [resp. m odd] by Deligne's solution to the Weil conjectures. For an arbitrary variety X, the formula becomes $\chi_m^c(X)$ = (#zeros - #poles) of the zeta function of $\bar{X}$ with absolute value $q^{-m/2}$. The varieties $C^n - \{0\}$ and a union of three lines in P^2 are easily computable examples of this.

ACKNOWLEDGEMENTS: I would like to thank Andrew Sommese for kindling my interest in this problem, and Josef Steenbrink for critically reading an earlier version of this paper and suggesting the present form of Theorem 2.

REFERENCES:

1. A. Bialynicki-Birula and A. Sommese, Quotients by C* and SL(2,C) actions, Trans. Amer. Math. Soc. 279 (1983), 773-800.
2. P. Deligne, Theorie de Hodge II, III, Publ. IHES 40 (1971), 5-58, and 44 (1975), 5-77.
3. P. Deligne, Poids dans la cohomologie des variétés algebriques, Acts du Congrès International des Mathematiciens, Vancouver, 1974, 79-85.
4. F. Kirwan, Cohomology of Quotients in Symplectic and Algebraic Geometry, Math. Notes 31, Princeton University Press, 1984.

The Lower Central Series of Generalized Pure Braid Groups

MICHAEL FALK AND RICHARD RANDELL / Department of Mathematics, University of Iowa, Iowa City, Iowa

Let W be a finite irreducible group generated by reflections across hyperplanes in a real vector space, and let M_W be the complexified complement of the union of the reflecting hyperplanes. Then (by definition) $H_W = \pi_1(M_W)$ is a generalized pure braid group. In this paper we will discuss a prospective formula for the ranks of the quotients of the lower central series of H_W in terms of the exponents of the group.

Let $H = \pi_1(M)$ and let $H_1 = H$, $H_{j+1} = [H_j, H]$, the group generated by commutators. (We henceforth drop the subscript W.) Let $Q_j = H_j/H_{j+1}$ be the j-th quotient. Then [12] Q_j is a finitely generated abelian group, so we may set $\varphi_j = \text{Rank } Q_j$. Finally, let

$$P_M(t) = \sum_{j=0}^{\infty} b_j(M) t^j$$

be the Poincaré polynomial of M.

CONJECTURE.

$$\prod_{j=1}^{\infty} (1-t^j)^{\varphi_j} = P_M(-t) = \prod_{k=1}^{\ell} (1-d_k t), \qquad (*)$$

where the d_k are the exponents of W.

COMMENTS. (i) The conjecture holds for $A_\ell, C_\ell, D_\ell, G_2$, and $I_2(p)$. (Notation of [1]), as we shall see.

(ii) The formula (*) is to be interpreted in $\mathbb{Z}[[t]]$, the formal power series ring. By expanding both sides of the first equality, then comparing coefficients, one obtains equations for computing the φ_j's from the b_j's, or vice versa. For example,

$$\varphi_1 = b_1$$

$$\varphi_2 = -b_2 + \binom{\varphi_1}{2}$$

$$\varphi_3 = +b_3 + \varphi_1\varphi_2 - \binom{\varphi_1}{3}$$

$$\varphi_4 = -b_4 + \varphi_1\varphi_3 + \binom{\varphi_2}{2} - \binom{\varphi_1}{2}\varphi_2 + \binom{\varphi_1}{4}$$

(iii) Another way of viewing (*) is that H has a lower central series whose quotients are the same as those for a product of ℓ free groups on $d_1, d_2, \cdots, d_\ell$ generators. In fact, in the cases mentioned in (i) for which (*) holds, except possibly D_ℓ, the Q_j are free, as they are for free groups.

(iv) The second equality of (*) is always true, and is due to Brieskorn in [3]. Much of the interest of the formula (*) is that the d_j, which are classically defined in terms of the degrees of generating invariant polynomials under the action of W, turn up again in $\pi_1(M)$.

EXAMPLE. $W = \Sigma_{\ell+1}$ (the symmetric group on $\ell+1$ letters). Then

$$M \times \mathbb{C} = \{(z_1, \cdots, z_{\ell+1}) \in \mathbb{C}^{\ell+1} | z_i \neq z_j, \forall i \neq j\}.$$

The exponents are $(1, 2, \cdots, \ell-1, \ell)$, and $\pi_1(M)$ is the classical pure (or "colored") braid group. Furthermore, $\pi_1(M/W)$ is the classical Artin braid group.

The simplest non-trivial such example occurs with $\ell = 2$. Then $P_M(t) = (1+t)(1+2t)$ and so $P_M(-t) = 1 - 3t + 2t^2$. Thus $\varphi_1 = 3$, $\varphi_2 = 1$, $\varphi_3 = 2$, $\varphi_4 = 3, \cdots$.

For these examples, Kohno [9] deduced (*) from a free resolution of $\mathbb{C}$ as trivial module over the holonomy Lie algebra of M. It was his work which aroused our interest.

Now the spaces M_W were considered in Brieskorn's paper [3]. In that paper he showed that for $W = A_\ell, C_\ell, D_\ell, F_4 G_2, I_2(p)$, M_W (and hence M_W/W) are $K(\pi,1)$ spaces. Subsequently, Deligne showed [4] that for all finite irreducible real reflection groups W, M_W is a $K(\pi,1)$. These results and Kohno's result for A_ℓ lead one to the conjecture.

Brieskorn obtained his results by an iterated fibering technique, so this was our first approach. In [6], we considered a specific type of arrangement in $\mathbb{C}^\ell$.

DEFINITION. An <u>ℓ-arrangement</u> $\mathscr{A}$ is a finite collection of n hyperplanes through the origin in $\mathbb{C}^\ell$.

Let M be the complement of the union of the hyperplanes.

DEFINITION. An arrangement $\mathscr{A}$ is <u>fiber-type</u> if there is a sequence of linear fiber bundle projections

$$M_{\mathscr{A}} = M_\ell \xrightarrow{p_\ell} M_{\ell-1} \xrightarrow{p_{\ell-1}} M_{\ell-2} \longrightarrow \cdots \longrightarrow M_2 \xrightarrow{p_2} M_1 = \mathbb{C} - \{\text{pt.}\}$$

in which each p_k has fiber a d_k-times punctured $\mathbb{C}$, and each M_k is the complement of a k-arrangement. The numbers $d_1, \cdots, d_\ell$ are called the <u>exponents</u>. It is a consequence of a result below that they are well-defined.

NOTE. A_ℓ, C_ℓ, G_2, $I_2(p)$ <u>are</u> fiber-type, whereas D_4 is not. There is a combinatorial (= based on intersection pattern of the hyperplanes) criterion called supersolvability [8], which enables one to determine precisely which arrangements are fiber type. This makes the next result more useful.

THEOREM [6]. The equations (*) hold for fiber-type arrangements.

Proof (Outline): One first uses a spectral sequence argument to show $P_M(t) = \Pi_{k=1}^{\ell}(1+d_k t)$. For this, one uses the linearity of the projections to find a section σ_k of each p_k. Also, one must show that $\pi_1(M_k)$ acts trivially on $H_1(p_{k+1}^{-1}(pt)) = H_1(F_{k+1})$.

Next one considers the long exact homotopy sequences of the fiber bundles, which inductively become

$$1 \longrightarrow \pi_1(F_{k+1}) \longrightarrow \pi_1(M_{k+1}) \underset{\sigma_{k+1}}{\overset{p_{k+1}}{\rightleftarrows}} \pi_1(M_k) \longrightarrow 1$$

By using the splitting, the trivial action of $\pi_1(M_k)$, and standard commutator identities, one obtains corresponding split short exact sequences of the quotients in the lower central series.

Finally, the use of induction and the fact that (*) holds for free groups [12] (i.e., $\pi_1(F_k)$ and $\pi_1(\mathbb{C}-\{0\})$) completes the proof.

COROLLARY. (i) (*) holds for the arrangements associated to $A_\ell, C_\ell, G_2, I_2(p)$.

(ii) (*) holds for the arrangements associated to the complex reflection groups $G(r,p,\ell)$, $p < r$.

(iii) (*) holds for arrangements with supersolvable lattice.

Proof: In all cases the arrangements are fiber-type.

References are (i) [3]; (ii) [13]; (iii) [16].

COMMENT. (i) As we have noted, (*) also holds for D_ℓ arrangements, even though D_4, for example, is not fiber-type (since it isn't supersolvable). The result for D_ℓ is due to Kohno [10]. The arrangement associated to D_ℓ does admit an iterated (non-linear) fibering with a section, but with non-trivial action.

(ii) A further corollary is that (*) holds for arrangements through 0 in $\mathbb{C}^2$, suggesting the study of (*) for classical links in S^3. In fact, T. Maeda [11] has shown that such a formula does hold when the link satisfies certain conditions. The Hopf links ((p,p) torus links) do satisfy these conditions, and are the links associated to $I_2(p)$ arrangements by intersecting with the unit 3-sphere in $\mathbb{C}^2$.

(iii) The conjecture and the partial results presented here are for the generalized pure braid groups. One is led to ask what happens for the generalized braid groups themselves.

PROPOSITION. (*) holds for M/W (the generalized braid group space) iff $W = A_\ell, D_\ell, E_6, G_2, I_2(p)$.

Proof: Brieskorn has computed $\pi_1(M/W)$ in [2]. Abelianizing, we find that $H_1(M/W) =$

$$\begin{cases} \mathbb{Z} & , W = A_\ell, D_\ell, E_6, E_7, E_8, H_3, H_4, I_2(p) \quad (p \text{ odd}) \\ \mathbb{Z} \oplus \mathbb{Z} & , W = C_\ell, F_4, G_2, I_2(p) \quad (p \text{ even}) \end{cases}$$

Furthermore, the Hurewicz map $\varphi : \pi_1(M/W) \to H_1(M/W)$ admits a section in every case. This is trivial if $H_1(M/W) \cong \mathbb{Z}$ and requires a certain $\mathbb{Z} \oplus \mathbb{Z}$ subgroup of $\pi_1(M/W)$ in the other cases. For example, consider the case $W = C_\ell$. Then

$$\pi_1(M/W) \cong \langle a_1, \cdots, a_\ell \mid a_1a_2a_1a_2 = a_2a_1a_2a_1;\ a_ia_{i+1}a_i = a_{i+1}a_ia_{i+1}, \\ i = 2, \cdots, \ell-1;\ [a_i, a_j], |i-j| > 1\rangle$$

and $H_1(M/W)$ is generated by $\bar{a}_1$ and $\bar{a}_2$, the classes of a_1 and a_2 under abelianization. Notice $\bar{a}_2 = \bar{a}_3$, since $a_2a_3a_2 = a_3a_2a_3$ in $\pi_1(M/W)$. The required section is thus obtained by mapping $\bar{a}_1$ to a_1 and $\bar{a}_2 = \bar{a}_3$ to a_3, and extending (since $[a_1,a_3] = 1$ in $\pi_1(M/W)$). The other cases are similar.

Setting $G = \pi_1(M/W)$ we have a central extension

$$1 \longrightarrow Q_2 = G_2/G_3 \longrightarrow G/G_3 \underset{\varphi'}{\overset{\longrightarrow}{\longleftarrow}} G/G_2 = Q_1 = H_1(M/W) \longrightarrow 1.$$

The existence of φ' (induced from the section) allows one to conclude that G/G_3 is abelian. Thus $[G,G] = G_2 \subset G_3$, and $G_3 \subseteq G_2$ always holds, so $G_2 = G_3$. Thus $G_4 = [G_3,G] = [G_2,G] = G_3$, etc., implying $G_n = G_2$, for all $n \geq 2$. Thus, (*) holds iff $P_{M/W}(-t) = (1-t)^{\varphi_1}$. One then checks $P_{M/W}(t)$ in [3] to reach the conclusion of the theorem.

Let us conclude by observing that there are arrangements with several pleasant properties for which (*) fails. Consider the 3-arrangement $\mathcal{A}$ of the seven lines $z = 0$, $x = \pm z$, $y = \pm z$, $x = \pm y$. Then it is known that

(i) $P_M(t) = (1+t)(1+3t)^2 = 1 + 7t + 15t^2 + 9t^3$.

(ii) $\mathcal{A}$ is a free arrangement [7].

(iii) M is a $K(\pi,1)$ [4],[7].

Now by constructing the Sullivan minimal model [14] of M as in [5] and [7], one may compute that $\varphi_3 = 12$. Since if (*) held, one would have $\varphi_3 = 16$, (*) does not hold here.

ACKNOWLEDGEMENT

The second author was partially supported by an NSF grant.

REFERENCES

1. Bourbaki, N., Groupes et Algèbrès de Lie, Chap. 4, 5 et 6, Herman, 1968.

2. Brieskorn, E., Die Fundamentalgruppe des Raumes der regulären Orbits einer endlichen komplexen Spiegelungsgruppe, Inventiones Math. 12 (1971), 57-61.

3. Brieskorn, E., Sur les groupes de tresses (d'après V.I. Arnold), Séminaire Bourbaki 24e année, 1971-72, Springer Lecture Notes No. 317, Springer Verlag, Berlin.

4. Deligne, P., Les immeubles des groupes de tresses généralisés, Inventiones math. 17 (1972), 273-302.

5. Falk, M., The Minimal model of the complement of an arrangement, preprint, 1985.

6. Falk, M., and Randell, R., The Lower Central Series of a Fiber-type arrangement, Inventiones math. 82 (1985), 77-88.

7. Falk, M., and Randell, R., On the Homotopy Theory of Arrangements, Proc. of the 1984 U.S.-Japan Singularities Seminar, to appear.

8. Jambu, M., and Terao, H., Free Arrangments of Hyperplanes and Supersolvable Lattices, Advances in Math. 52 (1984), 248-258.

9. Kohno, T., Serie de Poincaré-Koszul associee aux groupes de tresses pures, Inventiones Math. 82 (1985), 57-75.

10. Kohno, T., Poincare series of the Malcev completion of generalized pure braid groups, preprint, 1985.

11. Maeda, T., Ph.D. Thesis, University of Toronto, 1982.

12. Magnus, W., Karrass, A., and Solitar, D., Combinatorial Group Theory, John Wiley and Sons, 1966.

13. Orlik, P., and Solomon, L., Coxeter Arrangements, Proc. Symp. Pure Math., no. 40, Amer. Math. Soc., Providence, R.I., 1983.

14. Sullivan, D., Infinitesimal Computations in Topology, Publ. Math. IHES 47 (1977), 269-331.

15. Terao, H., Arrangements of Hyperplanes and their Freeness I, J. Fac. Sci. Univ. Tokyo Sec 1A Math. 27 (1980), 293-312.

16. Terao, H., Modular Elements of Lattices and Topological Fibrations, preprint.

Implication of the Geometrization Conjecture for the Algebraic K-Theory of 3-Manifolds

F. T. FARRELL / Department of Mathematics, Columbia University, New York, New York

L. E. JONES / Mathematics Department, State University of New York, Stony Brook, New York

In this note, the main results of [5], [6] (as announced in [4]) about the algebraic K-theory of hyperbolic manifolds are used to show Thurston's geometrization conjecture places strong constraints on the algebraic K-theory of three-dimensional manifolds. First, we recall Thurston's conjecture; cf. [13], then we state its consequence.

GEOMETRIZATION CONJECTURE. Let M be a closed, connected, orientable and irreducible three-dimensional manifold which is not sufficiently large but whose fundamental group contains infinitely many elements. Then, M is either a Seifert fibre space or M admits a hyperbolic structure.

THEOREM. If M is a compact, connected, orientable three dimensional manifold whose fundamental group is torsion-free and the geometrization conjecture is true, then

$$Wh_n(\pi_1 M) \otimes \mathbb{Z}\,[1/N] = 0$$

for all non-negative integers n, where $N = [(n+1)/2]!$ In particular, $\tilde{K}_0(\mathbb{Z}\pi_1 M)$, $Wh(\pi_1 M)$ and $Wh_2(\pi_1 M)$ vanish. In addition, $\pi_1 M$ is K-flat; i.e.,

$$Wh(\pi_1 M \oplus A) = 0$$

for any finitely generated free abelian group A; consequently,

$$K_n(\mathbb{Z}\pi_1 M) = 0$$

for all negative integers n.

Proof: We start by verifying the first sentence of the theorem and afterwards mention the modifications necessary to verify the second sentence.

A theorem of Knesser [7] shows M is the connected sum of a finite number of prime manifolds; i.e.,

1. $M = M_1 \# M_2 \# \cdots \# M_m$

where each M_i is prime. Let $LWh_n(G)$ denote $Wh_n(G) \otimes \mathbb{Z}[1/N]$ where $N = [(n+1)/2]!$ By a result of Waldhausen [15],

2. $LWh_n(\pi_1 M) \simeq \bigoplus_{i=1}^{m} LWh_n(\pi_1 M_i)$.

Recall a compact, orientable prime three-manifold with infinite fundamental group is either $S^2 \times S^1$ or is irreducible and aspherical. Consequently, for each i, one of the following is true.

3. (a) $\pi_1 M_i$ contains only one element,
   (b) $\pi_1 M_i$ is infinite cyclic, or
   (c) M_i is irreducible and aspherical.

In cases (a) and (b), $LWh_n(\pi_1 M_i) = 0$ by a result of Quillen [11]. In case (c), Waldhausen [15] has shown $LWh_n(\pi_1 M_i) = 0$ provided M_i is sufficiently large. Waldhausen also showed M_i is closed in case (c) when M_i is not sufficiently large. Consequently, we are left to handle the situation when M_i is a closed, connected, orientable and irreducible three-manifold which is not sufficiently large but whose fundamental group is infinite. So, by the geometrization conjecture, we are reduced to considering two subcases

4. (d) M_i is a closed Seifert fibre space, or
   (e) M_i is a closed hyperbolic manifold.

In subcase (d) and when n = 0 or 1, Plotnick [10] showed $LWh_n(\pi_1 M_i) = 0$. For certain special instances of (d); namely, when the fundamental group of the Seifert manifold is virtually poly-$\mathbb{Z}$, he used the result of Farrell and Hsiang [3]. In [8], Nicas and Stark showed $LWh_n(\pi_1 M_i) = 0$ when M_i is a closed Seifert fibre space except when $\pi_1 M_i$ is virtually poly-$\mathbb{Z}$. But these cases are handled by Quinn [12] and Nicas [9] (cf., also [2]).

When M_i is a closed hyperbolic manifold, we showed in Corollary 2 of [4] that $LWh_n(\pi_1 M_i) = 0$ completing the verification of the first sentence of our theorem. Earlier, Nicas and Stark [17], [18], and Nicas [19] had shown $LWh_n(\pi_1 M) = 0$ for certain interesting classes of non-sufficiently large hyperbolic three-manifolds M.

We now discuss the modifications in the above argument needed to show $\pi_1 M$ is K-flat. Let T be a torus such that $\pi_1 T$ is isomorphic to A. (A torus is a finite Cartesian product of circles.) Then, $\pi_1(M \times T)$ is isomorphic to $\pi_1 M \oplus A$ and (1) yields a splitting of $M \times T$ by codimension-one submanifolds homeomorphic to $S^2 \times T$. In particular, we obtain an amalgamated free product structure for $\pi_1 M \oplus A$ where the amalgamating subgroups are isomorphic to A. Hence by Waldhausen [15] and [1],

2'. $$Wh(\pi_1 M \oplus A) \cong \bigoplus_{i=1}^{m} Wh(\pi_1 M_i \oplus A).$$

In case (a) and (b), $Wh(\pi_1 M_i \oplus A) = 0$ by the Bass-Heller-Swan formula [1]. In case (c), Stark [14] showed $Wh(\pi_1 M_i \oplus A) = 0$ provided M_i is sufficiently large. Therefore, we again need only consider subcases (d) and (e). In subcase (d) Stark [14] extended Plotnick's argument to show $Wh(\pi_1 M_i \oplus A) = 0$; again using [3] to handle the cases where $\pi_1 M_i$ is virtually poly-$\mathbb{Z}$.

Finally, when $\pi_1 M_i$ is a closed hyperbolic manifold, we showed in Corollary 1 of [4] that $Wh(\pi_1 M_i \oplus A) = 0$. This completes the proof of Theorem.

We conclude with a remark. If the geometrization conjecture is true, then the above Theorem concatenated with the works of Waldhausen [16] would yield strong information about both the topological and smooth stable pseudo-isotopy spaces of compact, orientable, aspherical three-dimensional manifolds. An example of what we have in mind (which in fact does not depend on the geometrization conjecture) is the following special case of a result we prove in [6].

PROPOSITION. Let M be a compact orientable non-Haken hyperbolic three-manifold, then

$$\pi_i(P_s(M \times \mathbb{D}^n)) \otimes \mathbb{Q} \cong \begin{cases} \mathbb{Q} \text{ if } i \equiv 2 \text{ or } 3 \bmod 4 \text{ and } i > 2 \\ 0 \text{ otherwise} \end{cases}$$

provided $i < (n-7)/7$.

In this Proposition, $P_s(M \times \mathbb{D}^n)$ is the space of smooth pseudo-isotopies on $M \times \mathbb{D}^n$ where $\mathbb{D}^n$ denotes the closed ball of radius 1 in $\mathbb{R}^n$.

POSTSCRIPT. Andy Nicas points out the following relevant information. Culler and Shalen have shown that for <u>virtually</u> Haken manifolds without boundary the geometrization conjecture is true <u>up to homotopy equivalence</u>; cf. [20], theorem 4.2.2. Clearly, in our theorem it is only necessary that M satisfies the geometrization conjecture up to homotopy equivalence (this also suffices in stable psuedo-isotopy calculations).

ACKNOWLEDGEMENT

The authors were supported in part by the NSF.

REFERENCES

1. Bass, H., Heller, A., and Swan, R. G., The Whitehead Group of a Polynomial Extension, Publ. Math. I.H.E.S. 22 (1964), 61-79.
2. Farrell, F. T., and Hsiang, W. C., On the Rational Homotopy Groups of the Diffeomorphism Groups of Discs, Spheres and Aspherical Manifolds, Proc. of Symposia in Pure Math. 32 (1978), 325-337.
3. Farrell, F. T., and Hsiang, W. C., The Whitehead Group of Poly-(finite or cyclic) groups, J. London Math. Soc. (2) 24 (1981), 308-324.
4. Farrell, F. T., and Jones, L. E., Algebraic K-theory of Hyperbolic Manifolds, Bull. Amer. Math. Soc., 14 (N.S.) (1986), 115-119.
5. Farrell, F. T., and Jones, L. E., K-theory and Dynamics, I, (to appear).
6. Farrell, F. T., and Jones, L. E., K-theory and Dynamics, II, (submitted for publication).
7. Knesser, H., Greschlossene Flachen in Dreidimensionale Mannigfaltigkeiten, Jaresbericht der Deutsche Math. Verein, 38 (1929), 248-260.
8. Nicas, A. J., and Stark, C. W., Higher Whitehead Group of Certain Bundles over Seifert Manifolds, Proc. of Amer. Math. Soc. 91 (1984), 1-5.
9. Nicas, A. J., On the Higher Whitehead Groups of a Bieberbach Group, Trans. Amer. Math. Soc., 285 (1985), 853-859.
10. Plotnick, S., Vanishing of Whitehead Groups for Seifert Manifolds with Infinite Fundamental Group, Comment. Math. Helv. 55 (1980), 654-667.
11. Quillen, D., Higher Algebraic K-theory. I, Algebraic K-theory I, Lecture Notes in Math, vol. 341, Springer-Verlag, Berlin and New York, 1973, pp. 84-147.
12. Quinn, F., Algebraic K-theory of Poly-(finite or cyclic) Groups, Bull. Amer. Math. Soc. 12 (New Series) (1985), 221-226.

13. Scott, P., The Geometries of Three-manifolds, Bull. London Math. Soc., 15 (1983), 401-487.

14. Stark, C. W., Structure Sets Vanish for Certain Bundles over Seifert Manifolds, Trans. Amer. Math. Soc. 285 (1984), 603-615.

15. Waldhausen, F., Algebraic K-theory of Generalized Free Products, Annals of Math., 108 (1978), 135-256.

16. Waldhausen, F., Algebraic K-theory of Topological Spaces. I, Proc. of Symposia in Pure Math. 32 (1978), 35-60.

17. Nicas, A. J., and Stark, C. W., Whitehead Groups of Certain Hyperbolic Manifolds, Math. Proc. Camb. Phil. Soc. 95 (1984), 299-308.

18. Nicas, A. J., and Stark, C. W., Whitehead Groups of Certain Hyperbolic Manifolds II, (to appear).

19. Nicas, A. J., An Infinite Family of Non-Haken Hyperbolic Three-manifolds with Vanishing Whitehead Groups, (to appear).

20. Culler, M., and Shalen, P. B., Varieties of Group Representations and Splittings of Three-manifolds, Annals of Math., 117 (1983), 109-146.

On the Diffeomorphism Types of Certain Elliptic Surfaces

ROBERT FRIEDMAN AND JOHN W. MORGAN / Department of Mathematics, Columbia University, New York, New York

INTRODUCTION

The purpose of this note is to describe some recent results concerning certain closed, smooth 4-manifolds. These 4-manifolds have the structure of algebraic surfaces (more precisely, they are simply connected elliptic surfaces). Our results depend on a new invariant introduced by Donaldson in [3]. He used this invariant to prove that two well-known elliptic surfaces, which are homotopy equivalent (even homeomorphic) and hence h-cobordant, are not diffeomorphic. By using the formal properties he established for this invariant, we are able to extend his results to an entire class of elliptic surfaces. We show, in fact, that there are infinitely many simply connected 4-manifolds, all homotopy equivalent to one another (and hence homeomorphic), no two of which are diffeomorphic. The same techniques may be applied to study the group of self-diffeomorphisms of these elliptic surfaces. We show that, in the group of integral automorphisms of the cohomology ring, the subgroup of elements realized by self-diffeomorphisms may be characterized up to finite index and is of infinite index.

The basic ingredients needed in order to apply Donaldson's invariant in our situation are (a) the calculation of certain moduli spaces of stable holomorphic vector bundles over the algebraic surfaces in question, and (b) an analysis of the geometry of certain reflection groups associated to quadratic forms of hyperbolic type (i.e., those which have a maximal positive definite space of dimension one). The results described here are the simplest case of a series of more general results, described in full detail in [7]. Because of the length and technical nature of that paper, we felt that a short note, stating the most basic of these results, and giving an indication of the proof, would be a useful introduction to [7]. In this spirit, we have stated various results concerning stable vector bundles on elliptic surfaces, but have deferred the proofs to [7]. Likewise, the description of the chamber structure for a particular discrete reflection group acting on 9-dimensional hyperbolic space, a key part of the proof, will not be established here. This description follows from purely algebraic methods. It can also be quite easily established by studying the geometry of certain blowups of the projective plane; this approach is carried out in [7].

1. STATEMENT OF RESULTS

Let M be a closed oriented 4-manifold. Set $\Lambda(M) = H^2(M;\mathbb{Z})$. Define the self-intersection form

$$q_M : \Lambda(M) \to \mathbb{Z}$$

by $q_M(a) = \langle a \cup a, [M] \rangle$, where $[M]$ is the top cycle of M.

DEFINITION. A quadratic form $q : \Lambda \to \mathbb{Z}$, where Λ is a free $\mathbb{Z}$-module, is of <u>type</u> (r,s) if it is isomorphic, over $\mathbb{Z}$, to the form $q_{(r,s)}$ on $\mathbb{Z}^{r+s}$ given by

$$q_{(r,s)}(x_1, \cdots x_r, y_1, \cdots, y_s) = \Sigma_{i=1}^r x_i^2 - \Sigma_{j=1}^s y_j^2 .$$

We shall call M of <u>type</u> (r,s) if the self-intersection form q_M on $\Lambda(M)$ is of type (r,s).

Let us begin by describing the elliptic surfaces that we shall consider. Let $\mathbb{P}^2 = \mathbb{C}P^2$ denote the complex projective plane. Let C_0, C_∞ be generic cubic curves in $\mathbb{P}^2$, defined by homogeneous cubic polynomials $\{f_0 = 0\}$ and $\{f_\infty = 0\}$. Then C_0 and C_∞ define a <u>pencil</u> of cubics, i.e., a family $\{C_t : t \in \mathbb{P}^1\}$ of cubics, with equations

$$f_t(x) = f_0(x) + tf_\infty(x).$$

If C_0 and C_∞ are chosen to meet transversally at 9 points, the pencil $\{C_t\}$ defines the structure of an elliptic fibration on the algebraic surface X obtained from $\mathbb{P}^2$ by blowing up the 9 points $C_0 \cap C_\infty$. By this we mean that the map $\pi : X \to \mathbb{P}^1$ defined by $x \in C_t \mapsto t \in \mathbb{P}^1$ is a well-defined, holomorphic map with generic fiber a smooth elliptic curve. Since X is an algebraic surface, it has a canonical orientation, and is therefore in a canonical way a smooth oriented 4-manifold. As X is $\mathbb{P}^2$ blown up at 9 points, it is diffeomorphic, by an orientation-preserving diffeomorphism, to $\mathbb{C}P^2 \# 9\overline{\mathbb{C}P}^2$, where $\overline{\mathbb{C}P}^2$ denotes $\mathbb{C}P^2$ with the opposite orientation. Hence X is simply connected and of type (1,9).

Fix positive integers p,q. Choose two smooth fibers C_{t_0}, C_{t_1} of π, and let S(p,q) be the result of performing logarithmic transforms of orders p and q respectively at C_{t_0} and C_{t_1}. (See [8], p. 546 for a precise definition of logarithmic transforms.)

Hence S(p,q) is a complex analytic surface, and there is an induced map $S(p,q) \to \mathbb{P}^1$, also denoted π, which expresses S(p,q) as an elliptic surface. In fact, S(p,q) is always an algebraic surface. It is not difficult to show that the diffeomorphism type of S(p,q) depends only on the unordered pair (p,q).

THEOREM (Dolgachev [1]). If g.c.d.(p,q) = 1, then S(p,q) is a simply connected algebraic surface of type (1,9).

Thus, if g.c.d.(p,q) = 1, then S(p,q) is homotopy equivalent to, and hence h-cobordant to, X; moreover, by Freedman's results [6], S(p,q) is homeomorphic to X.

If either p or q is 1, one easily shows that S(p,q) is a rational surface, and hence diffeomorphic to X. We therefore make the following definition.

DEFINITION. A <u>Dolgachev surface</u> is an algebraic surface of the form S(p,q), where $p > 1$, $q > 1$ and g.c.d.(p,q) = 1.

Donaldson's remarkable example of two simply connected, h-cobordant 4-manifolds which are not diffeomorphic is the following.

THEOREM ([7]). S(2,3) and X are not diffeomorphic.

We extend this result as follows.

THEOREM 1. The function from unordered pairs of relatively prime integers both greater than one to diffeomorphism classes of 4-manifolds defined by

$$(p,q) \mapsto \text{diffeomorphism class of } S(p,q)$$

is finite-to-one. In particular, there is an infinite set of (p,q) such that no two of the associated surfaces S(p,q) are diffeomorphic.

There is a more precise result when $p = 2$.

THEOREM 2. If $S(2,q)$ and $S(2,q')$ are diffeomorphic Dolgachev surfaces, then $q = q'$.

The methods of proof also lead to restrictions on the self-diffeomorphisms of the $S(p,q)$. For a closed simply connected 4-manifold M, denote by $A(M)$ the automorphism group of the quadratic form on $H^2(M,\mathbb{Z})$. Any orientation-preserving self-homotopy equivalence of M induces an element of $A(M)$, and all elements of $A(M)$ arise in this way.

THEOREM 3. For every Dolgachev surface $S = S(p,q)$, the subgroup of $A(S)$ realized by self-diffeomorphisms is a subgroup of finite index inside

$$A_f(S) = \{\psi \in A(S) : \psi[C_t] = \pm[C_t]\}$$

where C_t is a general fiber of π and $[C_t]$ is the cohomology class Poincaré dual to its fundamental class. This subgroup is of infinite index in $A(S)$.

Theorems 1, 2, and 3 should be contrasted with the result in dimension ≥ 5 which says that the homotopy type and Pontrjagin classes of a simply connected C^∞-manifold determine its diffeomorphism class up to finitely many choices. (For a simply connected 4-manifold M, $p_1(M)$ is three times the signature of M and hence is a homotopy invariant.) Analogously, the group of self-diffeomorphisms of such a manifold maps onto a subgroup of finite index in the group of integral automorphisms of the cohomology ring which preserve the Pontrjagin classes. Theorem 3 should also be contrasted with the following result of C.T.C. Wall:

THEOREM [11]. Every element of $A(X)$ is realized by a self-diffeomorphism of X.

2. REVIEW OF THE DONALDSON INVARIANT

Throughout this section, M denotes a closed, oriented, simply connected, smooth 4-manifold of type $(1,n)$ for some $n \geq 1$. As before, we let $\Lambda(M)$ denote the lattice $H^2(M;\mathbb{Z})$ and $q_M : \Lambda(M) \to \mathbb{Z}$ the self-intersection form. Set

$$\Lambda_{\mathbb{R}}(M) = \Lambda(M) \otimes \mathbb{R}$$

and let $\bar{q}_M : \Lambda_{\mathbb{R}}(M) \to \mathbb{R}$ be the induced form. The associated inner product will be denoted $(\cdot)$.

The level set

$$\mathbb{H}(M) = \{x \in \Lambda_{\mathbb{R}}(M) : \bar{q}_M(x) = 1\}$$

is a hyperboloid of two sheets in $\Lambda_{\mathbb{R}}(M)$. Each sheet is a model of n-dimensional hyperbolic space. For every class $\alpha \in \Lambda(M)$ with $q_M(\alpha) = -1$, we define the wall associated to α:

$$W^{\alpha} = \{x \in \mathbb{H}(M) : x \cdot \alpha = 0\}.$$

The wall W^{α} has two components, each of which is a codimension-one, totally geodesic subspace of the component of $\mathbb{H}(M)$ in which it lies. Furthermore, each component is the negative of the other. Consider the collection

$$\mathcal{W}_{-1}(M) = \{W^{\alpha} : \alpha \in \Lambda(M), q_M(\alpha) = -1\}.$$

$\mathcal{W}_{-1}(M)$ is a locally finite set of walls, which partitions $\mathbb{H}(M)$ into chambers. Let $\mathcal{C}(M)$ denote the set of chambers. Since the walls W^{α} are invariant under multiplication by -1, if $C \in \mathcal{C}(M)$, then $-C \in \mathcal{C}(M)$ as well.

The Donaldson invariant is a function

$$\Gamma_M : \mathcal{C}(M) \to H^2(M;\mathbb{Z}).$$

Its basic properties (see [3]) are:

1) $\Gamma_M(-C) = -\Gamma_M(C)$ for all $C \in \mathcal{C}(M)$.

2) For every C_0 and $C_1 \in \mathcal{C}(M)$, if C_0 and C_1 meet along an open subset of a wall W^{α}, and α is chosen so that $\alpha \cdot C_0 \geq 0$ (i.e., so that $\alpha \cdot x \geq 0$ for every $x \in C_0$), then

$$\Gamma_M(C_1) = \Gamma_M(C_0) - 2\alpha.$$

3) If $f : M \to M'$ is an orientation-preserving diffeomorphism and $f^* : \mathcal{C}(M') \to \mathcal{C}(M)$ the induced map on chambers, then

$$\Gamma_M(f^*C) = f^*\Gamma_{M'}(C) \text{ for all } C \in \mathcal{C}(M').$$

There is a final property (4) which enables us to compute $\Gamma_M(C)$ in case M is an algebraic surface and C an appropriate chamber. To state it, we need some preliminary discussion. Suppose g is a Riemannian metric on M. Since M is of type (1,n), there is a unique harmonic two-form on M, up to non-zero scalar multiples, which is self-dual under the Hodge *-operator for g. Hence, there are exactly two such whose cohomology classes lie in $\mathbb{H}(M)$. They differ by a sign. Choose one such form and call it ω_g. Denote by $[\omega_g]$ its cohomology class.

In order to state the last property we shall make two assumptions about the metric g. To describe the first we need some notation. Let

$\Omega^2(M)$ be the vector bundle of C^∞ 2-forms on M. The *-operator for g acts on $\Omega^2(M)$ and $\Omega^2 M$ splits into eigenspaces $\Omega^2_\pm(M)$ for this operator. Each is oriented by the orientation on M, each is of rank 3, and ω_g is a section of $\Omega^2_+(M)$. Generically ω_g vanishes along a (real) 1-dimensional submanifold of M. Our first assumption on g is

ASSUMPTION (A): ω_g is nowhere vanishing.

If g is a Kähler metric on a complex surface, then ω_g is a multiple of the Kähler class. Hence, Assumption (A) holds in this case.

If g satisfies Assumption (A), then we define $k_g \in H^2(M;\mathbb{Z})$ to be the Euler class of the oriented two-plane bundle $(\omega)^\perp \subset \Omega^2_+(M)$.

Clearly, replacing ω_g by $-\omega_g$ reverses the orientation on $(\omega_g)^\perp$ and consequently replaces k_g by $-k_g$.

Our second assumption on g concerns the moduli space $\mathcal{M}$ of anti-self-dual connections (with respect to g) on the SU(2)-bundle P over M with $c_2 = 1$, modulo gauge equivalence. This moduli space is cut out of the space $\mathcal{A}$ of all irreducible connections modulo gauge equivalence by natural equations. The second assumption that we make is:

ASSUMPTION (B): The differentials of the natural defining equations for $\mathcal{M} \subset \mathcal{A}$ are of maximal rank.

Under Assumption (B), the Atiyah-Singer index theorem implies that $\mathcal{M} \subset \mathcal{A}$ will be a smooth 2-manifold, see [5], p. 49.

By a result of Uhlenbeck [10], if g satisfies Assumption (A), then $\mathcal{M}$ is compact. The choice of ω_g gives $\mathcal{M}$ an orientation, and changing the sign of ω_g reverses the induced orientation on $\mathcal{M}$.

There is a universal U(2) bundle with connection

$$\mathcal{P} \to M \times \mathcal{M},$$

by [4], which is tautological in an appropriate sense. In other words, $\mathcal{P}|M \times \{x\}$ "is" the bundle P with anti-self-dual connection corresponding to $x \in \mathcal{M}$. Set

$$m_g = (\pi_1)_* c_2(\mathcal{P}) \in H^2(M;\mathbb{Z})$$

where π_1 is the projection onto the first factor.

We can now state the last property of Γ_M as follows:

4) If g is a metric satisfying (A) and (B), and if $C \in \mathcal{C}(M)$ contains $[\omega_g]$, then

$$\Gamma_M(C) = k_g + 2m_g.$$

3. STABLE BUNDLES ON DOLGACHEV SURFACES

We define a Hodge metric g on a complex surface S to be a Kähler metric g whose Kähler class represents an integral cohomology class. According to the Kodaira embedding theorem, if g is a Hodge metric, there is an ample divisor L on S whose associated cohomology class is equal, in cohomology, to the Kähler class of g. By a fundamental result of Donaldson [2], there is a bijective correspondence.

$$\left\{\begin{array}{l}\text{irreducible anti-self-dual}\\ \text{connections on principal}\\ SU(2)\text{-bundles } P \text{ over } S\\ \text{modulo } C^\infty\text{- isomorphism}\end{array}\right\} \longleftrightarrow \left\{\begin{array}{l}L\text{-stable (in the sense of}\\ \text{algebraic geometry [2])}\\ \text{rank-two holomorphic vector}\\ \text{bundles } V \text{ over } S \text{ with } c_1(V)=0\\ \text{modulo algebraic isomorphism}\end{array}\right\}$$

In this correspondence, $c_2(P) = c_2(V)$. Hence, by (4) above, to calculate Γ_S, we need to study the moduli space of L-stable rank-two holomorphic vector bundles over S with $c_1 = 0$ and $c_2 = 1$. Moreover, if ω_g is the Kähler form associated to g, one checks easily that $k_g = [K_S]$, the canonical class of S viewed as an element of $H^2(S;\mathbb{Z})$.

Let $S = S(p,q)$ be a Dolgachev surface. Let f, F_p, F_q denote, respectively, a general fiber of the map π and the fibers of multiplicities p and q. Denote by [f], $[F_p]$, and $[F_q]$ the cohomology classes dual to the classes of f, F_p, F_q. One checks easily that $p[F_p] = q[F_q] = [f]$, and that $[f] = pq\kappa$, where $\kappa \in \Lambda(S)$ is a primitive class.

LEMMA 4. As divisor classes $K_S = -f + (p-1)F_p + (q-1)F_q$. Hence, as cohomology classes

$$[K_S] = (pq - p - q)\kappa.$$

To calculate m_g, we need an appropriate Hodge metric g. Since the surface S is projective algebraic, a Hodge metric exists. Choose one such, and suppose that it corresponds to the ample divisor L. If K_S is the canonical class, then since the divisor class K_S is a positive rational multiple of f, we have $K_S \cdot C \geq 0$ for all irreducible (holomorphic) curves

$C \subseteq S$. It follows easily that, for every $r \geq 0$, $L + rK_S$ is an ample divisor in S, corresponding to some Hodge metric.

THEOREM 5. If $r >> 0$, and m_g is the invariant described in §2, computed with respect to the Hodge metric g associated to $L + rK_S$, then

$$m_g = \varepsilon(p,q)\kappa \text{ for } \varepsilon(p.q) \text{ a positive integer depending only on p and q.}$$

Moreover, $\varepsilon(2,q) = \frac{q(q-1)}{2}$.

Strictly speaking, Theorem 5 is proved - and applied - when S is a generic Dolgachev surface in an appropriate sense. From this, the result follows easily for all Dolgachev surfaces.

Similar results have been obtained by Okonek and Van de Ven [9].

COROLLARY 6. If $C \in \mathcal{C}(S)$ and if $\mathbb{R}^+ \cdot C$ contains the class $[L + rK_S]$, for some r sufficiently large, then

$$\Gamma_S(C) = n(p,q)\kappa,$$

where $n(p,q)$ depends only on p and q,

$$n(p,q) \geq pq - p-q$$

and

$$n(2,q) = q^2 - 2.$$

4. ARITHMETIC OF FORMS OF TYPE (1,9).

Let $q : \Lambda \to \mathbb{Z}$ be a form of type (1,9). Setting $\Lambda_{\mathbb{R}} = \Lambda \otimes \mathbb{R}$, let $\bar{q} : \Lambda_{\mathbb{R}} \to \mathbb{R}$ be the natural extension of q. Let $(\cdot)$ denote the associated bilinear pairing $\Lambda_{\mathbb{R}} \otimes \Lambda_{\mathbb{R}} \to \mathbb{R}$. Define

$$\mathbb{H}(q) = \{x \in \Lambda_{\mathbb{R}} : \bar{q}(x) = 1\},$$

a hyperboloid of two sheets. There is a natural identification $\mathbb{H}(q) \cong \{x \in \Lambda_{\mathbb{R}} : \bar{q}(x) > 0\}/\mathbb{R}^+$. This leads to a compactification of $\mathbb{H}(q)$ given by $\overline{\mathbb{H}(q)} = (\{x \in \Lambda_{\mathbb{R}} : \bar{q}(x) \geq 0\} - \{0\})/\mathbb{R}^+$. $\overline{\mathbb{H}(q)}$ is the natural compactification of $\mathbb{H}(q)$ obtained by adding two spheres at infinity. For each $\alpha \in \Lambda$ with $q(\alpha) = -1$ define W^α analogously as in §2. Set

$$\mathcal{H}_{-1}(q) = \{W^\alpha \subseteq \mathbb{H}(q) : \alpha \in \Lambda,\ q(\alpha) = -1\},$$

$$R_{\alpha_i}(x) = x + 2(x \cdot \alpha_i)\alpha_i.$$

The reflections R_{α_i} act on $\mathbb{H}(S)$ and preserve the components and chamber structure of $\mathbb{H}(S)$. Hence $R_{\alpha_i}(C_{i-1}) = C_i$. Set

$$w = R_{\alpha_r} \circ R_{\alpha_{r-1}} \circ \cdots \circ R_{\alpha_1},$$

so that $wC = C'$.

CLAIM. $\Gamma_S(wC) - \Gamma_S(C) = \kappa_C - w\kappa_C$.

Proof. It suffices to show that $\Gamma_S(C_i) - \Gamma_S(C_{i-1}) = \kappa_{C_{i-1}} - \kappa_{C_i}$ and then sum over i. But, by Lemma 8 $\kappa_{C_i} = R_{\alpha_i}(\kappa_{C_{i-1}}) = \kappa_{C_{i-1}} + 2\alpha_i$, as $\alpha_i \cdot \kappa_{C_{i-1}} = 1$. So, the claim is an immediate consequence of property (2) for Γ_S.

To prove Theorem 10, we write

$$\Gamma_S(wC) = \Gamma_S(C) + \kappa_C - w\kappa_C$$

$$= (n(p,q) + 1)\kappa_C - w\kappa_C.$$

Hence $q_S(\Gamma_S(wC)) = -2(n(p,q) + 1)(\kappa_C \cdot w\kappa_C)$.

Since κ_C and $w\kappa_C$ lie in the same component of

$$(\{x \in \Lambda_{\mathbb{R}}(S) \mid \bar{q}_S(x) \geq 0\} - \{0\}),$$

we have $\kappa_C \cdot w\kappa_C \geq 0$. Thus, $q_S(\Gamma_S(wC)) \leq 0$. Suppose that $q_S(\Gamma_S(wC)) = 0$. Then $\kappa_C \cdot w\kappa_C = 0$. Since $q(\kappa_C) = q(w\kappa_C) = 0$, and a maximal dimensional isotropic subspace for $\bar{q}_S$ in $\Lambda_{\mathbb{R}}(S)$ is one dimensional, if $\kappa_C \cdot w\kappa_C = 0$, then κ_C and $w\kappa_C$ are multiples of each other. Since they lie in the same component, they are in fact positive multiples of each other. But both κ_C and $w\kappa_C$ are primitive in Λ. Hence, $\kappa_C = w\kappa_C$. Consequently $C = wC = C'$.

REMARK. One shows easily that for any Hodge metric g for the rational surface X, the invariant $m_g = 0$. Hence, for any chamber C in $\mathbb{H}(X)$ containing the class associated to a Hodge metric, we have $\Gamma_X(C) = -\kappa_C$. (i.e., $n(1,1) = -1$. The argument given in the proof of theorem 10 shows that $\Gamma_X(C) = -\kappa_C$ for all chambers $C \in \mathcal{C}(X)$.

Lemma 8 may be formulated in geometric terms as follows. If γ is a geodesic ray beginning at x and tending toward the "ideal point" $\kappa_C \in \overline{\mathbb{H}(q)}$, then there exists a $t_0 \geq 0$ such that, for all $t \geq t_0$, $\gamma(t) \in C$. In other words, the geodesic ray γ eventually enters C and remains there for all future time.

5. THE MAIN RESULT

We fix a Dolgachev surface $S = S(p,q)$. Let $\kappa = \kappa(p,q)$ be the primitive class defined in §3, so that $pq\kappa = [f]$ and $[K_S] = (pq - p - q)\kappa$.

LEMMA 9. κ satisfies (3) of theorem 7.

Proof. Since f is a fiber of π, $f \cdot f = 0$, so that $q_S(\kappa) = 0$. Since κ is a rational multiple of K_S, $(\kappa)^{\perp} = [K_S]^{\perp}$. But $[K_S]$ reduces modulo 2 to the second Stiefel-Whitney class of S, so that $q_S \,|\, [K_S]^{\perp} \cap \Lambda(S)$ is even.

Let C be the chamber corresponding to κ, i.e., $\kappa = \kappa_C$. Since κ is a positive multiple of [f] and L is ample $[L]\cdot\kappa > 0$. This means that L and κ, and hence [L] and C, lie in the same component of $\{x \in \Lambda_{\mathbb{R}}(S) \,|\, \bar{q}_S(x) \geq 0\} - \{0\}$. Hence, by lemma 8, for r sufficiently large, $[L + rK_S] \in \mathbb{R}^+ \cdot C$. By corollary 6, $\Gamma_S(C) = n(p,q)\kappa$.

THEOREM 10. C and -C are the unique chambers in $\mathcal{C}(S)$ such that $q_S(\Gamma_S(C)) = 0$. In fact, for all other chambers $C' \neq \pm C$, we have

$$q_S(\Gamma_S(C')) < 0.$$

Proof. Since $\Gamma_S(C) = n(p,q)\kappa$ and $\Gamma_S(-C) = -\Gamma_S(C)$, clearly

$$q_S(\Gamma_S(C)) = q_S(\Gamma_S(-C)) = n(p,q)^2 \cdot q_S(\kappa) = 0.$$

Let C' be any other chamber in $\mathcal{C}(S)$. Using Property (1) for Γ_S we may clearly assume that C' lies in the same component of $\mathbb{H}(S)$ as C. Choose a geodesic arc γ in $\mathbb{H}(M)$ joining C to C'. After perturbing γ, we may assume that it does not pass through the intersection of two walls in $\mathcal{W}_{-1}(q)$. Enumerate the chambers through which γ passes in order as $C_0 = C$, $C_1, \cdots, C_r = C'$, and suppose that the wall between C_{i-1} and C_i is W^{α_i}, with α_i chosen so that $\alpha_i \cdot C_{i-1} \geq 0$. It follows that $\alpha_i \cdot \kappa_{C_{i-1}} = 1$. To each α_i, we may associate the reflection R_{α_i} in the wall W^{α_i} (which is induced from an isometry of Λ):

and

$\mathcal{C}(q)$ = set of chambers for the walls $\mathcal{W}_{-1}(q)$.

Our main results concerning $\mathcal{C}(q)$ is the following:

THEOREM 7. 1) For each $C \in \mathcal{C}(q)$, there is a unique $\kappa_C \in \Lambda$ such that, for $\alpha \in \Lambda$ and $q(\alpha) = -1$, $\kappa_C \cdot \alpha = 1$ if and only if W^α is a wall of C and $\alpha \cdot C \geq 0$. Consequently, if φ is any integral automorphism of (Λ, q) then $\varphi(\kappa_C) = \kappa_{\varphi(C)}$.

2) The class κ_C is a primitive integral vector, $q(\kappa_C) = 0$, and $q|((\kappa_C)^\perp \cap \Lambda)$ is an even form.

3) Conversely, if $\kappa \in \Lambda$ satisfies the conditions in 2), $\kappa = \kappa_C$ for a unique chamber $C \in \mathcal{C}(q)$.

Theorem 7 is proved in [7] as follows. By the elementary algebraic geometry of rational surfaces, one checks that there exists a chamber C and class κ_C as described in 1) and 2) for $\Lambda = \Lambda(X)$. Here X is a generic rational elliptic surface as described in the introduction. In fact, C is the Kähler cone of X, intersected with $\mathbf{H}(X)$, and $\kappa_C = -[K_X]$, where $[K_X]$ denotes the cohomology class associated to the canonical divisor K_X of X. By transport of structure, 1) and 2) must hold for every chamber and every form isometric to $\Lambda(X) = \Lambda$. Finally, 3) is a simple exercise in quadratic forms.

We need one more result about κ_C.

LEMMA 8. If $x \in \mathbf{H}(q)$ is in the same component as C, then for all $N >> 0$,

$$x + N\kappa_C \in \mathbb{R}^+ \cdot C.$$

Proof. Let $A_C = \{\alpha \in \Lambda : q(\alpha) = -1 \text{ and } \kappa_C \cdot \alpha = 1\}$. Thus, A_C is the set of classes defining oriented walls of C.

CLAIM. $\{\alpha \in A_C : x \cdot \alpha < 0\}$ is a finite set.

Proof of the claim. Choose $y \in C$. Then $y \cdot \alpha \geq 0$ for all $\alpha \in A_C$. Let γ be a geodesic line segment in $\mathbf{H}(q)$ joining x to y. If $\alpha \in A_C$ and $x \cdot \alpha < 0$, then γ must cross the wall W^α. By local finiteness of $\mathcal{W}_{-1}(q)$, γ can meet only finitely many walls, and this proves the claim.

To complete the proof of Lemma 8, note that, since $x \cdot \alpha \geq 0$ for all but finitely many $\alpha \in A_C$ and $\kappa_C \cdot \alpha = 1$ for all $\alpha \in A_C$, if $N >> 0$, then $(x + N\kappa_C) \cdot \alpha > 0$ for $\alpha \in A_C$. In other words, if $N >> 0$, then $(x + N\kappa_C) \in \mathbb{R}^+ \cdot C$.

COROLLARY 11. Let $S = S(p,q)$ and $S' = S(p',q')$ be Dolgachev surfaces. Let $\varphi : S \to S'$ be a diffeomorphism. Let $C \in \mathcal{C}(S')$ be given with $q_S(\Gamma_S(C)) = 0 = q_{S'}(\Gamma_{S'}(C'))$. Then $\varphi^*C' = \pm C$.

Proof. Since S and S' are of type (1,9), every diffeomorphism $\varphi : S \to S'$ is automatically orientation preserving. Corollary 11 is now an immediate consequence of theorem 10 and property (3) for Γ_S.

Proof of Theorems 1, 2, and 3. Let $S = S(p,q)$ and $S' = S(p',q')$ be Dolgachev surfaces, and let $C \in \mathcal{C}(S)$ and $C' \in \mathcal{C}(S')$ be chambers for which $q_S(\Gamma_S(C)) = 0 = q_{S'}(\Gamma_{S'}(C'))$. Then $n(p,q)$ and $n(p',q')$ are the orders of divisibility in $\Lambda(S)$ and $\Lambda(S')$ of $\Gamma_S(C)$ and $\Gamma_{S'}(C')$. Suppose $\varphi : S \to S'$ is a diffeomorphism. Then by corollary 11, $\varphi^*(C') = \pm C$. By Properties (1) and (3) for the Donaldson invariant, it follows that

$$\varphi^*\Gamma_{S'}(C') = \pm \Gamma_S(C).$$

In particular, the orders of divisibility of $\Gamma_{S'}(C')$ in $\Lambda(S')$ and of $\Gamma_S(C)$ in $\Lambda(S)$ are equal, i.e., $n(p,q) = n(p',q')$. Theorems 1 and 2 are immediate from this and corollary 6.

Now suppose $\varphi : S \to S$ is a self-diffeomorphism. By Property (3) for the Donaldson invariant

$$q_S(\Gamma_S(\varphi^*C)) = q_S(\varphi^*(\Gamma_S(C))) = \varphi^*q_S(\Gamma_S(C)) = 0.$$

By corollary 11, this implies that $\varphi^*C = \pm C$. Hence $\varphi^*\kappa_C = \pm\kappa_C$. Since κ_C is a non-zero multiple of $[f]$, it follows that

$$\varphi^*[f] = \pm[f].$$

Hence, $\varphi^* \in A_f(S)$. Details of the proof that a subgroup of finite index in $A_f(S)$ is realized by self-diffeomorphisms may be found in [7].

Finally, we shall state the analogues of theorems 1 and 2 which hold for blown up Dolgachev surfaces. These results are technically much more difficult, but the philosophy of the proof is the same.

Let $\tilde{S}_r(p,q)$ = the blowup of $S(p,q)$ at r points. Note that there is an orientation preserving diffeomorphism from $\tilde{S}_r(p,q)$ to $S(p,q) \# r\overline{\mathbb{CP}}^2$.

THEOREM. If $\tilde{S}_r(p,q)$ is diffeomorphic to $\tilde{S}_r(p',q')$, then $n(p,q) = n(p',q')$, where $n(p,q)$ is the integer defined in corollary 6. In particular, for every r, there are infinitely many $\tilde{S}_r(p,q)$, no two of which are diffeomorphic.

There is also an analogue of theorem 3, which characterizes, up to finite index, the subgroup of those elements of $A(\tilde{S}_r(p,q))$ arising from self-diffeomorphisms. For proofs of these and other results, we refer to [7].

ACKNOWLEDGEMENT

The authors were supported by NSF grants.

REFERENCES

1. Dolgachev, I., Algebraic surfaces with $q = p_g = 0$, in Algebraic Surfaces, proceedings of 1977 CIME, Cortona, Liguori Napoli (1981) 97-215.
2. Donaldson, S., Anti-self-dual Yang-Mills Connections Over Complex Algebraic Surfaces and Stable Vector Bundles, Proc. London Math. Soc. 50 (1985), 1-26.
3. Donaldson, S., La Topologie Différentielle des Surfaces Complexes, C.R. Acad. Sci. 301 (1985), 317-320.
4. Donaldson, S., Connections, Cohomology and the Intersection Forms of 4-manifolds, preprint.
5. Freed, D., and Uhlenbeck, K., Instantons and Four-Manifolds. MSRI publications, New York, Springer, 1984.
6. Freedman, M., The Topology of Four-Dimensional Manifolds, J. Diff. Geom. 17 (1982), 357-454.
7. Friedman, R., and Morgan, J., On the Diffeomorphism Types of Certain Algebraic Surfaces, to appear.
8. Griffiths, P., and Harris, J., Principles of Algebraic Geometry. New York, John Wiley, 1978.
9. Okonek, C., and Van de Ven, A., preprint.
10. Uhlenbeck, Removable singularities in Yang-Mills fields, Comm. Math. Phys. 95 (1984), 345-391.
11. Wall, C.T.C., Diffeomorphisms of 4-manifolds, J. London Math. Soc. 39 (1964), 131-140.

Deformations of Flat Bundles Over Kähler Manifolds

WILLIAM M. GOLDMAN / Department of Mathematics, Massachusetts Institute of Technology, Cambridge, Massachusetts

JOHN J. MILLSON / Department of Mathematics, University of California, Los Angeles, California

The purpose of this paper is to describe a general group-theoretical property enjoyed by fundamental groups of compact Kähler manifolds. Let M be a compact Kähler manifold, and let $\pi = \pi_1(M)$ denote its fundamental group. Let G be a Lie group with Lie algebra g. We denote by Hom(π, G) the analytic variety consisting of all homomorphisms $\pi \to G$. The main result of this paper concerns the local structure of Hom(π, G) when G is a compact Lie group. Suppose that $\psi \in$ Hom(π, G). Then the Zariski tangent space to Hom(π, G) equals the space $Z^1(\pi; g_{Ad\psi})$ of 1-cocycles of π taking coefficients in the π-module $g_{Ad\psi}$ defined by the composition Ad ψ : $\pi \to G \to$ Aut(g). We shall call an element of the Zariski tangent space $Z^1(\pi; g_{Ad\psi})$ an <u>infinitesimal deformation</u>. Let [,] denote the bilinear operation $Z^1(\pi; g_{Ad\psi}) \to Z^1(\pi; g_{Ad\psi}) \to H^2(\pi; g_{Ad\psi})$ defined by cup product using the Lie bracket on g as a coefficient pairing. Our main result is the following:

THEOREM A. Let G be a compact Lie group and let $\psi \in$ Hom(π, G). Suppose that $u \in Z^1(\pi; g_{Ad\psi})$ is a infinitesimal deformation. Then u is tangent to an analytic path in Hom(π, G) if and only if $[u, u] = 0$.

Alternatively, Theorem A states that the tangent cone to Hom(π, G) is defined by the quadratic equations $[u, u] = 0$. From Theorem A we obtain a sufficient condition that ψ be a smooth point of Hom(π, G):

COROLLARY B. A representation ψ is a smooth point of Hom(π, G) if the pairing $Z^1(\pi; g_{Ad\psi}) \to Z^1(\pi; g_{Ad\psi}) \to H^2(\pi, g_{Ad\psi})$ is identically zero.

It is generally true (compare [10]) that if $H^2(\pi;g_{Ad\psi}) = 0$, then ψ is a smooth point of $Hom(\pi,G)$ (the converse is true if π is the fundamental group of a surface). It is tempting to conjecture that there is a neighborhood of ψ in $Hom(\pi,G)$ which is analytically equivalent to the quadratic cone in $Z^1(\pi;g_{Ad\psi})$ defined by $[u,u] = 0$. When M is a Riemann surface and G is compact, this follows from a general criterion of Arms-Marsden-Moncrief [1] dealing with level sets of momentum mappings of certain group actions. (We describe this in the appendix.) When M is a Riemann surface, G is a reductive algebraic group, and ψ is a representation whose image is Zariski dense in a reductive subgroup of G, then the analogue of Theorem A is proved in Goldman [7]. These results seem to suggest a more general result may hold.

Theorem A is proved by reducing the representation-theoretic problem involving group cohomology to an analytic problem involving connections and de Rham cohomology. Thus the first step is to reduce questions about homomorphisms of fundamental groups to questions about flat connections. The following lemma achieves this reduction:

LEMMA C. Let M be any manifold, G a compact Lie group, and P a principal G-bundle over M. Denote the space of flat connections on P by $\mathbf{F}(P)$. Let $Hom(\pi,G)_P$ denote the open subset of $Hom(\pi,G)$ consisting of representations whose associated principal bundle is isomorphic to P. Let $x_0 \in M$ be a basepoint and let $\pi = \pi_1(M,x_0)$ denote the fundamental group of M at x_0. Let hol : $\mathbf{F}(P) \to Hom(\pi,G)_P$ be the map which associates to a flat connection its holonomy representation at x_0. Let $\mathbf{G}_0$ denote the group of bundle automorphisms $P \to P$ which are the identity at the fiber over x_0. Then hol is a principal bundle with structure group $\mathbf{G}_0$.

This lemma relates the singularity of $Hom(\pi,G)$ at ψ to the singularity of the infinite-dimensional space $\mathbf{F}(P)$ at the corresponding flat connection with holonomy ψ. However, the sacrifice of finite-dimensionality is overcome by the ability to apply the techniques of Hodge theory on a Kähler manifold (with coefficients in a flat vector bundle). Thus, using Lemma C we shall reduce Theorem A to a result on the local structure of $\mathbf{F}(P)$. To state this result, let ad P denote the flat vector bundle associated to P via the adjoint representation. Then the Zariski tangent space to $\mathbf{F}(P)$ at a flat connection A is the space $Z^1(M, \text{ad } P)$ of ad P-valued 1-forms which are closed with respect to the covariant differential corresponding to A. Such a covariant closed ad P-valued 1-form we shall call an <u>infinitesimal</u>

deformation of the flat connection A. There is a quadratic function which we also denote by [,], on $Z^1(M, \text{ad } P)$ taking values in the de Rham cohomology group $H^2(M, \text{ad } P)$, given by exterior product of differential forms using the Lie bracket on coefficients. The analogue of Theorem A is the following:

THEOREM D. Let $\eta \in Z^1(M, \text{ad } P)$ be an infinitesimal deformation of a flat connection $A \in \mathbf{F}(P)$. Then η is tangent to an analytic path in $\mathbf{F}(P)$ if and only if $[\eta,\eta] = 0$.

Our original motivation for this work grew out of our earlier work on local rigidity of lattices in the unitary group U(n,1). In a previous paper [8], we showed that if ψ is the composition of the inclusion of a cocompact lattice $\pi \subset U(n,1)$ with the inclusion $U(n,1) \subset U(n+1,1)$, then the only deformations of ψ in $\text{Hom}(\pi,U(n+1,1))$ correspond to homomorphisms of π into the centralizer of U(n,1) in U(n+1,1). This local rigidity theorem is proved by identifying the Zariski normal space of $\text{Hom}(\pi,U(n+1,1))$ with a space of 1-cocycles of π with coefficients in the standard representation V of U(n,1), and showing that the corresponding cup product pairing $\mathbf{B} : H^1(\pi,V) \times H^1(\pi,V) \to H^2(\pi,u(n,1))$ is "positive definite," i.e., there exists a natural linear map $\kappa : H^2(\pi,u(n,1)) \to \mathbf{R}$ such that the composition $\kappa \circ \mathbf{B}$ is a positive definite quadratic form on $H^1(\pi,V)$. Although the Lie groups involved in this problem are noncompact, one can show (using Galois conjugation) that the analogue of Theorem A applies to deformations of homomorphisms of some cocompact arithmetic subgroups of U(n,1) which factor through representations of U(n,1).

In general, the condition that an infinitesimal deformation be tangent to an analytic path is equivalent (Artin [2]) to the existence of a formal power series deformation with the infinitesimal deformation as its leading term. If a k-jet of a deformation exists, the condition that this k-jet of a deformation extend to a (k+1)-jet of a deformation is given by the vanishing of a certain cohomology class, which depends on the previous k terms of the series. (Indeed, the coefficient of order k+1 is a cochain whose coboundary equals this cocycle.) Since the first obstruction is given by the cup product, the higher obstructions are secondary operations derived from the cup product, somewhat like power operations associated with Massey products, i.e., "Massey powers." Our proof of Theorems A and

C follows the general outline of the proof in Deligne-Griffiths-Morgan-Sullivan [4] that on a Kähler manifold Massey products vanish identically.

The paper is organized as follows. In the first section, Lemma C is proved, and using the canonical isomorphisms between de Rham cohomology and Eilenberg-MacLane cohomology, Theorem A is reduced to Theorem D. In the second section, the obstructions to realizing an infinitesimal deformation are developed. The third section brings in harmonic theory and Theorem D is proved. Finally, in the appendix, the arguments of Arms-Marsden-Moncrief [1] are used to show that when M is a Riemann surface, then **F**(P) and Hom(π,G) are locally equivalent (around any point) to the tangent cone, which is defined by a system of homogeneous quadratic equations.

1. FLAT CONNECTIONS AND REPRESENTATIONS OF THE FUNDAMENTAL GROUP

Let M be a manifold, G a Lie group and P a principal G-bundle over M. We shall assume that P admits a flat connection and the goal of this section is to understand the space **F**(P) of flat connections on P. For the definition and basic properties of connections on principal bundles, the reader is referred to [6]. Let **A**(P) denote the space of all connections on P. We shall always topologize spaces of connections by the Sobolev s-norms, where s is sufficiently large to ensure that the connections are at least C^2. We shall think of a connection in terms of its covariant differential operator as follows.

Let Ad P denote the g-bundle associated to P by the adjoint representation Ad : G $\to$ Aut(g) and let Ω^p(M, ad P) denote the space of all ad P-valued exterior p-forms on M. Exterior product combined with Lie bracket on the fibers of ad P defines a multiplication $[\ ,\ ] : \Omega^p(M, \text{ad } P) \otimes \Omega^q(M, \text{ad } P) \to \Omega^{p+q}(M, \text{ad } P)$. Under this multiplication, the complex Ω^*(M, ad P) is a graded Lie algebra, i.e., if $\alpha \in \Omega^p(M, \text{ad } P)$, $\beta \in \Omega^q(M, \text{ad } P)$, $\gamma \in \Omega^r(M, \text{ad } P)$, then

$$[\alpha,\beta] + (-1)^{p+q}[\beta,\alpha] = 0 \tag{1.1}$$

and

$$(-1)^{rp}[\alpha,[\beta,\gamma]] + (-1)^{qr}[\beta,[\gamma,\alpha]] + (-1)^{qr}[\gamma,[\alpha,\beta]] = 0. \tag{1.2}$$

(In particular if α is any ad P-valued form, then $[\alpha,[\alpha,\alpha]] = [[\alpha,\alpha],\alpha] = 0$.)

The covariant differential operator corresponding to a connection A is a map $d_A : \Omega^0(M, \text{ad } P) \to \Omega^1(M, \text{ad } P)$ which is linear over **R** and, for any smooth function f on M, satisfies $d_A(f\alpha) = f\, d_A(\alpha) + df \wedge \alpha$. It is easy to see that if A and A' are two connections, then the difference of the corresponding covariant differential operators is exterior multiplication by an ad P-valued 1-form. Thus the space **A**(P) of connections on P is an affine space whose vector space of translations is $\Omega^1(M, \text{ad } P)$. In particular, for each connection $A \in \mathbf{A}(P)$, the tangent space to **A**(P) at A is naturally identified with $\Omega^1(M, \text{ad } P)$. Given $\eta \in \Omega^1(M, \text{ad } P)$ and $A \in \mathbf{A}(P)$, the connection obtained by "translating" A by η is denoted by $A + \eta$, i.e., $d_{A+\eta}(\alpha) = d_A(\alpha) + [\eta,\alpha]$. Covariant differentiation and exterior multiplication are related by

$$d_A([\alpha,\beta]) = [d_A\alpha,\beta] + (-1)^p[\alpha,d_A\beta]. \tag{1.3}$$

Let $\eta \in \Omega^1(M, \text{ad } P)$ and let $A \in \mathbf{A}(P)$. The curvature of a connection A will be denoted F(A). Then we have:

$$F(A+\eta) - F(A) = d_A(\eta) + \frac{1}{2}[\eta,\eta] \tag{1.4}$$

from the formula for the curvature of a connection. Thus if A is a flat connection, then every other flat connection is given by $A + \eta$ where η satisfies

$$d_A(\eta) + \frac{1}{2}[\eta,\eta] = 0. \tag{1.5}$$

Suppose that A is a connection. Then given any path $\gamma(t)$ in M, there exists a parallel transport operator τ_γ which maps the fiber over $\gamma(0)$ to the fiber over $\gamma(1)$. If A is flat, then parallel translation τ_γ depends only on the homotopy class of γ. If γ is a closed loop in M based at x_0, then τ_γ is a transformation of the fiber over x_0, and hence can be identified with an element of G, the holonomy of the connection around γ. Denoting the fundamental group $\pi_1(M, x_0)$ by π, we obtain a homomorphism $\text{hol}(A) : \pi \to G$, called the holonomy homomorphism of the flat connection A. The main goal of this section is to discuss the holonomy map $\text{hol} : \mathbf{F}(P) \to \text{Hom}(\pi,G)$ and to show that it is a principal fibration having as structure group the group $\mathbf{G}_0$ of bundle automorphisms $P \to P$ which are the identity on the fiber over x_0 (Theorem D).

We shall break the proof of Theorem D into several lemmas:

LEMMA 1.1. Suppose that A and A' are two flat connections on P such that hol(A) = hol(A'). Then there exists a gauge tranformation $g \in \mathbf{G}_0$ such that g(A) = A'.

Proof: Let τ and τ' be the parallel transport operators for the connections A and A respectively. Let $\tilde{M}$ denote the universal covering space of M corresponding to the basepoint x_0, i.e., $\tilde{M}$ is defined as the space of homotopy classes of paths on M which begin at x_0. (The covering projection $p : \tilde{M} \to M$ associates to each path its endpoint in M.) Then for each such path γ, the operators τ_γ and τ'_γ map the fiber over x_0 to the fiber over $\gamma(1)$. Thus the product $\tau_\gamma \tau_\gamma'^{-1}$ is an inner automorphism of the fiber over $\gamma(1)$. In this way we construct, for every element $\tilde{x}$ of $\tilde{M}$, an inner automorphism $\tilde{g}(\tilde{x})$ of the fiber over $p(\tilde{x})$. The condition that hol(A) = hol(A') implies that $\tilde{g}(\tilde{x})$ depends only on $p(\tilde{x})$, i.e., that $\tilde{g}$ defines a gauge transformation g of P. Clearly g acts as the identity over x_0 and g(A) = A'.

LEMMA 1.2. $\mathbf{G}_0$ acts freely on **F**(P).

Proof: (Compare Atiyah-Jones [3].) Suppose that A is a flat connection and that $g \in \mathbf{G}_0$ fixes A. The condition that g fixes A is equivalent to g being parallel with respect to A. Thus g is a parallel section of the G-bundle Ad G associated to P by the action of G on itself by inner automorphisms. Since g is the identity transformation on the fiber of P over x_0, and M is connected, it follows that g must be the identity on each fiber, and hence g is the identity.

Since the map hol : $\mathbf{F}(P) \to \mathrm{Hom}(\pi, G)$ is clearly invariant under $\mathbf{G}_0$, to show that given any connection $A \in \mathbf{F}(P)$, there exists a local slice containing A for the action of $\mathbf{G}_0$. Such a slice is constructed in [3], to which we refer for details.

2. DEFORMATION THEORY OF FLAT CONNECTIONS

In this section we discuss the local theory of the space **F**(P) of flat connections on a principal bundle. Although this theory is a special case of the general theory of deformations of pseudogroup structures (compare Guillemin-Sternberg [9]) the special case we need is particulary neat, and we give a self-contained account here.

Let $A \in \mathbf{F}(P)$ be a flat connection; we are interested in the geometry of the space **F**(P) near A. Using A, we may identify the space **A**(P) of all connections with the space $\Omega^1(M, \text{ad } P)$ of ad P-valued 1-forms on M, namely to an ad P-valued 1-form ϕ on M we associate the connection $A + \phi$ on P. The condition that the connection $A + \phi$ is also a flat connection is precisely that

$$d_A(\phi) + \frac{1}{2}[\phi, \phi] = 0 . \tag{2.1}$$

Thus we may identify **F**(P) with the space of connections $A + \phi$ where ϕ satisfies (2.1); we henceforth identify **F**(P) with this subset of $\Omega^1(M, \text{ad } P)$. Suppose that $A + \phi(t)$ is a differentiable path of flat connections with tangent vector $\eta = \phi'(0) \in \Omega^1(M, \text{ad } P)$; then (2.1) is satisfied by $\phi = \phi(t)$ for each t, and by differentiating (2.1) with respect to t, we see that ϕ must satisfy the linear equation $d_A\eta = 0$. This equation represents the linearization of (2.1), and so (in analogy with finite-dimensional analytic varieties), we may say that the Zariski tangent space to **F**(P) at A is the space $Z'(M, \text{ad } P)$ of d_A-closed ad P-valued 1-forms on M. Such an ad P-valued 1-form will be called an infinitesimal deformation of the flat connection A. However, if **F**(P) is singular at A, η may very well not be tangent to a path of flat connections.

To determine whether η is actually tangent to a path of flat connections, we may assume that $\phi(t)$ is analytic at $t = 0$, and expand $\phi(t)$ in a power series:

$$\phi(t) = \sum_{n=1}^{\infty} \phi_n t^n \tag{2.2}$$

where $\phi_1 = \eta$. Then equation (2.1) is equivalent to the system of equations

$$d_A(\phi_k) = -\frac{1}{2} \sum_{i+j=k} [\phi_i, \phi_j] . \tag{2.3}$$

We shall denote the right hand side of (2.3) by Φ_k. Then we have the following basic lemma (compare Guillemin-Sternberg [9]):

LEMMA 2.1. Suppose that a sequence $\phi_1, \phi_2, \cdots, \phi_n$ of ad P-valued 1-forms exists satisfying (2.3) for $k = 1, \cdots, n$. Then the ad P-valued 2-form

$$\Phi_{n+1} = -\frac{1}{2} \sum_{i+j=n+1} [\phi_i, \phi_j]$$

is d_A-closed.

Proof: Let $\eta(t) = \Sigma_{i=1}^{n} t^i \phi_i$. The Jacobi identity (1.2) applied to $\eta(t)$ yields $[\eta(t),[\eta(t),\eta(t)]] = [[\eta(t),\eta(t)],\eta(t)] = 0$. Taking the coefficient of t^{n+1}, we obtain

$$\sum_{i+j+k=n+1} [\phi_i,[\phi_j,\phi_k]] = \sum_{i+j+k=n+1} [[\phi_i,\phi_j],\phi_k] = 0.$$

Now

$$d_A\phi_{n+1} = -\frac{1}{2} \sum_{i+j=n+1} ([d_A\phi_i,\phi_j] - [\phi_i,d_A\phi_j])$$

$$= \frac{1}{4} \sum_{i+j+k-n-1} ([[\phi_i,\phi_j],\phi_k] - [\phi_i,[\phi_j,\phi_k]]) = 0$$

With this the lemma is proved.

Thus given an infinitesimal deformation $\eta = \phi_1$, either there exists an infinite sequence $\phi_1, \cdots, \phi_n, \cdots$ satisfying all the conditions (2.3) for $n = 1,2,\cdots$ (in which case $\Sigma_{i=1}^{\infty} \phi_i t^i$ defines a formal power series deformation tangent to η), or there will eventually be some Φ_n which is not exact. In the latter case, the cohomology class of Φ_n in $H^2(M, \text{ad } P)$ is a nonzero obstruction to the realization of η as a tangent vector to a smooth (or analytic) path.

3. HODGE THEORY AND INFINITESIMAL DEFORMATIONS

The proofs of Theorems A and C involve the harmonic theory of M with coefficients in a flat vector bundle. The flat vector bundle we shall need is the complexification of the flat vector bundle ad P, and we shall denote this flat complex vector bundle by E. Since G is a compact group, the adjoint representation of G preserves a positive definite symmetric bilinear form on its Lie algebra. Consequently, the action of G on the complexification of its Lie algebra preserves a positive definite Hermitian form. Such a Hermitian form defines a Hermitian metric on the flat complex vector bundle E which is parallel with respect to the flat connection on E.

If M is a complex manifold, then the complex of E-valued differential forms on M splits:

$$\Omega^n(M,E) = \bigoplus_{p+q=n} \Omega^{p,q}(M,E)$$

and the covariant differential operator d_A decomposes as $\partial_A + \bar{\partial}_A$, where

$\partial_A : \Omega^{p,q}(M,E) \to \Omega^{p+1,q}(M,E)$ and $\bar{\partial}_A : \Omega^{p,q}(M,E) \to \Omega^{p,q+1}(M,E)$. The connection being flat implies that $\bar{\partial}_A\bar{\partial}_A = 0$ and $\partial_A\bar{\partial}_A = -\bar{\partial}_A\partial_A$.

We now suppose that M is a Kähler manifold. Then the Kähler metric on M together with the Hermitian metric on E defines Hermitian structures on all of the spaces of differential forms. In particular, the adjoints ∂_A^* and $\bar{\partial}_A^*$ of ∂_A and $\bar{\partial}_A$, respectively, are defined. One may form Laplacians in the usual manner and the Kähler condition implies that the Laplace operators associated to ∂_A and $\bar{\partial}_A$ are each exactly half the Laplace operator associated to d_A. Just as in the case of ordinary coefficients, there are the fundamental identities (compare Wells [13]), where Λ denotes interior multiplication by the Kähler form:

$$[\Lambda, \partial_A] = i\bar{\partial}_A^*$$

$$[\Lambda, \bar{\partial}_A] = -i\partial_A^* .$$

The following lemma will play a basic role in the proof of Theorem D:

LEMMA 3.1. (The $\bar{\partial}_A\partial_A$-lemma) Let ξ be a d_A-closed E-valued (p,q)-form which is ∂_A-exact (respectively, $\bar{\partial}_A$-exact, d_A-exact). Then there exists an E-valued (p-1,q-1)-form ζ such that $\xi = \partial_A\bar{\partial}_A(\zeta) = -\bar{\partial}_A\partial_A(\zeta)$. Explicitly, if G denotes the Green's operator for ∂_A (and for $\bar{\partial}_A$), then we may take

$$\zeta = \partial_A^* G \bar{\partial}_A^* G(\xi) .$$

REMARK. When E is a trivial coefficient system (i.e., E is a trivial complex line bundle over M), then Lemma 3.1 is the well-known "principle of two types." (See Deligne-Griffiths-Morgan-Sullivan [6]).) It is the main analytic tool used in [6] to prove the vanishing of Massey products. Similarly in the present work, Lemma 3.1 is the main analytic tool proving the vanishing of the secondary "obstructions" for realizing an infinitesimal deformation as tangent to an actual deformation. The proof of Lemma 3.1 is identical to the proof in the case of trivial coefficients (once the fundamental identities above are established), and therefore we omit it. It is worth remarking, however, that these fundamental identities are equivalent to the parallelism of the Hermitian structure, and although in certain cases of flat bundles over Hermitian locally symmetric spaces a Hodge decomposition exists without the existence of a parallel Hermitian structure on the fibers (see [8] and the references there), the first order Kähler identities will nonetheless fail.

COROLLARY 3.2. A d_A-exact form of pure type (p,0) or (0,q) is identically zero.

Using 3.1., we shall construct a map taking a neighborhood of 0 in $\Omega^1(M,E)$ to $\Omega^1(M,E)$ which maps a neighborhood of 0 in the subset of $Z^1(M,E)$ consisting of all η such that $[\eta,\eta]$ is d_A-exact to the set of all $\phi \in \Omega^1(M,E)$ satisfying (2.1). We denote by H the projection of $\Omega^2(M, \text{ad } P)$ onto the harmonic ad P-valued 2-forms. Given $\eta \in Z^1(M,E)$ satisfying $H[\eta,\eta] = 0$, $A + \phi(t\eta)$ will be an analytic path in $F(P_c)$ tangent to η. We shall construct this map $\phi(\eta)$ using the power series expansion (2.1).

We assume throughout that $[\eta,\eta]$ is d_A-exact. Writing $\eta = \eta^{(1,0)} + \eta^{(0,1)}$, we see that the (2,0)-part of $[\eta,\eta]$ equals $[\eta^{(1,0)},\eta^{(1,0)}]$, which is ∂_A-exact. Since a ∂_A-exact holomorphic (p,0)-form is identically zero, it follows $[\eta^{(1,0)},\eta^{(1,0)}] = 0$. Similarly $[\eta^{(0,1)},\eta^{(0,1)}] = 0$. Thus

$$[\eta,\eta] = 2[\eta^{(1,0)},\eta^{(0,1)}]$$

is an ad P-valued (1,1)-form.

Suppose that γ is an ad P-valued (1,1)-form which is d_A-exact. Then, denoting the projection onto the harmonic forms by H, we have $H(\gamma) = 0$ and $0 = d_A(\gamma) = \partial_A(\gamma) + \bar{\partial}_A(\gamma)$, whence the (2,1)-component $\partial_A(\gamma)$ and the (1,2)-component $\bar{\partial}_A(\gamma)$ are each 0. Thus

$$\Delta_{\bar{\partial}_A} G\gamma = \bar{\partial}_A\bar{\partial}_A G\gamma + \bar{\partial}_A\bar{\partial}_A^* G\gamma = \bar{\partial}_A\bar{\partial}_A^* G\gamma$$

The map $G\bar{\partial}_A^* = \bar{\partial}_A^* G : \Omega^{(1,1)}(M,E) \to \Omega^{(1,0)}(M,E)$ satisfies the following properties:

(i) $G\bar{\partial}_A^*$ is bounded linear with respect to the Sobolev s-norm on $\Omega^{(1,1)}(M, \text{ad } P)$ and the Sobolev (s+1)-norm on $\Omega^{(1,0)}(M, \text{ad } P)$;

(ii) If γ is d_A-exact, then $\gamma = \bar{\partial}_A(\bar{\partial}_A^* G)(\gamma)$;

(iii) If γ is d_A-exact, then $\bar{\partial}_A^* G(\gamma)$ is ∂_A-exact.

(iii) follows because

$$\bar{\partial}_A^* G(\gamma) = -i[\Lambda,\partial_A]G(\gamma) =$$
$$-i\Lambda G\partial_A(\gamma) + i\partial_A\Lambda G(\gamma) = \partial_A(i\Lambda G(\gamma))$$

(Since γ is a d_A-exact (1,1)-form, it is ∂_A-closed.) Thus we shall take for our choice of primitive of the square $[\eta,\eta]$ the (1,0)-form

$$\phi_2 = -G\bar{\partial}_A^* \frac{1}{2}[\eta,\eta] = -G\bar{\partial}_A^*[\eta^{(1,0)},\eta^{(0,1)}].$$

Then ϕ_2 is ∂_A-exact, and satisfies $\bar{\partial}_A\phi_2 = -[\eta^{(1,0)},\eta^{(0,1)}]$.

4. THE MAIN INDUCTION

We wish to find, given $\eta \in Z^1(M,E)$ such that $[\eta,\eta]$ is d_A-exact, a sequence $\phi_1(\eta) = \eta, \phi_2(\eta), \phi_3(\eta), \cdots$ of elements of $\Omega^1(M,E)$ satisfying (2.3) for all n. We shall show that we can inductively solve (2.3) under the additional hypothesis that each ϕ_n is a ∂_A-exact (1,0)-form for each $n > 1$.

Note first that if $\phi_2,\phi_3,\cdots,\phi_N$ are ∂_A-exact (1,0)-forms satisfying (2.3) for $n = 1,2,\cdots,N$, then

$$\sum_{i+j=N,i,j>1} [\phi_i,\phi_j]$$

and $[\eta^{(1,0)},\phi_{N-1}]$ are both ∂_A-exact (2,0)-forms. By Lemma 2.1, the ad P-valued 1-form

$$\Phi_N = -\frac{1}{2}\sum_{i+j=N}[\phi_i,\phi_j]$$

is d_A-closed and hence also its (2,0)-component

$$\Phi_N^{(2,0)} = -[\eta^{(1,0)},\phi_{N-1}] - \frac{1}{2}\sum_{i+j=N,i,j>1}[\phi_i,\phi_j]$$

is also d_A-closed. Hence by the $\partial_A\bar{\partial}_A$-lemma, $\Phi_N^{(2,0)}$ is identically zero. Hence Φ_N is an ad P-valued (1,1)-form and equals $-[\eta^{(0,1)},\phi_{N-1}]$. Thus we must show the following:

PROPOSITION 4.1. Suppose $\eta \in Z^1(M,E)$ and $[\eta,\eta]$ is d_A-exact. Let $\phi_i(\eta) = \eta$ and inductively define $\phi_n(\eta) = -G\bar{\partial}_A^*[\eta^{(0,1)},\phi_{n-1}]$ for $n = 2,3,\cdots$. Then $\phi_n(\eta)$ is ∂_A-exact for $n > 1$ and $\bar{\partial}_A\phi_n(\eta) = -[\eta^{(0,1)},\phi_{n-1}(\eta)]$.

LEMMA 4.2. $\partial_A\phi_n = 0$.

PROOF OF LEMMA 4.2. We have $\partial_A\phi_1 = 0$. Inductively assume that $\partial_A\phi_n = 0$. Then $\phi_{n+1} = -G\bar{\partial}_A^*[\eta^{(0,1)},\phi_n]$ implies that

$$\partial_A\phi_{n+1} = -G\partial_A\bar{\partial}_A^*[\eta^{(0,1)},\phi_n] = iG\partial_A\Lambda\partial_A\phi_n[\eta^{(0,1)},\phi_n]$$
$$= iG\partial_A\Lambda[\eta^{(0,1)},\partial_A\phi_n] = 0.$$

Thus $\partial_A\phi_n = 0$ for all $n > 0$.

PROOF OF LEMMA 4.3. We must show that the d_A-closed (1,1)-form $[\eta^{(0,1)},\phi_n]$ is exact. It will suffice to show that $[\eta^{(0,1)},\phi_n]$ is ∂_A-exact. Suppose inductively that $[\eta^{(0,1)},\phi_n]$ is ∂_A-exact. Then

$$\begin{aligned}[\eta^{(0,1)},\phi_{n+1}] &= -[\eta^{(0,1)},\bar{\partial}_A^*G[\eta^{(0,1)},\phi_n]] \\ &= i[\eta^{(0,1)},G(\Lambda\partial_A - \partial_A\Lambda)[\eta^{(0,1)},\phi_n]] \\ &= i[\eta^{(0,1)},G\partial_A\Lambda[\eta^{(0,1)},\phi_n]] \\ &= -i[\eta^{(0,1)},\partial_A(G\Lambda[\eta^{(0,1)},\phi_n])] \\ &= \partial_A(-i[\eta^{(0,1)},G\Lambda[\eta^{(0,1)},\phi_n]]).\end{aligned}$$

Since $[\eta^{(0,1)},\phi_{n+1}]$ is d_A-closed, ∂_A-exact and of pure type (1,1), there exists an ad P-valued 1-form θ such that

$$[\eta^{(0,1)},\phi_{n+1}] = \partial_A\bar{\partial}_A\theta = \tfrac{1}{2}(\partial_A + \bar{\partial}_A)(\bar{\partial}_A\theta - \partial_A\theta) = \tfrac{1}{2}\, d_A(\bar{\partial}_A\theta - \partial_A\theta)$$

is d_A-exact. Lemma 4.3 follows by induction.

PROOF OF PROPOSITION 4.1. Inductively we assume $\bar{\partial}_A\phi_n = -[\eta^{(0,1)},\phi_{n-1}]$. Then

$$\begin{aligned}\bar{\partial}_A\phi_{n+1} &= -\bar{\partial}_A\bar{\partial}_A^*G[\eta^{(0,1)},\phi_n] \\ &= -[\eta^{(0,1)},\phi_n] + H[\eta^{(0,1)},\phi_n] + \bar{\partial}_A^*\bar{\partial}_AG[\eta^{(0,1)},\phi_n] \\ &= -[\eta^{(0,1)},\phi_n] + G\bar{\partial}_A^*[\eta^{(0,1)},\bar{\partial}_A\phi_n] \\ &= -[\eta^{(0,1)},\phi_n] - G\bar{\partial}_A^*[\eta^{(0,1)},[\eta^{(0,1)},\phi_{n-1}]].\end{aligned}$$

Now $[\eta^{(0,1)},[\eta^{(0,1)},\phi_{n-1}]] = -\frac{1}{2}[\phi_{n-1},[\eta^{(0,1)},\eta^{(0,1)}]]$ (by the Jacobi identity), which equals zero because $[\eta^{(0,1)},\eta^{(0,1)}] = 0$. Thus $\bar{\partial}_A\phi_{n+1} = -[\eta^{(0,1)},\phi_n]$ as desired.

Therefore $[\eta^{(0,1)},\phi_n]$ is a $\bar{\partial}_A$-exact, d_A-closed (1,1)-form. It follows that $\phi_{n+1} = -\bar{\partial}_A^* G[\eta^{(0,1)},\phi_n]$ is d_A-exact.

Thus we have constructed a formal power series deformation of flat connections. To see this construction yields an analytic path of flat connections, we must prove the following:

PROPOSITION 4.4. There exists a neighborhood U of 0 in $Z^1(M,E)$ (with respect to the Sobolev s-norm) such that for $\eta \in U$, the series $\phi(\eta) = \Sigma_{n=1}^{\infty}\phi_n(\eta)$ converges absolutely.

Proof: The terms of the series satisfy the recursion

$$\phi_{n+1}(\eta) = - \bar{\partial}_A^* G[\eta^{(0,1)},\phi_n(\eta)]$$

whence there exist constants C, C' such that

$$\|\phi_{n+1}\| \leq C\|\eta\|_0\|\phi_n\|_{s-1} \leq C'\|\eta\|_0\|\phi_n\|_s .$$

Thus there exists $\varepsilon < 0$ and $k < 1$ such that if $\|\eta\|_0 < \varepsilon$, then $\|\phi_{n+1}\|_s < k\|\phi_n(\eta)\|_s$. It follows that $\phi(\eta)$ converges absolutely for η in a neighborhood of 0.

Writing ad μ for the linear transformation ad $\mu(\theta) = [\mu,\theta]$, we have the following expression for the map $\phi(\eta)$:

$$\phi(\eta) = \sum_{n=0}^{\infty} (-\bar{\partial}_A^* G \text{ ad } \eta^{(0,1)})^n(\eta) = (I + \bar{\partial}_A^* G \text{ ad } \eta^{(0,1)})^{-1}(\eta)$$

which defines an analytic embedding of a neighborhood of 0 in $Z^1(M,E)$ in $Z^1(M,E)$, which restricts to an embedding of the quadratic cone in $Z^1(M,E)$ defined by $H[\eta,\eta] = 0$ into a neighborhood of A in the space of flat connections.

REMARK. When M is a complex surface, $G = U(n)$ and $\psi \in \text{Hom}(\pi_1(M),G)$ is an irreducible representation, then it can be shown that Theorem A follows from the main result of Donaldson [5]. By using the correspondence between flat unitary structures on a holomorphic complex vector bundle ξ and stable holomorphic structures, one can reduce to the analogous problem for deformations of stable holomorphic bundles. The infinitesimal deformations of stable holomorphic structures form the space $H^1(M,o(\xi))$, which by the Dolbeault isomorphism, may be identified with a space of harmonic ξ-valued (0,1)-forms, in such a way that the corresponding obstruction maps agree. Suppose that η is an infinitesimal deformation of the flat unitary connection such that $H[\eta,\eta] = 0$. Then the corresponding infinitesimal deformation of the holomorphic vector bundle has a harmonic representative which is a ξ-valued (0,1)-form $\tilde{\eta}$ such that $H[\tilde{\eta},\tilde{\eta}] = 0$. By the $\partial\bar{\partial}$-lemma, the harmonic ξ-valued (0,2)-form $[\tilde{\eta},\tilde{\eta}]$ is identically zero. Since the holomorphic structures on a complex vector bundle form an affine space, the corresponding linear path with tangent vector $\tilde{\eta}$ is a path of holomorphic structures. Since the stable vector bundles are open in the space of holomorphic bundles, an open segment of this linear path will consist of stable holomorphic bundles. By Donaldson [5], there will be a corresponding path of irreducible flat unitary connections with tangent vector η.

APPENDIX: THE CASE OF A RIEMANN SURFACE

When M is a compact Riemann surface and G is a compact Lie group, then a much stronger theorem than Theorem A holds, due to the work of Arms-Marsden-Moncrief [1]. Namely in this case, the spaces $\mathrm{Hom}(\pi,G)$ and $\mathbf{F}(P)$ are actually locally analytically equivalent to cones defined by homogeneous systems of quadratic equations. This is much stronger than knowing that the tangent cones are quadratic, and it seems plausible that this stronger statement may hold in general for compact Kähler manifolds. The main result of [1] pertains to level sets of momentum mappings for Poisson actions of (possibly infinite-dimensional) Lie groups. In this appendix we discuss the application of the technique of Arms-Marsden-Moncrief in our special case. All of the ideas in this section are contained in [1] and we present them here as a simple application of their elegant general theory.

THEOREM. (Arms-Marsden-Moncrief) Let M be a compact Riemann surface and let G be a compact Lie group. Let P be a flat principal G-bundle over M and let $\mathbf{F}(P)$ denote the space of flat connections on P. Choose a flat connection $A \in \mathbf{F}(P)$ and let $Z^1(M, \mathrm{ad}\, P)$ be the space of infinitesimal deformations of A inside $\mathbf{F}(P)$. Then there exists an analytic diffeomorphism between a neighborhood of A in $\mathbf{F}(P)$ and a neighborhood of 0 in the subset of $Z^1(M, \mathrm{ad}\, P)$ consisting of ad P-valued 1-forms η such that $H[\eta,\eta] = 0$.

Using Lemma C we obtain as above an analogous statement for spaces of representations of fundamental groups of compact Riemann surfaces:

COROLLARY. Let M be a compact Riemann surface and let π be its fundamental group. Suppose that G is a compact Lie group and that $\psi \in \mathrm{Hom}(\pi, G)$. Then there exists a neighborhood of ψ in $\mathrm{Hom}(\pi, G)$ which is real-analytically diffeomorphic to a neighborhood of 0 in the subset of $Z^1(\pi; g_{\mathrm{Ad}\psi})$ consisting of all $\eta \in Z^1(\pi; g_{\mathrm{Ad}\psi})$ such that $[\eta,\eta] = 0$ in $H^2(\pi; g_{\mathrm{Ad}\psi})$.

PROOF OF THEOREM. As above, we identify the space F(P) of flat connections with the subset of $\Omega^1(M)$, ad P) satisfying (1.5). Let $* : \Omega^p(M, \mathrm{ad}\, P) \to \Omega^{2-p}(M, \mathrm{ad}\, P)$ be the Hodge star operator. Then the adjoint of the covariant differential operator d_A is given by $d_A^* = -*d_A*$. Let H denote harmonic projection and G the Green's operator as above. The following elementary fact may be proved by direct calculation in isothermic coordinates:

LEMMA. Let $\eta, \zeta \in \Omega^1(M, \mathrm{ad}\, P)$. Then $[*\eta, *\zeta] = [\eta, \zeta]$.

We shall work in the linear slice S for the action of the gauge group defined by $d_A^*(\phi) = 0$. (See Freed-Uhlenbeck [6], Atiyah-Jones [3], or Singer [12] for discussion.) Let $F : \Omega^1(M, \mathrm{ad}\, P) \to \Omega^1(M, \mathrm{ad}\, P)$ be the map defined by $F(\phi) = \phi + \frac{1}{2} G d_A^*[\phi,\phi]$ (the Kuranishi map). Since the derivative of F at 0 is the identity, the implicit function theorem implies that F maps a neighborhood U of 0 in $\Omega^1(M, \mathrm{ad}\, P)$ to another neighborhood V of 0 in $\Omega^1(M, \mathrm{ad}\, P)$. Clearly $F(S) = S$. We claim that F maps the intersection of U with S with **F**(P) bijectively to the intersection of V with the quadratic cone Q consisting of all $\eta \in Z^1(M, \mathrm{ad}\, P) \cap S$ such that $H[\eta,\eta] = 0$. To see this, we use the Hodge decomposition of $\Omega^2(M, \mathrm{ad}\, P)$ to rewrite (1.5) as a system of two equations, one whose solutions form a smooth manifold, and one which is homogeneous quadratic. Since M has real dimension 2, the identity map on $\Omega^2(M, \mathrm{ad}\, P)$ is the sum of two projections:

$$I = H + d_A d_A^* G$$

Thus equation (1.5) is equivalent to the two equations:

$$d_A\phi + \frac{1}{2} d_A d_A^* G[\phi,\phi] = 0 \tag{A.1}$$

$$\frac{1}{2} H[\phi,\phi] = 0 \tag{A.2}$$

Now (A.1) is equivalent to the equation

$$d_A F(\phi) = d_A(\phi + \frac{1}{2} d_A^* G[\phi,\phi]) = 0$$

which merely expresses the condition that $F(\phi)$ is d_A-closed. Thus $F^{-1}(F(P) \cap V)$ consists of all $\phi \in U$ such that $F(\phi) \in Z^1(M, \text{ad } P)$ and $H[\phi,\phi] = 0$. We claim that $H[\phi,\phi] = H[F(\phi),F(\phi)]$:

$$H[F(\phi),F(\phi)] - H[\phi,\phi] = H([\phi,d_A^* G[\phi,\phi]] + \frac{1}{4}[d_A^* G[\phi,\phi],d_A^* G[\phi,\phi]])$$

$$= H([-*\phi,-d_A*G[\phi,\phi]] + \frac{1}{4}[-d_A*G[\phi,\phi],-d_A*G[\phi,\phi]])$$

(by the lemma above)

$$= Hd_A([*\phi,\phi*G[\phi,\phi]] + \frac{1}{4}[*G[\phi,\phi],d_A*G[\phi,\phi]]) = 0$$

(because $d_A*\phi = 0$ as $\phi \in S$). Thus $\phi \in U \cap S \cap F(P)$ if and only if $F^{-1}(\phi)$ is a d_A-harmonic form η satisfying $H[\phi,\phi] = 0$. The proof of the theorem is complete.

REFERENCES

1. Arms, J., Marsden, J., and Moncrief, V., Symmetry and Bifurcation of Mappings, Comm. Math. Phys. 78 (1981), 455-478.
2. Artin, M., On the Solution of Analytic Equations, Inv. Math. 5 (1968), 277-291.
3. Atiyah, M. F., and Jones, J.D.S., Topological Aspects of Yang-Mills Theory, Comm. Math. Phys. 61 (1978), 97-118.
4. Deligne, P., Griffiths, P., Morgan, J., and Sullivan, D., Real Homotopy Theory of Kahler Manifolds, Inv. Math. 29 (1975), 245-274.
5. Donaldson, S., Anti Self-dual Connections over Complex Algebraic Surfaces and Stable Vector Bundles, Proc. London Math. Soc. 50 (1985), 1-26.
6. Freed, D., and Uhlenbeck, K., Instantons and Four-Manifolds, MSRI Publication 1 (1984), Springer Verlag, New York.
7. Goldman, W., Representations of fundamental groups of surfaces, Proceedings of the Special Year in Topology, University of Maryland 1983-1984, Lecture Notes in Mathematics, 1167 (1985), 95-117, Springer-Verlag, Berlin, Heidelberg.
8. Goldman, W., and Millson, W., Local Rigidity of Discrete Groups on Complex Hyperbolic Space (submitted).
9. Guillemin, V., and Sternberg, S., Deformation Theory of Pseudogroup Structures, Mem. A.M.S. 64 (1964)

10. Johnson, D., and Millson, J., Deformation Spaces Associated to Compact Hyperbolic Manifolds, in Discrete Groups in Geometry and Analysis, Proceedings of a conference held at Yale University in honor of G. D. Mostow on his sixtieth birthday, Birkhauser, Boston (to appear).

11. Palais, R., On the existence of slices for actions of noncompact Lie Groups, Ann. Math. 73 (1961), 295-323.

12. Singer, I. M., Some remarks on the Gribov Ambiguity, Comm. Math. Physics 60 (1978), 7-12.

13. Wells, R. O., Differential Analysis on Complex Manifolds, Graduate Texts in Mathematics, 65, Springer-Verlag, New York.

Imbeddings of Knot Groups in Knot Groups

F. GONZÁLES-ACUÑA / Instituto de Mathematicas, Universidad Nacional Autonoma de Mexico, Mexico City, Mexico

WILBUR WHITTEN / Department of Mathematics, University of Southwestern Louisiana, Lafayette, Louisiana

This paper deals with the question of how classical knot groups imbed in one another. The motivation for this question arises indirectly from the following result of the first author.

THEOREM [9]. There exists a smooth 3-knot (in S^5) whose group contains every finitely presented group.

In dimension 4, we do not know whether there exists a smooth 2-knot (in S^4) whose group contains all 2-knot groups, much less a larger class of groups. But the corresponding statement for classical knots is false.

PROPOSITION. No classical knot group contains all classical knot groups as subgroups.

Proof. Suppose that there exists a tame knot $K \subset S^3$ whose group contains all (classical) knot groups; clearly, K must be nontrivial. Let E(K) denote the exterior of K, and let M denote the double of E(K). Then, in particular, $\pi_1 E(K)$, and thus $\pi_1 M$, contains all torus-knot groups as subgroups. But this is a contradiction to [12; theorem 4.3, p. 163], which says that the group of a closed, irreducible, sufficiently large manifold cannot contain an infinity of torus-knot groups.

Now the class of classical knot groups is a varied and useful class whose members contain many subgroups that are not themselves knot groups. Hence, we pose the following problem.

Describe all imbeddings of classical knot groups in classical knot groups.

In this paper, we shall state some of our results on this problem and give indications of proofs of most of them, but complete proofs and many other results will appear elsewhere. Throughout, we deal only with the smooth category. Henceforth, all knots and knot groups belong to the classical case $(S^1 \subset S^3)$.

Our problem falls naturally into three parts, the first two of which overlap:

1) coverings, $E(K_1) \to E(K_2)$, of knot exteriors by knot exteriors;

2) knot groups as subgroups of <u>finite</u> index in knot groups;

3) knot groups as subgroups of <u>infinite</u> index in knot groups.

This paper is divided into three sections, a section covering each of 1), 2), and 3).

We wish to thank J. Simon and the referee for helpful remarks.

1. COVERINGS OF KNOT EXTERIORS BY KNOT EXTERIORS

To state our first theorem we recall some notation. If G is a group and if $S \subset G$, then $\langle S \rangle_G$ denotes the normal closure of S in G.

THEOREM 1.1. Let $\rho : E_1 \to E_2$ be a covering between compact orientable 3-manifolds with $\partial E_i \cong S^1 \times S^1$ $(i = 1,2)$. Assume that $\pi_1(\partial E_1)$ normally generates $\pi_1 E_1$. Then ρ is cyclic; that is, ρ is a regular covering and the group of covering transformations is cyclic.

Outline of proof. Let $G = \pi_1 E_2$, let $H = \rho_*(\pi_1 E_1)$, and let $P = i_*(\pi_1 \partial E_2)$, where $i : \partial E_2 \to E_2$ is the inclusion map. Then $P \cdot H = G$, because $\rho^{-1}(\partial E_2)$ is connected.

It is straightforward to show that $\langle P \cap H \rangle_G \subset H$, and the fact that $\pi_1(\partial E_1)$ normally generates $\pi_1 E_1$ implies that $H \subset \langle P \cap H \rangle_G$. Hence, $H = \langle P \cap H \rangle_G$ and therefore ρ is a regular covering.

Now, since $\langle P \rangle_G \supset P \cup \langle P \cap H \rangle_G = P \cup H$ and since $G = P \cdot H$, we have $\langle P \rangle_G = G$ and, therefore, $i_* : H_1(\partial E_2) \to H_1(E_2)$ is surjective. It follows that $H_1(E_2, \partial E_2) = 0$ and, using Poincaré duality and the universal coefficient theorem, we have

Torsion $H_1(E_2) \approx$ Torsion $H^2(E_2, \partial E_2) \approx$ Torsion $H_1(E_2, \partial E_2) = 0$. Also, $H_2(E_2) \approx H^1(E_2, \partial E_2) \approx \mathrm{Hom}(H_1(E_2, \partial E_2), Z) = 0$ and $H_2(E_2, \partial E_2) \approx H^1(E_2) \approx H_1(E_2)$, and so one has the exact sequence

$$0 \longrightarrow H_2(E_2, \partial E_2) \longrightarrow H_1(\partial E_2) \longrightarrow H_1(E_2) \longrightarrow 0.$$

Therefore, $H_1(E_2) \approx H_2(E_2, \partial E_2) \approx Z$. Since G/H $(=P \cdot H/H \approx P/P \cap H)$ is abelian, it is a quotient of G/G' $(\approx H_1(E_2))$ and, therefore, cyclic.

COROLLARY 1.2. If $\rho : E_1 \to E_2$ is a covering between knot exteriors, then ρ is cyclic.

A 3-manifold L is a <u>lens-like space</u>, if $\pi_1 L$ is cyclic and if the universal covering space of L is S^3. If K is a knot in S^3, then K(r/s) shall denote the manifold obtained by (r/s)-surgery on K.

THEOREM 1.3. Suppose that $p : E_1 \to E_2$ is an n-fold covering between knot exteriors $E_1(K_1)$ and $E_2(K_2)$, with $n \geq 2$ and with K_1 (and, therefore, K_2) nontrivial. Then

1) if K_2 is a nontorus knot, either (+n)-surgery of (-n)-surgery on K_2 yields a lens-like space; if K_2 is a (p,q)-torus knot, then there exists an integer d such that $n = dpq \pm 1$ and (n/d)-surgery on K_2 yields a lens space; in either case, (S^3, K_1) admits a free Z_n-action; and

2) K_1 has property P, and $K_2(1/m) \not\approx S^3$ for any $m \neq 0$; in fact, K_2 has property P, if it is a torus knot or if $n \geq 3$.

Outline of proof. Orient E_1 and E_2 so that p preserves orientation, and orient K_2. Now p is a cyclic covering, by Corollary 1.2. Let (μ, λ) be a standard, oriented meridian-longitude pair for K_2 on ∂E_2; we have $\mu \cdot \tilde{\lambda} = 1$. Let $(\mu_1, \tilde{\lambda})$ be an oriented meridian-longitude pair for K_1 on ∂E_1 such that $p(\tilde{\lambda}) = \lambda$, such that $p|\tilde{\lambda}$ preserves orientation, and such that $\mu_1 \cdot \tilde{\lambda} = 1$. Finally, let $\tilde{\mu} = p^{-1}(\mu)$, and orient $\tilde{\mu}$ so that $p_*([\tilde{\mu}]) = [\mu]^n$. (We take the point of intersection of $\tilde{\mu}$ and $\tilde{\lambda}$ as basepoint of $\pi_1 E_1$ and the point of intersection of μ and λ as basepoint of $\pi_1 E_2$.) Since $\tilde{\mu} \cdot \tilde{\lambda} = 1$ and $\mu_1 \cdot \tilde{\lambda} = 1$, there exists an integer d such that $[\mu_1] = [\tilde{\mu}][\tilde{\lambda}]^d$.

Now sew on a solid torus V to E_1 to obtain S^3 and thereby kill $[\mu_1]$. Let h be a generator of the covering-transformation group Z_n of p. Since $(h|\partial E_1)_*([\mu_1]) = [\mu_1]$, an application of the geometric loop theorem [16; theorem 6] shows that $p_*([\mu_1])$ $(= [\mu]^n[\lambda]^d)$ is represented by a simple closed curve on ∂E_1; that is, $(n,d) = 1$.

Thus, if $L = K_2(n/d)$, then S^3 $(= E_1(K_1) \cup_{\partial E_1} V)$ is the (unique) completion of $S^3 - K_1$, and we obtain an induced covering $S^3 \to L$. The cyclic action Z_n on $E_1(K_1)$ extends over V to S^3, and $L = S^3/h$. Moreover, by the Smith conjecture [15], the group Z_n acts freely on S^3, since $K_1 \neq 0$. Since $\pi_1 L \approx Z_n$, we have $d = \pm 1$ [2; Corollary 1], if K_2 is not a torus knot. If K_2 is a (p,q)-torus knot, then $n = dpq \quad 1$ [14]. Also, since h has no fixed points on (S^3,K_1), the knot K_1 has property p [2; corollary 7].

Suppose now that $K_2(1/m) \cong (S^3,K_2')$. If $m \neq 0$, then K_2 is not a torus knot, and it follows from [2; corollary 2] and the fact that we have a cyclic covering $E_1 \to E(K_2')$ $(\cong E_2)$ that there exists an integer d' such that both m and d' are in $\{-1,1\}$ and such that $(n/(m+d'))$-surgery on K_2 produces a lens-like space L'. Since $K_2(n/d) \cong L$, it follows from the cyclic surgery theorem [2] that either n or $3n \leq 1$, which is untrue. Hence, $K_2(1/m) \cong S^3$ only when $m = 0$.

Finally, suppose that $\pi_1 K_2(1/m) \cong \{1\}$, for some $m \neq 0$. Then, by [2; corollary 2] (since torus knots have property P), we have $m \in \{-1,1\}$. Furthermore, since $\pi_1 K_2(\pm n) \approx Z_n$, we also have $\Delta(1/m,\pm n) = \Delta(\pm 1,\pm n) \leq 1$, by the cyclic surgery theorem [2]. (Here, $\Delta(r/s,r'/s') = |rs' - r's|$.) Hence, $n = 2$, since $n \geq 2$. Thus, if $n \geq 3$, then K_2 has property P.

Our next result follows from Theorem 1.3, the cyclic surgery theorem [2], and a result of R. H. Fox [8].

COROLLARY 1.4. A knot exterior is covered by, at most, two distinct knot exteriors.

Proof. Let $p : E_1(K_1) \to E_2(K_2)$ be a (nontrivial) covering between knot exteriors. By Corollary 1.2 and Theorem 1.3, the covering p is cyclic, and we have the following "covering" diagram in which L is a lens-like space obtained by surgery on K_2:

$$\begin{array}{ccccc} B & \supset & E_1 & \subset & S^3 \\ \Big\downarrow{\scriptstyle 1} & & \Big\downarrow{\scriptstyle p} & & \Big\downarrow{\scriptstyle p_2} \\ S^3 & \supset & E_2 & \subset & L \end{array}$$

If K_2 is a nontorus knot, then by the cyclic surgery theorem [2], there are at most two ways to obtain a lens-like space (by surgery on K_2), each of which leads to a specific cyclic covering space of E_2.

If K_2 is a torus knot, then so is K_1 [13; lemma II.6.1, p.36]. It follows from Fox [8] that K_1 and K_2 are equivalent, and so $E_1 \cong E_2$.

COROLLARY 1.5. The exterior of a nontrivial knot covers only finitely many distinct knot exteriors.

Outline of proof. Let K be a nontrivial knot, and let $p : E(K) \to E(K')$ be a covering of index $n \geq 2$. By theorem 1.3, p is cyclic and the group of covering translations extends to a free Z_n-action on (S^3,K). If K is a torus knot, then it follows from [8] that the only knot exterior that E(K) can cover is itself.

Now assume that K is a nontorus knot. By [4], there are only a finite number of possible values of n. Henceforth, n is fixed.

We orient both S^3 and K, and we let $\mathrm{Diff}^*(S^3,K)$ denote the group of all diffeomorphisms of (S^3,K) preserving both orientations. Two Z_n-actions in $\mathrm{Diff}^*(S^3,K)$ are equivalent, if they are conjugate. Note that if $p_i : E(K) \to E(K_i)$ $(i = 1,2)$ are cyclic coverings (between knot exteriors) with equivalent groups of covering transformations, then $E(K_1) \cong E(K_2)$. To complete the proof, we apply either the main result of [1] or [18; theorem 5(2)], implying that there are only a finite number of equivalence classes of free Z_n-actions on (S^3,K).

If K is an amphicheiral hyperbolic knot, then (S^3,K) has no free (orientation-preserving) Z_n-actions [17]. It follows from theorem 1.3 that E(K) (properly) covers no knot exterior. An application of corollary 2.3 (below) and the fact that a Haken manifold that is homotopy equivalent to E(K) must be homeomorphic to E(K) will then complete the proof of our final result of this section.

COROLLARY 1.6. The exterior of an amphicheiral hyperbolic knot does not (properly) cover any knot exterior, and the group of an amphicheiral hyperbolic knot cannot be imbedded in a knot group as a (proper) subgroup of finite index.

2. KNOT GROUPS AS SUBGROUPS OF FINITE INDEX IN KNOT GROUPS

One can prove the following theorem case-by-case.

THEOREM 2.1. If M is a compact orientable 3-manifold homotopy equivalent to a nontrivial-knot exterior, then ∂M is an incompressible torus and $\pi_1 M/\langle \pi_1 \partial M\rangle$ is trivial.

Outline of proof. In three separate lemmas, we prove that if M is a Haken manifold homotopy equivalent to either a cable-knot exterior, a nontrivial torus-knot exterior, or a composite-knot exterior, then ∂M is a torus and $\pi_1 M/\langle \pi_1 \partial M\rangle$ is trivial (see lemma 2.2 below); in fact, if M is homotopy equivalent to a composite-knot exterior, then M is a composite-knot exterior. We also prove an easy proposition stating that a Haken manifold homotopy equivalent to a nontrivial-knot exterior has an incompressible torus as its boundary.

To complete the proof of theorem 2.1, we can assume that M is irreducible by replacing a fake 3-cell by a 3-cell, if necessary, and noting that M is homotopy equivalent to a knot exterior. Thus, since $H_1(M) \approx Z$, we can assume that M is Haken. The ∂M is thus an incompressible torus, by our proposition. By our three lemmas, we can assume that M is homotopy equivalent to a knot exterior containing no essential annuli. But then M is actually a knot exterior [3; theorem 10, p. 42].

As an example of one of the three lemmas mentioned in the last proof, we give the following.

LEMMA 2.2. Let M be a Haken manifold homotopy equivalent to a nontrivial torus-knot exterior. Then ∂M is a torus and $\pi_1 M/\langle \pi_1 \partial M \rangle \approx \{1\}$.

Proof. Let K be a nontrivial torus knot. Then E(K) is a union of two solid tori along an (essential) annulus. Since M is boundary irreducible (by our proposition mentioned in the last proof), we can obtain M by Dehn flips along these solid tori (in E(K)) [11; theorem 29.1, p.243]. Thus M is the union of two solid tori along an (essential) annulus.

Since M imbeds in a lens space and in a homology sphere (via Dehn surgery on M), there exists elements g_1 and g_2 in $\pi_1 M$ represented by simple closed curves on ∂M such that $\pi_1 M/\langle g_1 \rangle$ is finite cyclic and $\pi_1 M/\langle g_2 \rangle$ is perfect. Hence, $\pi_1 M/\langle g_1, g_2 \rangle$ is both cyclic and perfect, since a homomorph of a cyclic or perfect group is a cyclic or perfect group, respectively. Therefore, $\pi_1 M/\langle g_1, g_2 \rangle \approx \{1\} \approx \pi_1 M/\langle \pi_1 \partial M \rangle$.

COROLLARY 2.3. If G_1 and G_2 are knot groups, if $G_1 < G_2$, and if $[G_2 : G_1] < \infty$, then $G_1 \lhd G_2$ and G_2/G_1 is cyclic.

The proof of corollary 2.3 follows from straightforward applications of theorems 2.1 and 1.1.

THEOREM 2.4. The group of a composite knot contains no knot groups as proper subgroups of finite index.

Outline of proof. Let $E(K_2)$ be a composite-knot exterior, and suppose that there exists a knot K_1 with $\pi_1 E(K_1) < \pi_1 E(K_2)$ of finite index (≥ 2). Let $p : M \to E(K_2)$ denote the covering corresponding to $\pi_1 E(K_1)$. By an application of corollary 2.3, the covering p is finite cyclic (of index ≥ 2). Moreover, M is homotopy equivalent to $E(K_1)$, and ∂M is an incompressible torus.

Let $f : E(K_1) \to M$ be a homotopy equivalence, and set $g = pf : E(K_1) \to E(K_2)$. Of course, $g_* : \pi_1 E(K_1) \to \pi_1 E(K_2)$ is a monomorphism modulo basepoints. If $\beta(E(K))$ denotes the component of the characteristic submanifold of $E(K)$ meeting $\partial E(K)$, then by a deformation theorem (see 3.2) that we prove (in the spirit of Johannson [11]), we can assume that $g|\beta(E(K_1)) : \beta(E(K_1)) \to \beta(E(K_2))$. Since $(g|\beta(E(K_1)))_*$ is a monomorphism and since $\beta(E(K_2))$ is a composing space, it follows from a lemma of ours (not stated in this paper) that $\beta(E(K_1))$ is neither a cable space nor a torus-knot exterior. Hence, $\pi_1 M$ is neither a torus-knot group nor a cable-knot group.

In all other cases, we show that $M \cong E(K_1)$. It follows from theorem 1.3 that we can perform integral surgery on K_2 to obtain a lens-like space, $K_2(r)$. But nontrivial surgery on a composite knot always produces a Haken manifold [10; Lemma 7.1]. Thus, $\pi_1 K_2(r)$ is infinite, which is a contradiction.

A slight modification of the proof of theorem 2.4 yields the following.

COROLLARY 2.5. The group of a nontorus two-bridge knot or of any double of a nontrivial knot contains no knot groups as proper subgroups of finite index.

COROLLARY 2.6. If E is a knot exterior with a quadratic Alexander polynomial different from $t^2 - t + 1$, then

1) E is not covered (nontrivially) by a knot exterior and $\pi_1 E$ does not properly contain a knot group of finite index, and

2) E does not (nontrivially) cover a knot exterior and $\pi_1 E$ cannot be imbedded in a knot group as a proper subgroup of finite index.

Outline of proof. Suppose that one of the following four possibilities holds:

a) E is covered (nontrivially) by a knot exterior;

b) $\pi_1 E$ contains a knot group as a proper subgroup of finite index;

c) E (nontrivially) covers a knot exterior; or

d) $\pi_1 E$ imbeds in a knot group as a proper subgroup of finite index.

Then there exists a knot exterior E_2 of a knot K, a Haken manifold M_1 homotopy equivalent to a knot exterior, and a (nontrivial) cyclic covering $p : M_1 \to E_2$ (by Corollary 2.3). Let n be the covering index of p; by our assumptions we have $n \geq 2$.

Assuming that the Alexander polynomial of either $\pi_1 M$ or $\pi_1 E_2$ is quadratic and applying some old results of R. H. Fox, [5], [6], and [7], all related to the Alexander polynomial, we show that $\pi_1 M_1$ and $\pi_1 E_2$ have the same polynomial--namely, $t^2 - t + 1$. This contradicts our hypothesis.

REMARK. An example of a knot satisfying the hypothesis of corollary 2.6 is the figure-eight knot.

3. KNOT GROUPS AS SUBGROUPS OF INFINITE INDEX IN KNOT GROUPS

Our results in this section are, at present, mostly limited to cohopficity.

DEFINITION. A group is cohopfian, if it cannot be imbedded in itself as a proper subgroup.

REMARK. For contrast, recall that a group is Hopfian, if it is isomorphic to none of its proper factor groups, and that all knot groups are Hopfian, since knot groups are residually finite [19].

THEOREM 3.1. The only knots whose groups are not cohopfian are 1) torus knots, 2) composite knots, and 3) a certain class of iterated torus knots. Thus, the groups of the following classes of knots are cohopfian: a) hyperbolic knots, b) noncable, noncomposite, satellite knots, and c) all cable knots not included in 3).

To prove this theorem, one must consider each class of knots separately. In the proofs of these separate cases, we often use the following result, which is actually a corollary of a more general theorem. Recall that if M is a Haken manifold, then $\beta(M)$ denotes the components of the characteristic submanifold of M that meet the ∂M.

DEFORMATION THEOREM 3.2. Let E_1 and E_2 be exteriors of knots in S^3, and let $g : E_1 \to E_2$ be a map such that $g_* : \pi_1 E_1 \to \pi_1 E_2$ is a monomorphism. Then there exists a knot exterior $E \subset E_2$ and a map f homotopic to g such that

1) ∂E is incompressible in E_2,

2) f maps $(E_1, \beta(E_1), cl(E_1 - \beta(E_1)))$ into $(E, \beta(E), cl(E - \beta(E)))$, and

3) $f|cl(E_1 - \beta(E_1)) : cl(E_1 - \beta(E_1)) \to cl(E - \beta(E))$ is a covering map on each component of $cl(E_1 - \beta(E_1))$.

Furthermore, if $[\pi_1 E_2 : g_* \pi_1(E_1)] < \infty$, then E can be taken to be E_2.

ACKNOWLEDGEMENT

We wish to thank the University of Iowa for its hospitality while part of this work was done. The second author was partially supported by the ONR grant N0014-85-K-0099.

REFERENCES

1. Boileau, M., and Flapan, E., Uniqueness of Free Actions on S^3 Respecting a Knot, preprint.
2. Culler, M., Gordon, C., Leucke, J., and Shalen, P., Dehn Surgery on Knots, MSRI preprint.
3. Feustel, C. D., On the Torus Theorem and its Applications, Trans. Amer. Math. Soc. 217 (1976), 1-43.
4. Flapan, E., Infinitely Periodic Knots, Canad. J. Math., to appear.
5. Rox, R. H., The Homology Characters of the Cyclic Coverings of the Knots of Genus One, Ann. of Math. 71 (1960), 187-196.
6. Fox, R. H., Knots and Periodic Transformations, Topology of 3-Manifolds and Related Topics (Proc. The Univ. of Georgia Inst., 1961), Prentice-Hall, Englewood Cliffs, N.J., 1962, pp. 177-182.
7. Fox, R. H., On Knots Whose Points are Fixed Under a Periodic Transformation of the 3-Sphere, Osaka Math. J., 10 (1958), 31-35.
8. Fox, R. H., Two Theorems About Periodic Transformations of the 3-Sphere, Mich. Math. J., 14 (1967), 331-334.
9. González-Acuña, F., Gordon, C., and Simon, J., Unsolvable Problems About Knots and Related Groups, in preparation.
10. Gordon, C., Dehn Surgery and Satellite Knots, Trans. Amer. Math. Soc. 275 (1983), 687-708.
11. Johannson, K., Homotopy Equivalences of 3-Manifolds with Boundaries, Lecture Notes in Math. 761, Springer-Verlag; Berlin, Heidelberg, New York, 1979.
12. Jaco, W., and Myers, R., An Algebraic Determination of Closed Orientable 3-Manifolds, Trans. Amer. Math. Soc., 253 (1979), 149-170.
13. Jaco, W., and Shalen, P., Seifert Fibered Spaces in 3-Manifolds, Memoirs 21, Amer. Math. Soc., Providence, R.I., 1979.
14. Moser, L., Elementary Surgery Along Torus Knots, Pacific J. Math. 38 (1971), 41-53.
15. Morgan, J. W., and Bass, H., (editors), The Smith Conjecture, Academic Press, New York, 1984.

16. Meeks, III, W. H., and Yau, S.-T., The Equivariant Dehn's Lemma and Loop Theorem, Comm. Math. Helv. 56 (1981), 225-239.

17. Sakuma, M., Non-free-periodicity of Amphicheiral Hyperbolic Knots, preprint.

18. Sakuma, M., Uniqueness of Symmetries of Knots, preprint.

19. Thurston, W., The Geometry and Topology of 3-Manifolds, Lecture Notes, Princeton University, 1977.

Geometric Hopfian and Non-Hopfian Situations

JEAN-CLAUDE HAUSMANN / Section de Mathématiques, University of Geneva, Geneva, Switzerland

Let M^n be a closed oriented (smooth) manifold of dimension n. Let $f : M \to M$ be a map of degree d(f). In this paper we consider the following question:

Q : If $d(f) = \pm 1$, is f a homotopy equivalence?

This question naturally divides into two sub-questions:

Q1 : If $d(f) = \pm 1$, is $\pi_1 f : \pi_1(M) \to \pi_1(M)$ injective?

Observe that $\pi_1 f$ is always surjective (otherwise f would factor through a proper covering of M and hence could not be of degree ±1). Question Q1 was raised by H. Hopf around 1931 (see [10]) and gave rise to the concept of hopfian and non-hopfian groups (a group G is <u>hopfian</u> if any epimorphism of G onto itself is injective).

Q2 : If $d(f) = \pm 1$ and $\pi_1 f$ is injective (and hence bijective), is $f_* : H_*(M;\mathbb{Z}\pi) \to H_*(M;\mathbb{Z}\pi)$ an isomorphism? ($\pi = \pi_1(M)$).

The answer "yes" to both Questions Q1 and Q2 clearly implies the answer "yes" to Question Q.

Questions Q1 and Q2 may be illuminated by considering natural generalizations of them to maps of non-zero degree. Let $d = d(f)$ and denote by Λ the ring $\mathbb{Z}[1/d]$.

Q1a : If $d \neq 0$, is $\pi_1 f : \pi_1(M) \to \pi_1(M)$ injective?

Q2a : If $d \neq 0$ and $\pi_1 f$ is injective, is $f_* : H_*(M;\Lambda\pi) \to H_*(M;\Lambda\pi)$ an isomorphism?

Q2b : If $d \neq 0$ and $\pi_1 f$ is bijective, is $f_* : H_*(M;\Lambda\pi) \to H_*(M;\Lambda\pi)$ an isomorphism?

We present below a partial answer to Questions Q2 and Q2b and a surprising negative answer to question Q1a. The other questions remain open in general.

PARTIAL ANSWER TO QUESTIONS Q2 AND Q2b

PROPOSITION 1. The answer to Questions Q2 and Q2b is "yes":

a) when $\Lambda\pi$ is noetherian

or

b) when $n \leq 4$.

Proof: Let $f : M \to M$ be a map of degree $d \neq 0$ inducing an isomorphism on the fundamental group. Let B be a $\mathbb{Z}\pi$-module. We recall some consequences of Poincaré duality which are commonly used in the theory of surgery, from which we take the notations $K_i(f;B)$ and $K^i(f;B)$ for the relative terms of the homology and cohomology exact sequences for f. These sequences have then the following form:

$$\to K^{i-1}(f;B) \to H^i(M,B) \to H^i(M;B) \to K^i(f;B) \to$$

and

$$\to K_i(f;B) \to H_i(M;B) \to H_i(M:B) \to K_{i-1}(f,B) \to$$

Let $\Delta : H^i(M:B) \to H_{n-i}(M:B)$ be the Poincare duality isomorphism. The following formulas are classical (see [5, proof of Lemma 1.4]):

$$f_* \circ \Delta \circ f^* \circ \Delta^{-1}(a) = d \cdot a, \qquad a \in H_{n-i}(M:B)$$

$$\Delta^{-1} \circ f_* \circ \Delta \circ f^*(\alpha) = d \cdot \alpha, \qquad \alpha \in H^i(M;B)$$

Therefore, if B is a $\Lambda\pi$-module, the above sequences split:

$$H^i(M;B) \cong H^i(M;B) \oplus K^i(M;B), \qquad H_i(M;B) \cong H_i(M;B) \oplus K_i(M;B) \qquad (1)$$

The Poincaré duality isomorphism is known to preserve these splittings and gives an isomorphism $K^i(M;B) = K_{n-i}(M;B)$ (see [2, I.2.6]).

The chain complex $C_k(\tilde{M};\Lambda)$ ($\tilde{M}$ the universal covering of M), whose homology is $H_k(M;\Lambda\pi)$ can be taken finitely generated over $\Lambda\pi$. Therefore if $\Lambda\pi$ is noetherian, the splitting (1) for $B = \Lambda\pi$ implies $K_k(f;\Lambda\pi) = 0$ for all k, which proves a).

To prove b), observe that $K_i(f;\Lambda\pi) = 0$ for $i \leq k$ implies that $K^i(f;\Lambda\pi) = 0$ for $i \leq k$ and, then, $K_j(f;\Lambda\pi) = 0$ for $j \geq n-k$. Also, $K_i(f;\Lambda\pi) = 0$ for $i \leq 1$, since M is connected and $H_1(M;\Lambda\pi) = H_1(\tilde{M};\Lambda) = 0$. This implies b) for $n \leq 3$. When $n = 4$, one has $K_i(f;\Lambda\pi) = 0$ except for $i = 2$. An easy homological algebra argument shows that, as $K_i(f;\Lambda\pi) = 0$ for $i \leq 1$, then $K_i(f;B) = 0$ for $i \leq 1$ and any $\Lambda\pi$-module B. Therefore $K^3(M;B) = 0$ and one deduces, as in [12, Lemma 2.3], that $K_2(f;\Lambda\pi)$ is a finitely generated stably free $\Lambda\pi$-module. One then has:

$$K_2(f;\Lambda\pi) \oplus (\Lambda\pi)^r \cong (\Lambda\pi)^s \ . \qquad (2)$$

As $K_i(f;\Lambda\pi) = 0$ for $i \leq 1$, one has $K_2(f;\Lambda) = K_2(f;\Lambda\pi) \otimes_{\Lambda\pi} \Lambda$ (See [4, Lemma 1.4]). Since Λ is noetherian, Equation (1) for $B = \Lambda$ implies that $K_2(f;\Lambda) = 0$, and therefore $r = s$ by (2). This implies that $K_2(f;\Lambda\pi) = 0$ since, otherwise, $(\Lambda\pi)^r$ would be a proper summand of itself, which contradicts a theorem of Kaplanski [9].

COROLLARY 2. The answer to Question Q is "yes" if π is virtually nilpotent (i.e., if π contains a nilpotent subgroup of finite index).

Proof: Let π' be a nilpotent subgroup of π of finite index. A $\mathbb{Z}\pi$-module B is finitely generated over $\mathbb{Z}\pi$ if and only if it is finitely generated over $\mathbb{Z}\pi'$. As $\mathbb{Z}\pi'$ is noetherian (since π' is finitely generated and nilpotent, using [8 pp. 153-154]), one deduces that $\mathbb{Z}\pi$ is noetherian. Therefore, by Proposition 1, it is enough to prove that π is hopfian.

Let k be the index of π' in π and let Γ be the intersection of a the subgroups of π of index $\leq$ k. The group is of finite index in π and any epimorphism $\alpha : \twoheadrightarrow \pi$ satisfies $\alpha(\Gamma) = \Gamma$ (in particular, Γ is normal). One then has the following commutative diagram:

$$\begin{array}{ccccc}
\Gamma & \rightarrowtail & \pi & \twoheadrightarrow & \pi/\Gamma \\
\downarrow{\scriptstyle \alpha|\Gamma} & & \downarrow{\scriptstyle \alpha} & & \downarrow{\scriptstyle \bar{\alpha}} \\
\Gamma & \rightarrowtail & \pi & \twoheadrightarrow & \pi/\Gamma
\end{array}$$

The epimorphism $\bar{\alpha}$ is injective, since π/Γ is finite, and then $\alpha|\Gamma$ is surjective. But Γ is finitely generated (being of finite index in π) and nilpotent (being a subgroup of π'). Therefore, Γ is hopfian and then $\alpha|\Gamma$ is an isomorphism. By the five Lemma, α is an isomorphism, which shows that π is hopfian.

ANSWER TO QUESTION Q1a

Any manifold admits self-maps of degree 0 and 1. But the existence of self-maps of degree $\neq 0,1$ seems to be a strong condition on a manifold. We might then think that it would be harder to find strange self-maps for degree $\neq \pm 1$ then for degree ± 1. However, although Question Q1 remains completely open in general, we have a negative answer to Question Q1a:

THEOREM 3. For any $d \neq \pm 1$ and any $n \geq 6$, there exists a closed framed manifold M^n and a map $g : M \to M$ with $d(g) = d$, such that $\pi_1 g$ is surjective but not injective.

The proof of Theorem 3 requires some preliminaries. Let $\alpha : G \twoheadrightarrow Q$ be an epimorphism of groups and let $N = \ker\alpha$. Dividing out G by $[N,N]$ gives the following exact sequence:

$$H_1(N) \rightarrowtail G/[N,N] \xrightarrow{\alpha} Q$$

which is completely characterized by an element $H^2(Q;H_1(N))$ which will be called $c(\alpha)$. The following proposition permits us to realize geometrically some non-hopfian epimorphisms:

PROPOSITION 4. Let $d \in \mathbb{Z}$. Let $\alpha : G \twoheadrightarrow G$ be a self epimorphism of a group G such that:

1) BG ($= K(G,1)$) is homotopy equivalent to a finite 2-complex.
2) $F = \ker\alpha$ satisfies $H_2(F;\mathbb{Z}) = 0$.
3) $c(\alpha) \in H^2(G;H_1(F))$ satisfies $d \cdot c(\alpha) = 0$.

Then, there exists a closed framed manifold P of dimension 6 and a map $f : P \to P$ with $d(f) = d$, such that $\pi_1(P) = G$ and $\pi_1 f = \alpha$.

We shall prove Proposition 4 and then give examples of epimorphisms α, with non-trivial kernel, satisfying the hypotheses of Proposition 4 for $d \neq \pm 1$. This will prove Theorem 3 by taking $g = f\times id : P\times S^{n-6} \to P\times S^{n-6}$. We were just not able to find such examples (with non-trivial kernel) for $d = \pm 1$.

Proof of Proposition 4: Let P be the boundary of a regular neighborhood of an embedding of the 2-dimensional complex BG into $\mathbb{R}^7$. The retraction of this regular neighborhood onto BG produces a map $\tau : P \to BG$ inducing an isomorphism on the fundamental group (this is the identification of $\pi_1(P)$ with G). We denote also by $\alpha : BG \to BG$ the map inducing the epimorphism α and we considered α and τ as a Serre fibrations (the fiber of α being BF). Form the pull-back diagram:

$$\begin{array}{ccccc} BF & \longrightarrow & X & \xrightarrow{\tilde{\alpha}} & P \\ \big\downarrow{=} & & \big\downarrow{\bar{\tau}} & & \big\downarrow{\tau} \\ BF & \longrightarrow & BG & \xrightarrow{\alpha} & BG \end{array}$$

This gives rise to a morphism of the Serre spectral sequence:

$$\begin{array}{ccc} E^2_{p,q}(\tilde{\alpha}) = H_p(P;H_q(BF)) & \Longrightarrow & H_{p+q}(X) \\ \big\downarrow{\bar{\tau}_*} & & \big\downarrow{\tau_*} \\ E^2_{p,q}(\alpha) = H_p(BG;H_q(BF)) & \Longrightarrow & H_{p+q}(X) \end{array}$$

Whose E^2-terms look as follows:

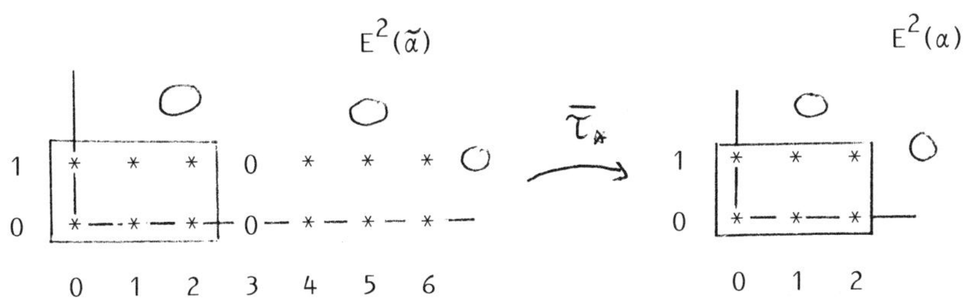

(This uses the fact that BF is a 1-dimensional complex (since $H_2(F; \mathbb{Z}) = 0$), that BG is a 2-dimensional complex and that P has a handle decomposition without handles of index 3.) By construction of P, the fiber of τ is 3-connected, and therefore the homomorphism $\bar{\tau}_* : E^2_{p,q}(\hat{\alpha}) \to E^2_{p,q}(\alpha)$ is an isomorphism for $p \leq 2$ (inside the rectangles in the above picture).

As $E^2_{p,q}(\tilde{\alpha})$ has only two non-zero lines, it gives rise to the long exact sequence:

$$\longrightarrow H_k(X) \xrightarrow{\tilde{\alpha}_*} H_k(P) \xrightarrow{d_2} H_{k-2}(P;H_1(BF)) \longrightarrow H_{k-1}(X) \longrightarrow$$

We shall use the following affirmation:

AFFIRMANTION. The differential d_2 coincides (at least up to sign) with the cap product with $c(\alpha) \in H^2(P;BF)$, the first obstruction for $\tilde{\alpha}$ to admit a section.

This affirmation was presented as a conjecture in [5], proven only in special cases. Since then, the author realized that this result is considered as "folklore" by the specialists. For the Hochschild-Serre spectral sequence, it is announced in [1, last theorem] and the proof of the corresponding result for the cohomology spectral sequence can be found in [6, theorem 4]. Still, no proof of the affirmation has been published (techniques are tedious). (If one is reluctant to use the affirmation, the same proof, with the techniques of [5, proof of theorem 1.5], permits us to prove a weaker version of proposition 4 and theorem 3, in which the sentence "with d(f) = d" in the conclusions has to be replaced by "with d(f) invertible in $\mathbb{Z}[1/d]$".)

By naturality of the first obstruction, one has $\tau^*(c(\alpha)) = c(\tilde{\alpha})$ and then $d \cdot c(\tilde{\alpha}) = 0$. The first consequence of the affirmation is then that $d \cdot [P] \in H_6(P)$ is in the image of $\tilde{\alpha}_*$. The second is that $H_1(P;H_1(BF)) = 0$. Indeed, the isomorphism $\bar{\tau}_* : E^2_{p,q}(\tilde{\alpha}) \to E^2_{p,q}(\alpha)$ (for $p \leq 2$), produces the following commutative diagram:

$$\begin{array}{ccccccccc}
0 \to & H_1(P;H_1(BF)) & \longrightarrow & H_2(X) & \xrightarrow{\tilde{\alpha}_a} & H_2(P) & \xrightarrow{d_2} & H_0(P;H_1(BF)) \\
& \downarrow\simeq & & \downarrow\simeq & & \downarrow\simeq & & \downarrow\simeq \\
0 \to & H_1(BG;H_1(BF)) & \longrightarrow & H_2(BG) & \xrightarrow{\alpha_a} & H_2(BG) & \xrightarrow{d_2} & H_0(BG;H_1(BF))
\end{array}$$

As $d \cdot c(\alpha) = 0$, the image of d_2 is finite and therefore α_* is injective, which gives $H_1(P;H_1(BF)) = H_1(BG;H_1(BF)) = 0$. We also have $H_2(P;H_1(BF)) = H_2(BG;H_1(BF)) = 0$, since $H_3(BG) = 0$. We then have obtained that $H_*(P,X) = 0$ for $* \leq 6$, except if $* = 2$ or 6. The E^2-term of the Atiyah-Hirzebruch spectral sequence for computing the framed bordism $\Omega^{fr}_*(P,X)$ looks as follows:

$$E^2_{p,q} = H_p(P,X;\Omega^{fr}_q(pt))$$

$$\begin{array}{c|ccccccc}
2 & 0 & 0 & * & 0 & 0 & 0 & * \\
1 & 0 & 0 & * & 0 & 0 & 0 & * \\
0 & 0 & 0 & * & 0 & 0 & 0 & * \\
\hline
 & 0 & 1 & 2 & 3 & 4 & 5 & 6
\end{array}$$

which, together with the classical fact that $\Omega^{fr}_4(pt) = \pi^s_4 = 0$, implies that the homomorphism $\Omega^{fr}_6(P,X) \to H_6(P,X)$ is injective. Using the diagram:

$$\begin{array}{ccccc}
\Omega^{fr}_6(X) & \longrightarrow & \Omega^{fr}_6(P) & \longrightarrow & \Omega^{fr}_6(P,X) \\
\downarrow & & \downarrow & & \downarrow \\
H_6(X) & \longrightarrow & H_6(P) & \longrightarrow & H_6(P,X)
\end{array}$$

one deduces that $d \cdot [P] \in H_6(P)$ (which is in the image of the framed bordism, since P is framed) can be represented by a composed map $T \xrightarrow{h} X \longrightarrow P$, where T^6 is a framed manifold.

We now turn our attention to the composed map $T \circ h : T \to BG$. This map represents a class in $\Omega^{fr}_6(BG)$. Using that BG is 2-dimensional and that $\Omega^{fr}_4(pt) = \Omega^{fr}_5(pt) = 0$, we obtain from the Atiyah-Hirzebruch spectral sequence for $\Omega^{fr}_6(BG)$ that $\Omega^{fr}_6(BG) \cong \Omega^{fr}_6(pt)$. Let $(\overline{T},\overline{h})$ be such that $\overline{T} = \overline{T} \# (-T)$, $\overline{h}|T$ homotopic to h and $\overline{h}|T)$ is a constant map. As $\overline{T}$ represents 0 in $\Omega^{fr}_6(pt)$, the above argument shows that there exists a framed manifold W of dimension 7 with boundary $\partial W = \overline{T}$ and an extension $\beta : W \to BG$ of $\tau \circ h$. By low dimensional framed surgeries, which are possible by [7, lemma 6.2], one may assume that $\pi_1\beta$ is an isomorphism and that $\pi_2(W) = 0$ (namely, β is a 3-connected map).

Since BG is 2-dimensional, the map β admits a homotopy section $\gamma : BG \to \text{int}W$ which induces an isomorphism on the fundamental groups. One may suppose that γ is an embedding. Let V be the compact manifold obtained from W by removing the interior of a regular neighborhood R of $\gamma(BG)$.

Since P is framed, it coincides with the regular neighborhood of an embedding of BG into $\mathbb{R}^7$ (one uses, for instance, the classification of the stable thickenings of Wall, see [11, proposition 5].) Therefore, $V = T \amalg P$. By general position, one gets that $\pi_i(P) = \pi_i(V)$ for $i \leq 2$. By Poincaré duality, one has $H^i(V,\overline{T};B) = 0$ when $i \geq 5$, for any $\mathbb{Z}\pi_1(V)$-module B. The above mentioned handle decomposition of P shows that $\pi_j(P) = 0$ for $j \leq 3$. Therefore, there is no obstruction to extend $\tilde{\alpha} o h : \overline{T} \to P$ to a map $F : V \to P$. One checks easily that $f = F|P : P \to P$ satisfies $\pi_1 f = \alpha$ and $d(f) = d$.

We give now examples of non-injective epimorphisms satisfying the conditions of proposition 4. They are straightforward generalizations of the first non-hopfian epimorphism for finitely presented groups, discovered by B.H. Neumann [10]. Let R be the group given by presentation:

$$R = \langle z,t \mid tzt^{-1} = z^d \rangle$$

Let R_i $(i = 1,2)$ denote two copies of R, with generators z_i and t_i. Form the free products with amalgamation:

$$\tilde{G} = R_1 * R_2/\{z_1^d = z_2^d\}$$

and

$$G = R_1 * R_2/\{z_1 = z_2\}$$

There is an obvious epimorphism $\tilde{\alpha} : G \to G$ extending the identities on R_1 and R_2. On the other hand, one checks easily that the correspondence $t_i \mapsto t_i$ and $z_i \mapsto t_i z_i t_i^{-1}$ gives rise to an isomorphism fro G to $\tilde{G}$. Call $\alpha : G \to G$ the epimorphism obtained by composing this isomorphism with α. If $d \neq \pm 1$, the element $z_1 z_2^{-1} \in \tilde{G}$ represent a non-trivial element of $\ker\alpha$ (by well-known techniques on free products with amalgamation), and therefore $\ker\alpha \neq \{1\}$. Let us now check the conditions of proposition 4.

1. It is classical that the finite two-dimensional complex associated to the above presentation of G is aspherical and therefore has the homotopy type of BG.

2. The intersection of $F = \ker\tilde{\alpha}$ with R_i is the trivial group. As F is a normal subgroup, it intersects trivially any conjugate of R_i. It follows that F is free, by the theorem of Kurosh, and therefore $H_2(F;\mathbb{Z}) = 0$.

3. To check that $d \cdot c(\alpha) = 0$, one writes the Mayer-Vietoris sequences for G and G [3, corollary 7.7]. The vertical arrows are induced by $\tilde{\alpha}$ and the coefficients are $H_1(F)$. We denote by $\tilde{C}$ the subgroup of G generated

by z_1^d and by C the subgroup of G generated by z_1.

$$\begin{array}{ccccccccc}
H^1(R_1)\oplus H^1(R_2) & \longrightarrow & H^1(C) & \longrightarrow & H^2(G) & \longrightarrow & H^2(R_1)\oplus H^2(R_2) & \longrightarrow & 0 \\
\downarrow \simeq & & \downarrow \tilde{\alpha}^1 & & \downarrow & & \downarrow & & \\
H^1(R_1)\oplus H^1(R_2) & \longrightarrow & H^1(\tilde{C}) & \xrightarrow{\lambda} & H^2(\tilde{G}) & \longrightarrow & H^2(R_1)\oplus H^2(R_2) & \longrightarrow & 0
\end{array}$$

As $d \cdot (c(\alpha)) = 0$, a diagram chasing argument shows that $c(\alpha) = \lambda(a)$, with $a \in H^1(C)$ satisfying $\tilde{\alpha}^{-1}(a) = 0$. But α^1 is induced by a d-sheeted self-covering of $BG = S^1$. By transfer, one deduces that $d \cdot a = 0$, and therefore $d \cdot c(\alpha) = 0$.

REMARK. By [5, lemma 1.4], the condition $d \cdot c(\alpha) = 0$ is necessary, in order to obtain the conclusion of Proposition 4.

REFERENCES

1. Andre, M., Le d_2 de la suite spectrale en cohomologie des groupes. C.R. Acad. Sc. Paris, 260 (1965), 2669-2671.
2. Browder, W., Surgery on Simply Connected Manifolds. Springer-Verlag, 1972.
3. Brown, K., Cohomology of Groups. Springer-Verlag, 1982.
4. Cappell, S., and Shaneson, J., The Codimension Two Placement Problem and Homology Equivalent Manifolds. Annals of Math. 99 (1974), 277-348.
5. Hausmann, J-Cl., Fundamental Group Problems Related to Poincaré Duality. Canad. Math. Soc. Conf. Proc. Vol 2, Part 2), (1982) (Conf. London, Ontario, 1981) 327-336.
6. Hochschild, G., and Serre, J-P., Cohomology of Groups Extensions. Trans. AMS 74 (1953), 110-134.
7. Kervaire, M., and Milnor, J., Groups of Homotopy Spheres. Ann. of Math. 77 (1963), 504-537.
8. Lambeck, J., Lectures on Rings and Modules. Blaisdell Publ. Co., 1966.
9. Montgomery, S., Left and Right Inverses in Group Algebras. Bull. AMS 75 (1969), 539-540.
10. Neumann, B. H., On a Problem of Hopf. J. of London Math. Soc. 28 (1953), 351-353.

11. Wall, C.T.C., Classification Problems, IV (Thickenings). Topology 5 (1966), 73-94.

12. Wall, C.T.C., Surgery on Compact Manifolds. Academic Press, 1970.

Deformations of Totally Geodesic Foliations

DAVID L. JOHNSON / Department of Mathematics, Lehigh University, Bethlehem, Pennsylvania

0. INTRODUCTION

Various differential-geometric structures have been imposed on either the leaves or the transverse directions of foliations in the past several years; among the most interesting being totally geodesic foliations (each leaf being a totally geodesic submanifold of the ambient manifold) and Riemannian foliations (the leaves remain a constant distance apart in some transverse metric, or in the ambient Riemannian manifold) [1], [8], [9], [13], [14]. The theorems proved about such structures tend to be either topological obstructions on the characteristic algebra or classifications of all manifolds admitting such foliations (again, usually topological classifications). Recently, A.M. Naveira has developed a geometric rationalization for which geometric structures are natural [2], [11], in that they appear as invariant subspaces of a certain representation space for tensors naturally associated to the foliations. Fortunately for those of us who have worked on such structures, all of the typical geometric notions which have been attached to foliations appear in Naveira's classification, but still there is some question of which geometric structures are natural from a more classical viewpoint.

One criterion of naturality would seem to be a sort of stability; a geometric structure on a foliation is <u>stable</u> if all nearby foliations also possess that structure. Those would correspond to geometric conditions which reflect some topological property of the foliation. Conversely, instable structures reflect some property of the C^∞ structure of the foliation. The purpose of this article was to have been to show that geodesibility is a stable condition in this sense; interesting enough, this is not the case except in codimension-one. This phenomenon appears even for such simple examples as the Hopf fibration of S^3. To a degree this result extends the result of Gluck [4], which illustrated a class of foliations near the Hopf fibration which are not geodesible (although the present result concerns only deformations of the initial metric), and to the same degree answers a question posed by Gluck in [4], originally due to Thurston.

The condition that a foliation be Riemannian for some metric is very easily seen to be an instable notion. The only foliations of the 2-dimensional torus T that are Riemannian are conjugate to foliations by either skew lines or circles; but a generic deformation of such a foliation has limit cycles, and so cannot be Riemannian. It seems reasonable to conjecture that the more general condition, due to Blumenthal and Hebda [1], that the foliation admit an Ehresmann connection (a transverse distribution with a unique horizontal path-lifting property) will be a stable notion in this sense. While not a classical differential-geometric condition, the existence of an Ehresmann connection is essentially geometric, and is guaranteed by both the existence of a Riemannian foliation structure and a totally geodesic structure. It also implies many of the structure theorems and topological obstructions of both.

In the light of the results of this note, it seems unlikely that any of Naveira's special geometric classes of foliations will be stable in the sense used here, except in special cases. These geometric conditions will more likely represent special elements in some (which?) topological classes of foliations, analogous to harmonic maps within a homotopy class of maps.

I would like to thank Karsten Grove for suggesting to me the possibility that geodesibility might be an instable conditions. Herman Gluck pointed out an example which showed that an earlier version of this article was flawed; I wish to thank him for pointing out the error of my ways and for many useful discussions which were invaluable in finding the correct description of the phenomena of deformations of the Hopf fibration.

1. DEFORMATIONS OF FOLIATIONS

Foliations F and F' on an n-dimensional manifold M are conjugate if they differ only by the action of a smooth diffeomorphism of M. The set of smooth conjugacy classes of codimension-q foliations F on M, denoted $F^q(M)$, has a natural topology inherited from the C -topology on the space of sections of the Grassmann bundle of (n-q)-planes in $T_*(M)$. A codimension-q foliation can be regarded as such a section, and Diff(M) acts smoothly on $\Gamma(G(n-q,T_*M))$; $F^q(M)$ is the resulting quotient space [7]. The topology is natural in that <u>deformations</u> F_t of a foliation $F \in F^q(M)$ - i.e., families F_t of foliations, t in some connected parameterizing manifold T, with $F \simeq F_0$ for some $0 \in T$ - correspond to maps into $F^q(M)$. Even though $F^q(M)$ is not usually a manifold, there is a sense in which maps $T \to F^q(M)$ are smooth, corresponding to smoothly varying families F_t. Clearly, foliations F and F' are smoothly homotopic if and only if they are in the same path-component of $F^q(M)$.

The infinitesimal structure of $F^q(M)$ is governed by a parabolic cohomology theory introduced by J. Heitsch [6] (cf. also [7]). For a given foliation F, let H be a complementary distribution, and let $P : T_*(M) \to T_*(M)$ be the <u>almost product structure</u> defined by F and H, given by $P^2 = I$, $P_{|F} = I$, $P_{|H} = -I$ (conversely, such an automorphism determines F and H). In terms of such a structure, the <u>Heitsch cohomology</u> $H^*(M,F)$ can be defined as the cohomology of the complex

$$\cdots \xrightarrow{d_P} \Gamma(\Lambda^k(F) \otimes H) \xrightarrow{d_P} \Gamma(\Lambda^{k+1}(F) \otimes H) \xrightarrow{d_P} \cdots$$

where, for $\omega \in \Gamma(\Lambda^k(F) \otimes H)$ a vertical, H-valued k-form, $d_P(\omega)$ is given by

$$d_p(\omega)(A_1, \cdots, A_{k+1}) = (I - P)d\omega(A_1, \cdots, A_{k+1})$$

$$= \sum_{j=1}^{k+1} (I - P)(-1)^{j+1}[A_j, \omega(A_1, \cdots, \hat{A}_j, \cdots, A_{k+1})]$$

$$+ \sum_{i<j} (-1)^{i+j+1} \omega([A_i, A_j], A_1, \cdots, \hat{A}_i, \cdots, \hat{A}_j, \cdots, A_{k+1})$$

for $A_1, \cdots, A_{k+1}$ tangent to F. Integrability of F implies that this is a complex, that is, $(d_p)^2 = 0$. Any one-parameter deformation F_t, $t \in (-\varepsilon, \varepsilon)$, of $F = F_0$ has H as a complementary distribution for t near 0, and the associated almost-product structures P_t satisfy $\frac{d}{dt} P_{t|0} = h \in \Gamma(\Lambda^1(F) \otimes H)$, with $d_p(h) = 0$; so h represents an element of $H^1(M,F)$ [7]. Thus $H^1(M,F)$ is the space of <u>infinitesimal deformations</u> of F. There are, as in complex deformation theory, obstructions lying in $H^2(M,F)$ to an infinitesimal deformation's being integrable to a deformation of F. Also, $H^0(M,F)$ reflects the local topological properties of $F^q(M)$, being locally Hausdorff and embeddable in $H^1(M,F)$ when $H^0(M,F)$ is minimized.

The following results are stated to provide examples of the structure of $F^q(M)$ in a few specific examples. Their proofs are contained in [7].

THEOREM 1.1. Let T be the 2-dimensional torus. There is an open, dense subset of the path-component of $F^1(T)$ consisting of foliations without Reeb components which is an infinite disjoint union of infinite-dimensional manifolds, consisting of structurally stable foliations F with a finite number of compact leaves. These foliations satisfy $H^0(T,F) = 0$.

THEOREM 1.2. Let $SF^1(M)$ be the set of codimension-one foliations F on a compact manifold M satisfying the following conditions:

1. F is structurally stable.
2. There are a finite number $L_1, \cdots, L_k$ of compact leaves.
3. In the leaf space M/F, the closure of any noncompact leaf L contains at least two compact leaves.
4. $H^0(M,F)$ is minimal (either 0- or 1-dimensional).

Either $SF^1(M,F)$ is empty, or is an open, Hausdorff subspace of $F^1(M)$ which locally embeds into $H^1(M,F)$.

2. DEFORMATIONS AND METRIC STRUCTURES

Let F be a codimension-q foliation on a Riemannian manifold M with metric $\langle , \rangle$. The metric determines a specific choice of complementary distribution H, $H = F^{\perp}$, which will be assumed in the sequel. Also, the almost-product structure P associated to F will be assumed to have H as its (-1)-eigenspace. A vector field will be called vertical if it is tangent to F at each point, and horizontal if it is tangent to H at each point.

F is totally geodesic if each leaf is a totally geodesic submanifold of M, or, equivalently, for X,Y vertical, Z horizontal, $\langle \nabla_X Y, Z\rangle = 0$. F is Riemannian if the leaves locally are a constant distance apart; that is $\langle \nabla_Z W, X\rangle = -\langle \nabla_W Z, X\rangle = \frac{1}{2}\langle [Z,W], X\rangle$ for Z,W horizontal and X vertical. In a geometric sense these two notions are essentially dual, in that, if H is integrable, then one foliation is totally geodesic if and only if the other is Riemannian [8]. A horizontal vector field Z is basic if for any local submersion $f_U : U \to \mathbb{R}^q$ defining F, Z is f_U-related to a vector field on $\mathbb{R}^q$. Equivalently, Z is basic if and only if [Z,X] is vertical whenever X is vertical [9], [12]. If F is Riemannian, then basic vector fields have constant length. Basic vector fields will always exist locally, but globally the set of basic vector fields is precisely $H^0(M,F)$, which may be 0 [7].

A one parameter deformation $\langle , \rangle_t$ of the Riemannian metric $\langle , \rangle$ of M is determined by a family A_t of positive-definite symmetric endomorphisms of $T_*(M)$, by $\langle \cdot , \cdot \rangle_t = \langle A_t \cdot , \cdot \rangle$. Since

$$2\langle \nabla_X Y, Z\rangle = X\langle Y,Z\rangle + Y\langle X,Z\rangle - Z\langle X,Y\rangle + \langle [X,Y],Z\rangle + \langle [Z,X],Y\rangle + \langle [Z,Y],X\rangle$$

the covariant derivative is then deformed according to:

$$2\langle A_t(\nabla_t)_X Y, Z\rangle = X\langle A_t Y,Z\rangle + Y\langle A_t X,Z\rangle - Z\langle A_t X,Y\rangle + \langle A_t[X,Y],Z\rangle + \langle A_t[Z,X],Y\rangle + \langle A_t[Z,Y],X\rangle .$$

Now let Y = X be vertical and Z be horizontal. If $A_0 = I$, $\nabla_0 = \nabla$, $\frac{d}{dt}\big|_0 A_t = A'$, and $\frac{d}{dt}\big|_0 \nabla_t = \nabla'$, then:

$$2\langle \nabla'_X X, Z\rangle = 2\langle \nabla_X A'X, Z\rangle - 2\langle A'\nabla_X X, Z\rangle - \langle \nabla_Z A'X, X\rangle + \langle \nabla_Z X, A'X\rangle .$$

Let F be totally geodesic. In order for F_t to be totally geodesic in some deformed metric $\langle A_t \ ,\rangle$, it is necessary and sufficient that

$$\langle A_t(\nabla_t)_{X_t} X_t, Z_t\rangle = 0$$

for X_t in F_t and Z_t perpendicular to F_t with respect to the deformed metric. Setting $X_t = \frac{1}{2}(I + P_t)(X)$ for X an arbitrary F-vertical vector field, and $Z_t = \frac{1}{2} A^{-1}(I - P_t^*)Z$ (where P_t^* is the adjoint of P_t with respect to the metric $\langle \ , \ \rangle$), it is easy to see that X_t will be F_t-vertical and Z_t will be perpendicular to F_t with respect to the metric $\langle A_t \ ,\rangle$ when t is near 0. Also, any vertical and horizontal vectors in the deformed structures can be represented in this form. Note that H, the (-1)-eigenspace of P_t by the assumptions of section 1, is not necessarily orthogonal to F_t in the deformed metric.

PROPOSITION 2.1. Let F be a codimension-q totally geodesic foliation on a Riemannian n-manifold M, and let F_t, $t \in (-\varepsilon,\varepsilon)$, be a deformation of F. If a deformation $\langle A_t \ ,\rangle$ of the metric makes each F_t totally geodesic, then

$$0 = \langle \nabla_X A'X, Z\rangle - \frac{1}{2}\langle \nabla_Z A'X, X\rangle + \frac{1}{2}\langle \nabla_Z X, A'X\rangle + \frac{1}{2}\langle \nabla_{hX} X, Z\rangle + \frac{1}{2}\langle \nabla_X hX, Z\rangle - \langle \nabla_X X, A'Z\rangle - \frac{1}{2}\langle \nabla_X X, h^* Z\rangle ,$$

where $A' = \frac{d}{dt}\big|_0 A_t$, $h = \frac{d}{dt}\big|_0 P_t$, and h^* is the $\langle \ , \ \rangle$ - adjoint of h.

Proof: Differentiating the equation $0 = \langle A_t(\nabla_t)_{X_t} X_t, Z_t\rangle$ for F_t to be totally geodesic, where $\frac{d}{dt}\big|_0 X_t = \frac{1}{2}hX$ and $\frac{d}{dt}\big|_0 Z_t = -A'Z - \frac{1}{2}h^* Z$ from the above (note that $A_0 = I$), immediately gives

$$0 = \langle \nabla_X A'X, Z\rangle + \langle \nabla'_X X, Z\rangle + \frac{1}{2}\langle \nabla_X hX, Z\rangle + \frac{1}{2}\langle \nabla_X hX, Z\rangle - \langle \nabla_X X, A'Z\rangle$$
$$- \frac{1}{2}\langle \nabla_X X, h^* Z\rangle .$$

Substituting in the expression for $\langle \nabla'_X X, Z\rangle$ above, and noting that A' is symmetric,

$$0 = \langle \nabla_X A'X, Z\rangle - \frac{1}{2}\langle \nabla_Z A'X, X\rangle + \frac{1}{2}\langle \nabla_Z X, A'X\rangle + \frac{1}{2}\langle \nabla_{hX} X, Z\rangle + \frac{1}{2}\langle \nabla_X hX, Z\rangle$$
$$- \langle \nabla_X X, A'Z\rangle - \frac{1}{2}\langle \nabla_X X, h^* Z\rangle ,$$

as claimed.

Since any foliation is locally a product, there is no local obstruction to finding a family A_t which keeps F_t totally geodesic. The question then becomes how to ensure that local deformations can be pieced together, or, equivalently, to construct the infinitesimal deformation A' globally. The two constructions are indeed equivalent, since there is no obstruction to integrating an infinitesimal family A' to a family A_t of metrics (for t not too large).

Consider first the case where F is of codimension-one. Then, as hX is horizontal for X vertical, if Z is any nonzero horizontal vector field, $hX = \lambda Z$, where λ is a function of X. Thus $\nabla_{hX} X = \lambda\nabla_Z X$ and $\nabla_X hX = \lambda\nabla_X Z + X(\lambda)Z$, which simplifies the expression above.

THEOREM 2.2. Let F be a codimension-one totally geodesic foliation on a Riemannian n-manifold M. Then, for any sufficiently small deformation F_t of F, there is a family $\langle A_t\ ,\ \rangle$ of metrics for which each F_t is totally geodesic.

Proof: It suffices to solve the above infinitesimal equation, and to show that such a solution can be integrated to a deformation A_t of the metric. However, setting $A' = -\frac{1}{2}(h + h^*)$, which in codimension-one gives $A'X = -\frac{1}{2}\lambda Z$, the equation above becomes:

$$-\frac{1}{2}\langle \nabla_X \lambda Z, Z\rangle + \frac{1}{4}\langle \nabla_Z \lambda Z, X\rangle - \frac{1}{4}\langle \nabla_Z X, \lambda Z\rangle + \frac{1}{2}\langle \nabla_{\lambda Z} X, Z\rangle + \frac{1}{2}\langle \nabla_X \lambda Z, Z\rangle +$$

$$+ \frac{1}{2}\langle\nabla_X X, h^* Z\rangle - \frac{1}{2}\langle\nabla_X X, h^* Z\rangle$$

$$= \frac{1}{4}\lambda\langle\nabla_Z Z, X\rangle + \frac{1}{4}Z(\lambda)\langle Z, X\rangle - \frac{1}{4}\lambda\langle\nabla_Z X, Z\rangle + \frac{1}{2}\lambda\langle\nabla_Z X, Z\rangle$$

$$= \frac{1}{4}\lambda Z\langle Z, X\rangle = 0.$$

Thus, if A_t satisfies the differential equation $A_t' = \frac{1}{2}(h_t + h_t^*)$, where $h_t = \frac{d}{dt}P_t$, then A_t will be (for t sufficiently small) a positive-definite symmetric matrix at each point, and will then satisfy the stated condition.

3. DEFORMATIONS OF THE HOPF FIBRATION

H. Gluck has constructed two examples of deformations of the foliation of S^3 by the circles of the Hopf fibration, one which is not geodesible [5], and another which is geodesible by a deformation of the round metric. The nongeodesible flow is a deformation of the Hopf foliation to one with only two closed linked closed orbits, one repelling and the other attracting. A theorem of Sullivan [15] is then used to show that this foliation is not geodesible. The geodesible deformations of the Hopf foliation are described using an observation of Gluck [personal communication] that, for any Riemannian manifold M, the flow on $T_*(M)$ of unit tangent vector fields to geodesics on M is a geodesic flow on $T_*(M)$ with the standard induced metric. The unit tangent flow to the geodesics of S^2 is then a geodesic flow on $T_1(S^2) = \mathbb{R}P^3$ which lifts to the Hopf fibration on the round S^3. Any deformation of the round metric on S^2 generates a deformation of the Hopf foliation, along with a deformation of the metric keeping the leaves totally geodesic.

The main result of this section places these examples in a more general, although infinitesimal, contex.

View S^3 as SU(2), specifically as the $\frac{1}{2}$-unit quaternions. Define orthonormal, left-invariant vector fields X, Y, and Z on S^3 by, $X(v) = 2v\cdot i$, $Y(v) = 2v\cdot j$, and $Z(v) = 2v\cdot k$, $v \in S^3$. Then $[X,Y] = 2Z$, $\nabla_X Y = Z$, $\nabla_Y X = -Z$, etc., permuting cyclically. These are the conventions for which S^3 has the metric of the symmetric space SU(2) with curvature 4, radius $\frac{1}{2}$. Let $F = \mathrm{Span}[X]$ be the distribution tangent to the usual Hopf fibration $\pi : S^3 \to S^2$, making S^3 an S^1-principal bundle.

Let $h \in \Lambda^1(F) \otimes F^\perp$, $h(X) = a(X)Y + b(X)Z = aY + bZ$. It will be convenient to consider complex multiplication on $F^\perp$ on the right, so that

$Y \cdot i = -Z$, and $h(X) = Y\,(a - ib)$. h is exact if there is a $\xi \in \Gamma(F^{\perp})$, $\xi = 1/2(fY + gZ)$, for which $h(X) = d_p(\xi) = (I - P)[X,\xi] = (-2g + X(F))Y + (2f + X(g))Z = Y\cdot(X(f - ig) - 2i(f - ig))$; thus h is exact if and only if

$$a + ib = X(f + ig) + 2i(f + ig).$$

LEMMA 3.1. h is <u>harmonic</u> in $\Lambda^1(F) \otimes F^{\perp}$ if and only if

$$X(a + ib) + 2i(a + ib) = 0.$$

Proof: By fiber integration, $\int_{S^3} = \int_{S^2} \int_{S^1}$. Thus, if $\int_{S^1} \langle h, d_F\xi\rangle = \int_{S^1} \langle h(X), d_F(\xi)(X)\rangle dx = 0$, where x is the fiber parameter of S^1, $\langle h, d_F(\xi)\rangle = 0$ in the L^2-metric on $\pi^1(F) \otimes F^{\perp}$. Conversely, considering ξ which have support in $\pi^{-1}(U)$, where U is a small neighborhood of a point $p \in S^2$, clearly h harmonic (orthogonal to $d_p(\Gamma(F^{\perp}))$ implies that $\int_{S^1} \langle h(X), d_p(\xi)\rangle dx = 0$ for all fibers.

Thus, h is harmonic if and only if

$$0 = \int_0^{\pi} \langle h(X), d_p\xi(X)\rangle dx$$

$$= \int_0^{\pi} \langle a + ib, 2i(f + ig) + f' + ig)\rangle dx$$

$$= \int_0^{\pi} \langle 2i(a + ib) + (a' + ib'),\ f + ig\rangle dx\ ,$$

noting that all functions are periodic with period π, and using integration by parts, where the prime denotes differentiation with respect to the fiber parameter x (representing the fiber as $\frac{1}{2}\{e^{2ix}, x \in [0,\pi]\}$). Since this must hold for all $(f + ig)$, $(a + ib)' + 2i(a + ib) = 0$, or

$$X(a) = 2b$$
$$X(b) = -2a\ .$$

COROLLARY 3.2. $H^1(S^3,F) = \Gamma(T_*(S^2))$.

Proof: In a local trivialization $\pi^{-1}(U) = U \times S^1$, $a + ib = Ge^{-2ix}$ for some complex-valued function G on $U \subset S^2$. $h_{|\pi^{-1}(U)}$ is then given by

$$h(X)_{(p,x)} = Y\cdot Ge^{2ix}\ ,$$

defining a tangent vector $G\bar{Y}$ on $T_*(U)$, where $\bar{Y} = \pi_*(Y_{|(p,0)})$. Note that here we do not change the sign when shifting from multiplication on the left to the right. The conventions used in defining a + ib and G were chosen so that h(X) would be right-invariant under the action of S^1, and so define a vector field $H = G\overline{Y}$ on S^2. Since h(X) is defined independently of local trivialization, clearly H is a globally-defined vector field on S^2. That the correspondence between the set of all harmonic h and all vector fields on S^2 is an isomorphism is not obvious.

REMARK: This generalizes the work of Girbau, Haefliger, and Sundararaman [3] to general deformations, rather than deformations of transversely holomorphic foliations. They show, in particular, that the set of transversely holomorphic deformations of the transversely holomorphic structure of the Hopf foliation is given by $H^0(S^2,T_*(S^2)) \simeq H^0(\mathbb{P}^1,O(2)) \simeq \mathbb{C}^3$. Their characterization will be used in the proof of the main result of this section.

THEOREM 3.3. $h \in H^1(S^3,F) \simeq \Gamma(T_*(S^2))$ is infinitesimally geodesible if and only if H is orthogonal to the set of holomorphic vector fields on S^2, i.e., $H \in H^0(\mathbb{P}^1,O(2))^{\perp}$. Thus the set of infinitesimally geodesible deformations is of codimension-3 in the set of all deformations.

Proof: By Proposition 2.1., an infinitesimal deformation A' of the metric must satisfy the following equation in order to infinitesimally preserve geodesibility: $0 = \langle\nabla_X A'X,E\rangle - \frac{1}{2}\langle\nabla_E A'X,X\rangle + \frac{1}{2}\langle\nabla_E X,A'X\rangle + \frac{1}{2}\langle\nabla_{h(X)}X,E\rangle + \frac{1}{2}\langle\nabla_X h(X),E\rangle - \langle\nabla_X X,A'E\rangle - \frac{1}{2}\langle\nabla_X X,h^*(E)\rangle$, where E = Y or Z. Since X is a geodesic flow field, the last two terms vanish. Set $A' = \alpha X + \beta Y + \gamma Z$, $h(X) = aX + bY$ as before, and assume that h is harmonic. Then, with E = Y,

$$0 = \langle\nabla_X(\alpha X + \beta Y + \gamma Z),Y\rangle - \frac{1}{2}\langle\nabla_Y(\alpha X + \beta Y + \gamma Z),X\rangle + \frac{1}{2}\langle\nabla_Y X,\alpha X + \beta Y + \gamma Z\rangle$$
$$+ \frac{1}{2}\langle\nabla_{aY + bZ}X,Y\rangle + \frac{1}{2}\langle\nabla_X(aY + bZ),Y\rangle$$
$$= -2\gamma + X(\beta) - \frac{1}{2}Y(\alpha) + \frac{1}{2}X(a),$$

since X,Y and Z are orthonormal. Similarly, with E = Z, $0 = 2\beta + X(\gamma) - \frac{1}{2}Z(\alpha) + \frac{1}{2}X(b)$. In complex form, this yields the single equation

$$\frac{1}{2}(Y + iZ)(\alpha) - \frac{1}{2}X(a + ib) = X(\beta + i\gamma) + 2i(\beta + i\gamma) .$$

These functions must also be periodic in x (the S^1 parameter) when restricted to a locally trivial neighborhood $\pi^{-1}(U) \simeq U \times S^1$, so all functions can be written in terms of their fourier series in x,

$$\alpha = \sum_n \alpha_n e^{2inx}, \quad \beta = \sum_n \beta_n e^{2inx}, \quad \gamma = \sum_n \gamma_n e^{2inx}, \quad a = a_{-1}e^{-2ix} + a_1 e^{2ix},$$

and

$$b = b_{-1}e^{-2ix} + b_1 e^{2ix},$$

where, for α, β, and γ, the sum runs over all $n \in \mathbb{Z}$, but for a and b the sum runs over only -1 and +1, since $X(a + ib) + 2i(a + ib) = 0$. This equation also implies that $b_{-1} = -ia_{-1}$ and $b_1 = ia_1$. All fourier coefficients satisfy $\bar{\alpha}_{-n} = \alpha_n$, etc., since all the functions are real-valued, and all coefficients are functions on $U \subset S^2$.

Given any function α, for $n \neq \pm 1$ the nth terms of the fourier coefficients of β and γ satisfy the functional equations (not differential equations)

$$\begin{pmatrix} \frac{1}{2}Y(\alpha_n) \\ \frac{1}{2}Z(\alpha_n) \end{pmatrix} = \begin{pmatrix} 2in & 2 \\ -2 & 2in \end{pmatrix} \begin{pmatrix} \beta \\ \gamma \end{pmatrix},$$

so $\beta_n = \frac{1}{4 - 4n^2}(inY(\alpha_n) + Z(\alpha_n))$, $\gamma_n = \frac{1}{4 - 4n^2}(-Y(\alpha_n) + inZ(\alpha_n))$, which satisfies $\bar{\beta}_{-n} = \beta_n$, $\bar{\gamma}_n = \gamma_n$, and the fourier series will converge for any α_n. Similarly, for $n = 1$, $\frac{1}{2}(Y + iZ)(\alpha_1) = -2\gamma_1 + 2i\beta_1$. The $n = -1$ term of the equation, however, becomes

$$\frac{1}{2}(Y + iZ)(\alpha_{-1}) + 2ia_{-1} = 0,$$

since the right-hand side vanishes (this must occur to avoid resonance in the ordinary differential equation in x, to ensure that $\beta + i\gamma$ be periodic).

Conjugating the above equation yields

$$\frac{1}{2}(Y - iZ)(\alpha_1) = 2ia_1 = iG,$$

since $a - ib = Ge^{2ix} = 2a_1 e^{2ix}$.

Thus the vector field $H = G\bar{Y}$ on S^2 must satisfy

$$\bar{\partial}^*(\alpha_1 d\bar{z}\bar{Y}) = -\bar{*}\bar{\partial}\bar{*}(\alpha_1 d\bar{z}\bar{Y}) = -\bar{*}\bar{\partial}\bar{\alpha}_1 \wedge dz\bar{Y}^* = -\bar{*}\frac{1}{2}(Y + iZ)(\bar{\alpha}_1)d\bar{z} \wedge dz\bar{Y}^*$$

$$= \frac{1}{2}(Y - iZ)(\alpha_1)\bar{Y} = iH,$$

where $\bar{Y}^* \in \Gamma(T^*(S^2)) \simeq \overline{\Gamma(T_*(S^2))}$ is dual to $\bar{Y}$, and dz is dual to $\frac{1}{2}(\bar{Y} - iZ) = \partial/\partial z$, with horizontal lift $\frac{1}{2}(Y - iZ)$ to $U \times X^1$.

Thus, in order for a suitable α to exist, H must be orthogonal to the subspace of holomorphic vector fields on S^2. On the other hand, if H is orthogonal to the holomophic vector fields, H will be $\bar{\partial}$ - exact, and so an α satisfying the necessary equation will exist. In that case, β and γ can be determined, completing the proof of the Theorem.

A similar proof establishes the following corollary, showing that deformations of the Hopf foliations of S^{2n+1} are always infinitesimally geodesible except for a finite-dimensional subspace.

COROLLARY 3.4. Let F be the Hopf foliation of S^{2n+1}. $h \in H^1(S^{2n+1}, F) \simeq \simeq \Gamma(T_*(\mathbb{CP}^n))$ will be infinitesimally geodesible if and only if h is orthogonal to the space of holomorphic vector fields on $\mathbb{CP}^n$.

REMARK. The non-geodesible foliation of S^3 described by Gluck [5] can be constructed to have infinitesimal deformation h corresponding to a holomorphic vector field H on $S^2 = \mathbb{C} \cup \{\infty\}$ with a source at the south pole and a sink at the north pole given by $H = x\partial/\partial x + y\partial/\partial y$ on $\mathbb{C}$.

ERRATUM

The conventions used in the discussion of the Hopf flow on S^3 are not as stated. The radius for which the covariant derivatives are as indicated is 1, not 1/2. This necessitates systematic changes in the periods of the functions discussed in §3, but does not change the form of the differential equation for harmonic forms, nor the form of the solutions. The critical equation identifying a harmonic $h \in \Lambda^1(F) \otimes F^\perp$ as a holomorphic vector field on S^2 is correct as stated (with the factor of 2), since such h(X) are right-invariant as vector fields on S^3. Also, due to the change in period, all fourier series must run over all n, not just even n. This change is completely inconsequential, and the conclusions remain as stated.

REFERENCES

1. Blumenthal, R. and J. Hebda, Ehresmann Connections for Foliations, Indiana U. Math J., 33 (1984), 597-611.
2. Gil-Medrano, O, On the Geometric Properties of Some Classes of Almost-product Structures, Rend. Circ. Mat. Palermo, 32 (1983), 315-329.
3. Girbau, J., Haefliger, A., and Sundararaman, D., On Deformations of Transversely Holomorphic Foliations, Journal fuer die reine und angw. Math., 346 (1983), 122-147.
4. Gluck, H., Can Space be Filled by Geodesics, and if so, How?, manuscript.
5. Gluck, H., Dynamical Behavior of Geodesic Fields, Global Theory of Dynamical Systems, Proc. Northwestern Univ., 1979, Springer Lecture Notes in Math., 819 (1980), 190-215.
6. Heitsch, J., A Cohomology for Foliated Manifolds, Comment. Math. Helvetici, 50 (1975), 197-218.
7. Johnson, D. L., Families of Foliations, to appear.
8. Johnson, D. L., and Naveira, A. M., Obstructions to Geodesibility of a Foliation of Odd Dimension, Geometria Dedicata, 9 (1981), 347-352.
9. Johnson, D. L., and Whitt, L. B., Totally Geodesic Foliations, J. Diff. Geo., 15 (1980), 225-235.
10. Miguel, V., Some Examples of Riemannian Almost-product Structures, Pac. J. Math. 111 (1984), 163-178.
11. Naveira, A. M., A Classification of Riemannian Almost-product Structures, Rendiconti di Matematica, 3 (1983), 577-592.
12. O'Neill, B., The Fundamental Equations of a Submersion, Michigan Math. J., 13 (1966), 459-469.
13. Pasternack, J., Classifying Spaces of Riemannian Foliations, Proc. Sym. in Pure Math., 27 (1975), 303-310.
14. Reinhart, B., Foliated Manifolds with Bundle-like Metrics, Ann. of Math. (2), 69 (1959), 119-131.
15. Sullivan, D., A Foliation of Geodesics in Characterized by having no Tangent Homologies, J. Pure and Applied Algebra, 13 (1978), 101-104.

Automorphisms of Punctured-Surface Bundles

DARRYL McCULLOUGH / Department of Mathematics, University of Oklahoma, Norman, Oklahoma

0. INTRODUCTION

Let S be a connected orientable surface of genus $g \geq 1$ with one boundary component, and let ϕ be a self-homeomorphism of S. The mapping torus $M(\phi) = S \times I/((x,0) \sim (\phi(x),1))$ is a 3-manifold which fibers over S^1 with S as fiber. Such manifolds arise naturally as complements of fibered knots in closed orientable 3-manifolds. In fact, any closed orientable 3-manifold contains a fibered knot [22],[14] whose complement $M(\phi)$ has a hyperbolic structure [13].

When $g = 1$, $M(\phi)$ is a punctured-torus bundle. In the orientable case these manifolds have been studied in [12], [11], and [24]; in particular the incompressible surfaces in $M(\phi)$ can be completely understood. In [12] it is observed that the mapping class groups $\Gamma(M(\phi))$ can be computed. In the first three sections of the present work, we carry this out in both the orientable and nonorientable cases. The results are summarized in section 4.

In section 5, still in the punctured-torus case, we analyze the effects of mapping classes on the boundary of $M(\phi)$. The fundamental group of $\partial M(\phi)$ is generated by the boundary of a fiber, called c, and a section to the fibering on $\partial M(\phi)$, called t. While c is determined up to orientation, a choice of t, called a framing of $M(\phi)$ in [11], must be made. All homeomorphisms must carry c to $c^{\pm 1}$ (since c generates the kernel of $H_1(\partial M) \to H_1(M)$) and hence must carry t to $t^{\pm 1}c^k$ for some $k \in Z$. In section 5 we come to a complete understanding of the restriction homomorphism $\Gamma(M(\phi)) \to \Gamma(\partial M(\phi))$.

When a solid torus is attached to (an orientable) $M(\phi)$ so that the meridian of the solid torus is attached along t, the result is a closed orientable 3-manifold $N(\phi)$, and the core of the solid torus is a fibered knot K in $N(\phi)$ whose genus equals the genus of S. When $M(\phi)$ has a homeomorphism carrying t to $t^{\pm 1}c^k$, it is clear that $\pm 1/k$ surgery on $N(\phi)$ along k yields a manifold homeomorphic to $N(\phi)$. In particular, if $N(\phi) = S^3$ and such a homeomorphism exists, then K would be a nontrivial knot in S^3 which does not have property P [3]. Since it is known [10] that this could only happen if $k = \pm 1$, we focus on this case. Motivated by the analysis of the punctured-torus case in section 5, we give in section 6 constructions of many $M(\phi)$ having homeomorphisms carrying t to $(tc)^{\pm 1}$, so that +1 surgery along K yields $N(\phi)$. For $g = 3$ we find some examples where $N(\phi)$ is a Z-homology 3-sphere. However, we also prove that our constructions can never yield $N(\phi) = S^3$.

The reader whose main interest is the surgery examples will have no difficulty starting in section 6 after a cursory reading of section 1.1 and lemma 1.2.2.

I wish to thank Vaughan Jones, Andy Miller, and Ulrich Oertel for helpful discussions and correspondence.

1. PRELIMINARIES

We work in the PL category without explicit mention.

1.1. NOTATION. The mapping class group $\Gamma(M(\phi))$ is the group of isotopy classes of homeomorphisms of $M(\phi)$. For a fixed fibering of $M(\phi)$ over S^1, we denote by $FP(M(\phi))$ the subgroup of $\Gamma(M(\phi))$ consisting of all classes containing a representative that carries fibers to fibers. The group of homotopy classes of self-homotopy-equivalences of $M(\phi)$ is denoted $E(M(\phi))$; since $M(\phi)$ is aspherical, $E(M(\phi))$ is isomorphic to the outer automorphism group $\mathrm{Out}(\pi_1(M))$.

For $A \subseteq X$, the notation $\Gamma(X \text{ rel } A)$ refers to isotopy classes of homeomorphisms that restrict to the identity on A, by isotopies that restrict to the identity on A.

It is clear that $M(\phi)$ depends up to homeomorphism only on the mapping class of ϕ in $\Gamma(S)$; however, as we are interested in the effect of homeomorphisms on $\partial M(\phi)$, it is convenient to work relative to a basepoint b_0 in the boundary of S. Let c be an element of $\pi_1(S, b_0)$ represented by a loop which runs once around ∂S. Define

$$\mathrm{Aut}^+(\pi_1(S,b_0)) = \{\phi \in \mathrm{Aut}(\pi_1(S,b_0)) \mid \phi(c) = c\}$$

$$\mathrm{Aut}^\pm(\pi_1(S,b_0)) = \{\phi \in \mathrm{Aut}(\pi_1(S,b_0)) \mid \phi(c) = c^{\pm 1}\}.$$

Classical methods show

PROPOSITION 1.1.1. The function sending $\langle f\rangle$ to $f_\#$ defines isomorphisms $\Gamma(S \text{ rel } \partial S) \to \mathrm{Aut}^+(\pi_1(S,b_0))$ and $\Gamma(S \text{ rel } b_0) \to \mathrm{Aut}^\pm(\pi_1(S,b_0))$.

Throughout the present work, we use the letters ϕ, γ, and τ to denote elements of these groups, moving freely between the interpretation as a mapping class or as an automorphism. We denote by ϕ_* the induced automorphism on $H_1(S; Z)$; since we will usually have a fixed basis for $H_1(S; Z)$, we usually regard ϕ_* as a $2g\times 2g$ matrix.

Under the correspondence of proposition 1.1.1, a Dehn twist about ∂S corresponds to the automorphism that sends x to cxc^{-1}. We denote this automorphism by $\mu(c)$.

Fix a standard generating set $\{a_1, b_1, a_2, \cdots, a_g, b_g\}$ for $\pi_1(S,b_0)$ so that the boundary curve c equals $\Pi_{i=1}^g [a_i, b_i]$. The basepoint of $M(\phi)$ will be $[(b_0,0)]$, denoted by b_0. The inclusion of S onto $[S \times \{0\}]$ induces an imbedding of $\pi_1(S,b_0)$ into $\pi_1(M,b_0)$ as a normal subgroup.

Let t be the element of $\pi_1(M(\phi),b_0)$ represented by the path sending s to $[(b_0,s)]$ for $0 \le s \le 1$. Then $\pi_1(M(\phi),b_0)$ is an HNN extension with presentation $\langle t, a_i, b_i \mid ta_it^{-1} = \phi(a_i), tb_it^{-1} = \phi(b_i), 1 \le i \le g\rangle$. Note that $\pi_1(\partial M(\phi),b_0)$ is the subgroup $\langle t, c \mid tct^{-1} = \phi(c)\rangle$.

We use the symbol C_n to denote the cyclic group of order $n \le \infty$, and D_n to denote the dihedral group of order n.

1.2. HOMEOMORPHISMS OF PUNCTURED-SURFACE BUNDLES. By theorem 7.1 of [27], or using corresponding results from [15] and [19] if $M(\phi)$ is nonorientable, we have

LEMMA 1.2.1. Homotopic homeomorphisms of $M(\phi)$ are isotopic. Equivalently, $\Gamma(M(\phi)) \to E(M(\phi)) \simeq \mathrm{Out}(\pi_1(M(\phi),b_0))$ is injective.

The next lemma is well-known in case $k = 0$. In the lemma, ε is 1 or -1.

LEMMA 1.2.2. The following are equivalent. (a) There is a fiber-preserving homeomorphism $H : M(\phi) \to M(\phi)$ such that $H_\#(t) = c^k t^\varepsilon$ and $H_\# |\pi_1(S,b_0) = \gamma$.

(b) There is an automorphism $\Theta : \pi_1(M(\phi),b_0) \to \pi_1(M(\phi),b_0)$ that preserves the subgroup $\pi_1(S,b_0)$, with $\Theta(t) = c^k t^\varepsilon$ and $\Theta|_{\pi_1(S,b_0)} = \gamma$.

(c) There is a $\gamma \in \mathrm{Aut}^{\pm}(\pi_1(S,b_0))$ such that $\gamma\phi\gamma^{-1} = \mu(c^k)\phi^\varepsilon$.

Proof: (a) $\Rightarrow$ (b) is immediate.

(b) $\Rightarrow$ (c) Let $x \in \pi_1(S,b_0)$. Applying Θ to $txt^{-1} = \phi(x)$, we find $\gamma\phi(x) = c^k t^\varepsilon \gamma(x) t^{-\varepsilon} c^{-k} = \mu(c^k)\phi^\varepsilon\gamma(x)$.

(c) $\Rightarrow$ (a) Let f and g be representatives for ϕ and γ respectively. If $\varepsilon = 1$, then $(fgf^{-1}g^{-1})_\# = \mu(c^{-k})$, so there is an isotopy j_s from 1_S to $j_1 = fgf^{-1}g^{-1}$ with trace(j_s) (i.e., the loop sending s to $j_s(b_0)$) representing c^k. Define H by $H([(x,s)]) = [(j_s(g(x)),s)]$. If $\varepsilon = -1$, then $(f^{-1}gf^{-1}g^{-1})_\# = \mu(c^{-k})$. If $\phi(c) = c^\delta$, we may choose j_s with $j_0 = f$, $j_1 = gf^{-1}g^{-1}$, and trace(j_s) representing $c^{\delta k}$. This time, H is defined by $H([(x,s)]) = [(j_s(g(x)), 1 - s)]$.

DEFINITION 1.2.3. We use the symbol $H(\gamma,\varepsilon)$ to denote a basepoint-preserving fiber-preserving homeomorphism of $M(\phi)$ such that $H(\gamma,\varepsilon)_\#(t) = c^k t^\varepsilon$ and $H(\gamma,\varepsilon)_\#|_{\pi_1(S,b_0)} = \gamma$. From lemma 1.2.2, we know that k is determined by γ and ε, hence $H(\gamma,\varepsilon)_\#$ is determined by γ and ε. Therefore, by lemma 1.2.1, the isotopy class of $H(\gamma,\varepsilon)$ in $\Gamma(M(\phi))$ is determined by γ and ε .

LEMMA 1.2.4. Let h be a homeomorphism of $M(\phi)$. The following are equivalent.

(a) h is isotopic to some $H(\gamma,\varepsilon)$

(b) h is isotopic to a fiber-preserving homeomorphism

(c) $h_*(\mathrm{image}(H_1(S) \to H_1(M(\phi)))) \subseteq \mathrm{image}(H_1(S) \to H_1(M(\phi)))$.

Proof: (a) $\Rightarrow$ (b) and (b) $\Rightarrow$ (c) are obvious. Suppose (c) holds. For any $x \in \pi_1(S,b_0)$, the homology class of $h_\#(x)$ is in the image of $H_1(S) \to H_1(M(\phi))$, which implies that the exponent of t in $h_\#(x)$ is zero. But this shows $h_\#(x) \in \pi_1(S,b_0)$. Therefore there is a morphism of extensions

$$\begin{array}{ccccccccc}
1 & \longrightarrow & \pi_1(S,b_0) & \longrightarrow & \pi_1(M(\phi),b_0) & \longrightarrow & C_\infty & \longrightarrow & 1 \\
 & & \downarrow{\scriptstyle h_\#|\pi_1(S,b_0)} & & \simeq\downarrow{\scriptstyle h_\#} & & \downarrow{\scriptstyle \overline{h}_\#} & & \\
1 & \longrightarrow & \pi_1(S,b_0) & \longrightarrow & \pi_1(M(\phi),b_0) & \longrightarrow & C_\infty & \longrightarrow & 1
\end{array}$$

By commutativity of the diagram $\overline{h}_\#$ is surjective, hence is an isomorphism.

Therefore $h_{\#}|_{\pi_1(S,b_0)}$ is an isomorphism, which we denote by γ. Now $\gamma(c) \in \pi_1(S,b_0) \cap \pi_1(\partial M(\phi),b_0)$, so $\gamma(c) = c^{\pm 1}$. Since $h|_{\partial M(\phi)}$ is a homeomorphism, $h_{\#}(t) = c^k t^{\varepsilon}$ for some $k \in Z$ and $\varepsilon = 1$ or -1. For all $x \in \pi_1(S,b_0)$, we have $\gamma\phi(x) = h_{\#}(\phi(x)) = h_{\#}(txt^{-1}) = c^k t^{\varepsilon}\gamma(x)t^{-\varepsilon}c^{-k} = \mu(c^k)\phi^{\varepsilon}\gamma(x)$, so $\gamma\phi\gamma^{-1} = \mu(c^k)\phi^{\varepsilon}$. By lemma 1.2.1, h is isotopic to $H(\gamma,\varepsilon)$.

The last two propositions in this section follow from the work in [6], but we give self-contained arguments for our special cases.

PROPOSITION 1.2.5. If $\det(I - \phi_*) \neq 0$, then $\Gamma(M(\phi)) = FP(M(\phi))$.

Proof: Abelianizing the presentation for $\pi_1(M(\phi),b_0)$ shows that $H_1(M(\phi)) \simeq Z \oplus (Z^{2g}/\text{image}(I - \phi_*))$, with Z generated by t and the other summand equal to image $(H_1(S) \to H_1(M(\phi)))$. When $\det(I - \phi_*) = 0$, this summand is the torsion subgroup of $H_1(M(\phi))$, so the proposition follows from lemma 1.2.4.

In the following proposition, we work in $\Gamma(S)$ rather than our usual $\Gamma(S \text{ rel } b_0)$. For $\phi \in \Gamma(S)$, let $\text{Norm}(\phi) = \{\gamma \in \Gamma(S) \mid \gamma\phi\gamma^{-1} = \phi^{\pm 1}\}$. This contains the cyclic normal subgroup (ϕ) generated by ϕ.

PROPOSITION 1.2.6. (a) If ϕ^2 is not isotopic to 1_S, then $FP(M(\phi)) \simeq \text{Norm}(\phi)/(\phi)$.

(b) If ϕ^2 is isotopic to 1_S, then $FP(M(\phi)) \simeq C_2 \times \text{Norm}(\phi)/(\phi)$, with $H(1_S,-1)$ representing the C_2 factor.

Proof: By lemma 1.2.4, if $\langle h\rangle \in FP(M(\phi))$ then h is isotopic to some $H(\gamma,\varepsilon)$. Define $\Phi : FP(M(\phi)) \to C_2 \times \text{Norm}(\phi)/(\phi)$ by $\Phi(\langle h\rangle) = (\varepsilon,\gamma(\phi))$. Isotopy only changes γ by multiples of ϕ and by Dehn twists about ∂S, which do not change the class of γ in $\text{Norm}(\phi)/(\phi)$. Therefore Φ is well-defined. If $\Phi(\langle h\rangle) = (1, (\phi))$, then h is isotopic to $H(1,\phi^k)$ for some k, which is isotopic to the identity. Therefore Φ is injective. Lemma 1.2.2 shows that the composite $FP(M(\phi)) \to C_2 \times \text{Norm}(\phi)/(\phi) \to \text{Norm}(\phi)/(\phi)$ is surjective. Also, lemma 1.2.2 shows that the element $(-1, (\phi))$ is in the image of Φ if and only if $\phi = \phi^{-1}$ in $\Gamma(S)$, so the proposition is proved.

1.3. AUTOMORPHISMS OF PUNCTURED-TORUS BUNDLES. Throughout this section we assume genus(S) = 1. The next proposition is quite special for this case. We fix a standard generating set {a, b} for $\pi_1(S,b_0)$ such that $[a, b] = c$. This determines an identification of $\text{Aut}(H_1(M(\phi)))$ with $GL(2,Z)$.

PROPOSITION 1.3.1. (a) The homomorphism $\Gamma(S) \to E(S)$ is an isomorphism.

(b) The homomorphism $\Gamma(S) \to GL(2,Z)$ that sends ϕ to ϕ_* is an isomorphism.

Proof: (b) follows from (a) since $E(S) \cong Out(\pi_1(S,b_0)) \cong Aut(H_1(S)) \cong GL(2,Z)$. For (a), recall that $GL(2,Z)$ is generated by the matrices

$$\begin{bmatrix} 1 & 1 \\ 0 & 1 \end{bmatrix}, \begin{bmatrix} 1 & 0 \\ 1 & 1 \end{bmatrix}, \text{ and } \begin{bmatrix} 0 & 1 \\ 1 & 0 \end{bmatrix}.$$

Since these are induced by homeomorphisms of S, $\Gamma(S) \to E(S)$ is surjective. Injectivity follows from a classical result of Baer [16, theorem 13.1].

Because of the previous proposition, we can strengthen lemma 1.1.4 in the punctured-torus case.

LEMMA 1.3.2. Let $H : M(\phi) \to M(\phi)$ be a homotopy equivalence of a punctured-torus bundle. The following are equivalent.

(a) h is homotopic to a fiber-preserving homeomorphism.

(b) $h_*(\text{image}(H_1(S) \to H_1(M(\phi)))) \subseteq \text{image}(H_1(S) \to H_1(M(\phi)))$.

Proof: (a) $\Rightarrow$ (b) is obvious. Assuming (b), an argument as in lemma 1.2.4 shows $h_\#(\pi_1(S,b_0)) = \pi_1(S,b_0)$. By proposition 1.3.1, h is homotopic to a map taking c to $c^{\pm 1}$, so assume $h_\#(c) = c^{\pm 1}$. Now write $h_\#(t) = xt^\varepsilon$ with $x \in \pi_1(S,b_0)$ and $\varepsilon = 1$ or -1, and let $\phi(c) = c^\delta$ with $\delta = 1$ or -1. Then $c^{\pm 1} = h_\#(c) = h_\#(tc^\delta t^{-1}) = xt^\varepsilon c^{\pm\delta} t^{-\varepsilon} x^{-1} = xc^{\pm 1}x^{-1}$. Since the centralizer of c in the free group $\pi_1(S,b_0)$ is the subgroup generated by c, we have $x = c^k$ for some $k \in Z$. The argument can now be completed as in lemma 1.2.4.

We denote the determinant of a matrix P by det(P), and its trace by tr(P).

PROPOSITION 1.3.3. Let $M(\phi)$ be a punctured-torus bundle. If ϕ_* satisfies neither of the exceptional cases $\det(\phi_*) = 1$ and $\mathrm{tr}(\phi_*) = 2$ or $\det(\phi_*) = -1$ and $\mathrm{tr}(\phi_*) = 0$, then $FP(M(\phi))) = \Gamma(M(\phi))$, and $FP(M(\phi)) \to E(M(\phi))$ is an isomorphism.

Proof: Since $\phi_* \in GL(2, Z)$, we have $\det(I - \phi_*) = 1 + \det(\phi_*) - \mathrm{tr}(\phi_*)$. Under the hypothesis, this will be nonzero, so as in the proof of proposition 1.2.5, $\text{image}(H_1(S) \to H_1(M(\phi)))$ is the torsion subgroup of $H_1(M(\phi))$. The conclusion now follows from lemmas 1.3.2 and 1.1.1.

2. CALCULATION OF NORM(P)/(P) FOR $P \in GL(2, Z)$

From propositions 1.2.6 and 1.3.1, we know that calculation of $FP(M(\phi))$ in the punctured-torus case amounts to the calculation of $Norm(P)/(P)$ for $P \in GL(2, Z)$. In this section we will calculate these groups by very elementary methods. We will omit many of the details.

When we say matrices are conjugate, we mean conjugate in $GL(2, Z)$ unless otherwise stated. Let

$$I = \begin{bmatrix} 1 & 0 \\ 0 & 1 \end{bmatrix}, \quad J = \begin{bmatrix} 0 & 1 \\ 1 & 0 \end{bmatrix}, \quad A = \begin{bmatrix} 1 & 1 \\ -1 & 0 \end{bmatrix}, \text{ and } B = \begin{bmatrix} 0 & 1 \\ -1 & 0 \end{bmatrix}.$$

It is well-known [20, p. 47] that $SL(2, Z)$ has the structure of a free product with amalgamation $C_6 *_{C_2} C_4$, where A generates C_6, B generates C_4, and $A^3 = B^2 = -I$ generates C_2.

2.1. MATRICES WITH SMALL TRACE.

The characteristic polynomial of $P \in GL(2, Z)$ is $\lambda^2 - tr(P)\lambda + det(P)$. Using elementary linear algebra, we have

PROPOSITION 2.1.1. Suppose $det(P) = 1$. Then

(a) P has order 2 if and only if $P = -I$

(b) P has order 3 if and only if $tr(P) = -1$

(c) P has order 4 if and only if $tr(P) = 0$

(d) P has order 6 if and only if $tr(P) = 1$.

Otherwise, $P = \pm I$ or P has infinite order.

Similarly, we have

PROPOSITION 2.1.2. Suppose $det(P) = -1$. Then P has order 2 if and only if $tr(P) = 0$. Otherwise, P has infinite order.

The next propositions give the conjugacy classes of matrices with small trace. For $n \in Z$, let P_n denote the matrix

$$\begin{bmatrix} 1 & n \\ 0 & 1 \end{bmatrix}.$$

PROPOSITION 2.1.3. Suppose $det(P) = 1$.

(a) If $tr(P) = -2$ then P is conjugate to $-P_n$ for exactly one $n \geq 0$

(b) If $tr(P) = -1$ then P is conjugate to A^2

(c) If $tr(P) = 0$ then P is conjugate to B

(d) If $tr(P) = 1$ then P is conjugate to A

(e) If $tr(P) = 2$ then P is conjugate to P_n for exactly one $n \geq 0$.

This is easily proved by examining the effects of conjugating by

$$\begin{bmatrix} 1 & 0 \\ 1 & 1 \end{bmatrix}, \begin{bmatrix} 1 & 1 \\ 0 & 1 \end{bmatrix}, \text{ and } \begin{bmatrix} 1 & 0 \\ 0 & -1 \end{bmatrix}.$$

Similarly, we have

PROPOSITION 2.1.4. Suppose $det(P) = -1$. If $tr(P) = 0$, then P is conjugate to exactly one of J or BJ.

2.2. CENTRALIZERS OF ELEMENTS OF GL(2, Z). We first determine the centralizer of an element $P \in GL(2, Z)$, which we will denote by Cent(P). This is easily accomplished using the structure of SL(2, Z) as a free product with amalgamation, and theorem 4.5 of [20]. Alternatively, the method described in [8] can be used. We leave details of the next proposition to the reader.

PROPOSITION 2.2.1. Let $P \in GL(2, Z)$.

(a) If $P = \pm I$, then $Cent(P) = GL(2, Z)$

(b) If $det(P) = \pm 1$ and $tr(P) = \pm 1$, then $Cent(P) \simeq C_6$ generated by a conjugate of A

(c) If $det(P) = 1$ and $tr(P) = 0$, then $Cent(P) \simeq C_4$ generated by a conjugate of B

(d) If $det(P) = 1$ and $|tr(P)| \geq 2$, and $P \neq \pm I$, then $Cent(P) \simeq C_2 \times C_\infty$, with the C_2 factor generated by -I

(e) If $det(P) = -1$ and $tr(P) = 0$, then $Cent(P) \simeq C_2 \times C_2$ generated by -I and P

(f) If $det(P) = -1$ and $tr(P) \neq 0$, then $Cent(P) \simeq C_2 \times C_\infty$, with the C_2 factor generated by -I.

DEFINITION 2.2.2. Suppose P has infinite order. By proposition 2.2.1, $Cent(P) \simeq C_2 \times C_\infty$ with the C_2 factor generated by -I. Since $P \in Cent(P)$, the other factor is generated by a matrix T so that $P = \pm T^n$ for n maximal. If we further require that $tr(T) > 0$, then T is uniquely determined by this equation. We call n the <u>divisibility</u> of P and denote it by n(P). We say P is <u>special</u> if $det(T) = -1$; clearly P is special if and only if its centralizer contains a matrix with determinant -1, and if and only if P or -P is a power of a matrix with determinant -1.

As a first approximation to Norm(P)/(P), we will compute the quotient Cent(P)/(P).

PROPOSITION 2.2.3. Let $P \in GL(2, Z)$. If P has infinite order, let $n = n(P)$.

(a) $Cent(I)/(I) \simeq GL(2, Z)$

(b) $Cent(-I)/(-I) \simeq PGL(2, Z)$

(c) If $\det(P) = 1$ and $tr(P) = 0$, then $Cent(P)/(P) \simeq \{1\}$

(d) If $\det(P) = 1$ and $tr(P) = 1$, then $Cent(P)/(P) \simeq \{1\}$

(e) If $\det(P) = 1$ and $tr(P) = -1$, then $Cent(P)/(P) \simeq C_2$

(f) If $\det(P) = 1$, $tr(P) \geq 2$, and $P \neq I$, then $Cent(P)/(P) \simeq C_2 \times C_n$

(g) If $\det(P) = 1$, $tr(P) \leq -2$, and $P \neq -I$, then $Cent(P)/(P) \simeq C_{2n}$

(h) If $\det(P) = -1$ and $tr(P) = 0$, then $Cent(P)/(P) \simeq C_2$

(i) If $\det(P) = -1$ and $tr(P) \neq 0$, then $Cent(P)/(P) \simeq C_{2n}$.

Proof: In view of proposition 2.2.1, only (f), (g), and (i) require explanation. Let T be as in definition 2.2.2. If $tr(P) > 0$ then $P = T^n$ so $Cent(P)/(P) \simeq C_2 \times C_n$, while if $tr(P) < 0$ then $P = -T^n$ so $Cent(P)/(P) \simeq C_{2n}$. When $\det(P) = -1$, n must be odd, so $C_2 \times C_n \simeq C_{2n}$ and the separate cases are not needed.

2.3. CALCULATION OF NORM(P)/(P)

PROPOSITION 2.3.1. Let $P \in GL(2, Z)$. If P is not conjugate to P^{-1}, or if $P = P^{-1}$, then Norm(P)/(P) = Cent(P)/(P).

Proof: In these cases, Norm(P) = Cent(P).

PROPOSITION 2.3.2. If $\det(P) = -1$, then Norm(P)/(P) = Cent(P)/(P).

Proof: When $\det(P) = -1$, $QPQ^{-1} = P^{-1}$ implies $tr(P) = tr(P^{-1}) = -tr(P)$ so $tr(P) = 0$. But then, $P = P^{-1}$ by proposition 2.1.2.

The remaining cases are covered by the next proposition. Let $C_n \propto C_4$ denote the semidirect product $\langle x,y \mid x^n = y^4 = 1, yxy^{-1} = x^{-1}\rangle$, let $\langle 2,2,n\rangle$ denote the dicyclic group $\langle x,y \mid x^2 = y^2 = (xy)^n\rangle$, let $\langle n,2 \mid 2; 2\rangle$ denote the group $\langle x,y \mid y^4 = 1, x^n = y^2, (xy)^2 = 1\rangle$ (see [9, p.71]), and let $(4,n \mid 2,2)$ denote the group $\langle x,y \mid x^n = y^4 = (yx)^2 = (y^{-1}x)^2 = 1\rangle$ (see [9, p.109]). The groups $\langle 2,2,n\rangle$, $\langle n,2 \mid 2; 2\rangle$, and $(4, n \mid 2,2)$ have order 4n.

PROPOSITION 2.3.3. Suppose $\det(P) = 1$. If P has infinite order, let $n = n(P)$.

(a) If $tr(P) = 0$, then $Norm(P)/(P) \simeq C_2$

(b) If $tr(P) = 1$, then $Norm(P)/(P) \simeq C_2$

(c) If $tr(P) = -1$, then $Norm(P)/(P) \simeq C_2 \times C_2$

(d) If $tr(P) = 2$ and $P \neq I$, then $Norm(P)/(P) \simeq C_2 \times D_{2n}$

(e) If $tr(P) = -2$ and $P \neq -I$, then $Norm(P)/(P) \simeq D_{4n}$

(f) If $tr(P) \geq 3$, P is not special, and P is conjugate to P^{-1} in $SL(2, Z)$, then $Norm(P)/(P) \simeq C_n \propto C_4$

(g) If $tr(P) \geq 3$, P is not special, and P is conjugate to P^{-1} in $GL(2,Z)$ but not in $SL(2, Z)$, then $Norm(P)/(P) \simeq C_2 \times D_{2n}$

(h) If $tr(P) \geq 3$, P is special, and P is conjugate to P^{-1} in $SL(2, Z)$, then $Norm(P)/(P) \simeq (4, n \mid 2, 2)$

(i) If $tr(P) \leq -3$, P is not special, and P is conjugate to P^{-1} in $SL(2, Z)$, then $Norm(P)/(P) \simeq \langle 2, 2, n \rangle$

(j) If $tr(P) \leq -3$, P is not special, and P is conjugate to P^{-1} in $GL(2, Z)$ but not in $SL(2, Z)$, then $Norm(P)/(P) \simeq D_{4n}$

(k) If $tr(P) \leq -3$, P is special, and P is conjugate to P^{-1}, then $Norm(P)/(P) \simeq \langle n, 2 \mid 2; 2 \rangle$.

Proof: When P is conjugate to P^{-1} and $P \neq P^{-1}$, we have an extension

$$1 \to Cent(P)/(P) \to Norm(P)/(P) \to C_2 \to 1. \qquad (*)$$

Thus (a) and (b) are immediate from proposition 2.2.4. If $QPQ^{-1} = P^{-1}$ with $P \neq I$, then a simple calculation on the entries of these matrices shows that $tr(Q) = 0$. Therefore $Q^2 = -I$ when $det(Q) = 1$ and $Q^2 = I$ when $det(Q) = -1$. Given P we choose a coset representative Q for the nonzero coset of $Cent(P)/(P)$ in $Norm(P)/(P)$ so that $det(Q) = 1$ when P is conjugate to P^{-1} in $SL(2, Z)$ (in particular, whenever P is special). Otherwise, we have $det(Q) = -1$. We will determine the groups $Norm(P)/(P)$ from the presentations obtained from the extension (*). To simplify notation, we will write S in place of the coset S(P).

(c) When $tr(P) = -1$, P is conjugate to A^2 by proposition 2.1.3. It is easy to check that A^2 is conjugate to A^{-2} in $GL(2, Z)$ but not in $SL(2, Z)$, so $det(Q) = -1$. Therefore, if $P = XA^2X^{-1}$, we have
$Norm(P)/(P) \simeq \langle XAX^{-1}, Q \mid (XAX^{-1})^2 = I, Q^2 = I, Q(XAX^{-1})Q^{-1} = (XAX^{-1})^{-1} \rangle$
$\simeq C_2 \times C_2$.

(d) and (e) are essentially the same as (g) and (j) respectively, so we omit their arguments.

For (f) through (k), note the following. We have $Cent(P)/(P) \simeq C_2 \times C_\infty$ generated by $-I$ and T, with $P = T^n$ if $tr(P) > 0$ and $P = -T^n$ if $tr(P) < 0$. Moreover $(QTQ^{-1})^n = QT^nQ^{-1} = \pm QPQ^{-1} = \pm P^{-1} = (T^{-1})^n$; since

QTQ^{-1} and T^{-1} lie in Cent(P), this implies $QTQ^{-1} = \pm T^{-1}$ lie in Cent(P), this implies $QTQ^{-1} = \pm T^{-1}$. Considering the trace shows $QTQ^{-1} = T^{-1}$ if P is not special and $QTQ^{-1} = -T^{-1}$ if P is special.

(f) $\mathrm{Norm}(P)/(P) \simeq \langle -I, T, Q \mid (-I)^2 = I, [-I, T] = [-I, Q] = I, QTQ^{-1} = T^{-1}, Q^2 = -I, T^n = I\rangle \simeq \langle T, Q \mid T^n = Q^4 = I, QTQ^{-1} = T^{-1}\rangle \simeq C_n \propto C_4$.

(g) $\mathrm{Norm}(P)/(P) \simeq \langle -I, T\, Q \mid (-I)^2 = I, [-I, T] = [-I, Q] = I, QTQ^{-1} = T^{-1}, Q^2 = I, T^n = I\rangle \simeq C_2 \times \langle T, Q \mid T^n = Q^2 = I, QTQ^{-1} = T^{-1}\rangle \simeq C_2 \times D_{2n}$.

(h) $\mathrm{Norm}(P)/(P) \simeq \langle -I, T, Q \mid (-I)^2 = I, [-I, T] = [-I, Q] = I, QTQ^{-1} = -IT^{-1}, Q^2 = -I, T^n = I\rangle \simeq \langle T, Q \mid [Q^2, T] = I, T^n = Q^4 = I; QTQ^{-1} = Q^2T^{-1}\rangle \simeq \langle T, Q \mid [Q^2, T] = T^n = Q^4 = I, (QT)^2 = I, (Q^{-1}T)^2 = I\rangle \simeq \langle T, Q \mid T^n = Q^4 = (QT)^2 = (Q^{-1}T)^2 = I\rangle \simeq (4, n \mid 2, 2)$.

(i) $\mathrm{Norm}(P)/(P) \simeq \langle -I, T, Q \mid (-I)^2 = I, [-I, T] = [-I, Q] = I, QTQ^{-1} = T^{-1}, Q^2 = -I, T^n = -I\rangle \simeq \langle T, Q \mid T^n = Q^2 = (QT)^2, Q^4 = I\rangle \simeq \langle 2, 2, n\rangle$.

(j) $\mathrm{Norm}(P)/(P) \simeq \langle -I, T, Q \mid (-I)^2 = I, [-I, T] = [-I, Q] = I, QTQ^{-1} = T^{-1}, Q^2 = I, T^n = -I\rangle \simeq \langle T, Q \mid T^{2n} = Q^2 = I, QTQ^{-1} = T^{-1}\rangle \simeq D_{4n}$.

(k) $\mathrm{Norm}(P)/(P) \sim \langle -I, T, Q \mid (-I)^2 = I, [-I, T] = [-I, Q] = I, QTQ^{-1} = -IT^{-1}, Q^2 = -I, T^n = -I\rangle \simeq \langle T, Q \mid Q^4 = I, T^n = Q^2, QTQ^{-1} = Q^2T^{-1}\rangle \simeq \langle T, Q \mid Q^4 = (TQ)^2 = I, T^n = Q^2\rangle \simeq \langle n, 2 \mid 2; 2\rangle$.

2.4. SOME COMPUTATIONS AND EXAMPLES. In [12], the authors construct a "strip" Σ_P for $P \in SL(2, Z)$ with $|tr(P)| \geq 3$. Explicitly, Σ_P consists of the (ideal) triangles in the "diagram for PSL(2, Z)" in the hyperbolic disc H which meet the geodesic whose endpoints in ∂H are the slopes of the eigenvectors of P. This geodesic is the invariant axis of P when P is considered as a Möbius transformation of H, hence Σ_P is left invariant by P. Therefore Σ_P is periodic and the pattern of triangles in Σ_P can be described, as in [12], by a sequence of positive integers $(a_1, a_2, \cdots, a_{2n})$, so that P translates each block of 2n triangles in Σ_P onto the succeeding block.

LEMMA 2.4.1. Suppose $P, Q \in SL(2, Z)$ with $|tr(P)| \geq 3$ and $|tr(Q)| \geq 3$. If Σ_P is combinatorially isomorphic to Σ_Q, then there is an element of GL(2, Z) that moves Σ_P onto Σ_Q.

Proof: Consider three successive triangles of Σ_P and their corresponding triangles of Σ_Q. The action of GL(2, Z) is transitive on triangles of the diagram, and can effect any permutation of the vertices of a triangle. There is an element of SL(2, Z) that moves the first two triangles in Σ_P onto their corresponding triangles in Σ_Q. An element of

GL(2, Z) - SL(2, Z) might now be needed to move the third triangle onto the corresponding third triangle, but then the rest of Σ_P must be carried isomorphically onto Σ_Q.

Let

$$U = \begin{bmatrix} 1 & 1 \\ 0 & 1 \end{bmatrix} \quad \text{and} \quad L = \begin{bmatrix} 1 & 0 \\ 1 & 1 \end{bmatrix} .$$

One checks easily that

LEMMA 2.4.2. The strip for $U^{a_1}L^{a_2}U^{a_3}\cdots L^{a_{2n}}$ is described by the sequence $(a_1, a_2, a_3, \cdots, a_{2n})$.

Therefore by lemma 2.4.1 we deduce

PROPOSITION 2.4.3. Suppose Σ_P is described by $(a_1, a_2, \cdots, a_{2n})$. Then Σ_P is conjugate in GL(2, Z) to $\varepsilon U^{a_1}L^{a_2}\cdots L^{a_{2n}}$, where ε is the sign of tr(P).

Since any element of Norm(P) must leave Σ_P invariant, it follows, as observed in [12], that Norm(P)/(-I,P) is the group of combinatorial symmetries of the triangulated cylinder Σ_P/P. The next three easy propositions relate these symmetries to the conditions of proposition 2.3.3. In these propositions, $|tr(P)| \geq 3$ and $(a_1, a_2, \cdots, a_{2n})$ is a sequence describing Σ_P, and the subscripts of the a_i are to be interpreted (mod 2n).

PROPOSITION 2.4.4. The following are equivalent.

(a) P is special.

(b) There is an odd k so that $a_i = a_{i+k}$ for $1 \leq i \leq 2n$.

(c) There is an orientation-reversing automorphism of the triangulated cylinder Σ_P/P that interchanges its boundary components.

PROPOSITION 2.4.5. The following are equivalent.

(a) P is conjugate to P^{-1} in SL(2, Z).

(b) There is an even k so that $a_{i+k} = a_{2n-i+1}$ for $1 \leq i \leq 2n$.

(c) There is an orientation-preserving homeomorphism of the triangulated cylinder Σ_P/P that interchanges its boundary components.

PROPOSITION 2.4.6. The following are equivalent.

(a) P is conjugate to P^{-1} by an element of $GL(2, Z) - SL(2, Z)$.

(b) There is an odd k so that $a_{i+k} = a_{2n-i+1}$ for $1 \leq i \leq 2n$.

(c) There is an orientation-reversing automorphism of the triangulated cylinder Σ_P/P that takes each boundary component to itself.

PROOF OF PROPOSITION 2.4.4. (a) $\Rightarrow$ (c) Suppose $P = \pm Q^n$ with $\det(Q) = -1$. Then $QPQ^{-1} = P$ so Q must leave Σ_P invariant. Since $\det(Q) = -1$, Q must interchange the boundary components of Σ_P, and (c) follows.

(c) $\Rightarrow$ (b) is obvious.

(b) $\Rightarrow$ (a) Let $J = \begin{bmatrix} 0 & 1 \\ 1 & 0 \end{bmatrix}$, so that $J^2 = I$, $JUJ^{-1} = L$, and $JLJ^{-1} = U$.

By proposition 2.4.3, P is conjugate to

$$\varepsilon(U^{a_1}L^{a_2} \cdots U^{a_k}L^{a_1} \cdots L^{a_k})^m$$

$$= \varepsilon(U^{a_1}L^{a_2} \cdots U^{a_k}JJL^{a_1}U^{a_2} \cdots L^{a_k}J^{-1}J)$$

$$= \varepsilon(U^{a_1}L^{a_2} \cdots U^{a_k}J)^{2m}$$

hence P is special.

PROOF OF PROPOSITION 2.4.5. Similar to proposition 2.4.4. For (b) $\Rightarrow$ (a), note that

$$\begin{bmatrix} 0 & 1 \\ -1 & 0 \end{bmatrix} P^t \begin{bmatrix} 0 & -1 \\ 1 & 0 \end{bmatrix} = P^{-1}.$$

But $P^t = U^{a_{2n}}L^{a_{2n-1}} \cdots U^{a_2}L^{a_1}$.

PROOF OF PROPOSITION 2.4.6. Similar to proposition 2.4.5.

REMARK 2.4.7. From the strip Σ_P, one can easily discern the divisibility $n(P)$. For if $P = \pm T^m$, then T must be a translation with the same invariant axis as P, so $a_i = a_{i+k}$ for $1 \leq i \leq 2n$ with $k = 2n/m$.

REMARK 2.4.8. Although we shall not make use of it here, [1] contains some interesting computational methods for determining conjugacy in $GL(2, Z)$.

EXAMPLE 2.4.9. $P = \phi_* = \begin{bmatrix} 2 & 1 \\ 1 & 1 \end{bmatrix}$.

In this case, it is well-known that $M(\phi)$ is the complement of the figure-8 knot in S^3. Since $P = UL$, Σ_P is described by the sequence (1, 1). By propositions 2.4.4 and 2.4.5, P is special and is conjugate to P^{-1} in $SL(2, Z)$, and by remark 2.4.7, $n(P) = 2$. By proposition 2.3.3(h), $\Gamma(M(\phi)) \simeq (4, 2 \mid 2, 2) \simeq D_8$. This result appears in [25].

EXAMPLE 2.4.10. $P = \phi_* = \begin{bmatrix} -2 & -1 \\ -1 & -1 \end{bmatrix}$.

Again, Σ_P is described by the sequence (1, 1). By proposition 2.3.3(k), $\Gamma(M(\phi)) \simeq \langle 2, 2 \mid 2; 2\rangle \simeq C_2 \times C_4$.

EXAMPLE 2.4.11. $P = \phi_* = -\begin{bmatrix} 13 & 5 \\ 5 & 2 \end{bmatrix}^2$

Here, $P = -(U^2LUL^2)^2$ so Σ_P is determined by the sequence (2, 1, 1, 2, 2, 1, 1, 2). By propositions 2.4.4 and 2.4.5, P is not special but is conjugate to P^{-1} in $SL(2, Z)$. By proposition 2.3.3(i), $\Gamma(M(\phi)) \simeq \langle 2, 2, 2\rangle \simeq Q_8$, the quaternion group of order 8.

EXAMPLE 2.4.12. $P = \phi_* = -\begin{bmatrix} 18 & 7 \\ 5 & 2 \end{bmatrix}^n$

Here, Σ_P is described by the sequence (2, 1, 1, 3, 2, 1, 1, 3, $\cdots$, 2, 1, 1, 3), so P is not conjugate to P^{-1}. By propositions 2.3.1 and 2.2.3(g), $\Gamma(M(\phi)) \simeq C_{2n}$.

3. THE EXCEPTIONAL CASES

3.1. THE CASE OF $\det(\phi_*) = 1$ AND $\mathrm{tr}(\phi_*) = 2$

By proposition 2.1.3, ϕ_* is conjugate to $\begin{bmatrix} 1 & n \\ 0 & 1 \end{bmatrix}$ for exactly one $n \geq 0$. Define $\phi_n \in \mathrm{Aut}^+(\pi_1(S, b_0))$ by $\phi_n(a) = a$ and $\phi_n(b) = ba^n$, so $M(\phi) = M(\phi_n)$.

Suppose first that $n \neq 0$. Let χ be the automorphism of $\pi_1(M(\phi), b_0)$ defined by $\chi(a) = a$, $\chi(b) = bt$, and $\chi(t) = t$. It is easy to construct a homeomorphism f of $M(\phi_n)$ which is the identity on $\partial M(\phi_n)$ and induces χ (f is a "Dehn twist" about an imbedded torus transverse to all the fibers). Since $\chi^k(b) = bt^k$, $\langle f \rangle$ generates an infinite cyclic subgroup of $\Gamma(M(\phi_n))$ and $E(M(\phi_n))$, which intersects $FP(M(\phi_n))$ only in the identity element.

Let δ, $\beta \in \mathrm{Aut}^{\pm}(\pi_1(S, b_0))$ be defined by $\delta(a) = ca^{-1}$, $\delta(b) = b^{-1}c$, $\beta(a) = bab^{-1}$, and $\beta(b) = b^{-1}$. From proposition 2.3.3(d), we know $N(\phi_*)/(\phi_*) \simeq C_2 \times D_{2n}$, where the C_2 factor is generated by $\delta_* = -I$ and the D_{2n} by $(\phi_1)_*$ and

$$\beta_* = \begin{bmatrix} 1 & 0 \\ 0 & -1 \end{bmatrix}.$$

By lemma 1.2.2., we have corresponding fiber-preserving homeomorphisms $H(\delta, 1)$, $H(\phi_1, 1)$, and $H(\beta, -1)$. A calculation shows that up to inner automorphism, $H(\delta, 1)_\# \chi H(\delta, 1)_\#^{-1} = \chi^{-1}$, $H(\phi_1, 1)_\# \chi H(\phi_1, 1)_\#^{-1} = \chi$, and $H(\beta, -1)_\# \chi H(\beta, 1)_\#^{-1} = \chi$. Therefore $\langle f \rangle$, $\langle H(\delta, 1) \rangle$, $\langle H(\phi_1, 1) \rangle$, and $H(\beta, -1)$ generate a subgroup of $\Gamma(M(\phi_n))$ which is isomorphic to $(C_\infty \propto C_2) \times D_{2n}$.

PROPOSITION 3.1.1. The homotopy classes of f, $H(\delta, 1)$, $H(\phi_1, 1)$, and $H(\beta, -1)$ generate $E(M(\phi_n))$.

Proof: Let g be a self-homotopy-equivalence of $M(\phi_n)$. There is a "vertical" torus T that carries the elements a and t of $\pi_1(M(\phi_n))$. The complement of an open regular neighborhood of T is a submanifold Σ which is the product of a disc-with-two-holes with S^1, with $\pi_1(\Sigma)$ generated by a, bab^{-1}, and t. It is clear that Σ is a characteristic submanifold in the sense of [17] or [18]. By the Mapping Theorem of [17] or the Enclosing Theorem of [18], g(T) is homotopic into Σ. Now $H_1(M(\phi_n)) \simeq Z \oplus Z \oplus Z/n$ generated by $\bar{t}$, $\bar{b}$, and $\bar{a}$. Since g(T) is homotopic into Σ, we have $g_*(\bar{t}) = \bar{t}^{\pm 1} + p\bar{a}$ for some p. Since g induces an automorphism on $H_1(M(\phi_n))$/torsion, it follows that $g_*(\bar{b}) = \bar{b}^{\pm\varepsilon} + q\bar{t} + r\bar{a}$ for some q and r and $\varepsilon = 1$ or -1. Therefore $(f^{-\varepsilon q}g)_*(\bar{b}) \in \mathrm{image}(H_1(S) \to H_1(M(\phi_n)))$. Since $\bar{a}$ has finite order, the same holds for $(f^{-\varepsilon q}g)_*(\bar{a})$. By lemma 1.3.2, $f^{-\varepsilon q}g$ is homotopic to a fiber-preserving homeomorphism. Since $H(\delta, 1)$, $H(\phi_1, 1)$, and $H(\beta, -1)$ generate $FP(M(\phi_n))$, the proposition is proved.

Using proposition 3.1.1 and lemma 1.2.1, we have

COROLLARY 3.1.2. For $n > 0$, $\Gamma(M(\phi_n)) \simeq (C_\infty \propto C_2) \times D_{2n}$, and $\Gamma(M(\phi_n)) \to E(M(\phi_n))$ is an isomorphism.

When $n = 0$, $M(\phi_0) = S \times S^1$ so $\mathrm{Out}(\pi_1(M(\phi_0))$ is known [6, p. 66] to be a semidirect product $(\mathrm{Hom}(\pi_1(S, b_0), Z) \propto GL(1, Z)) \propto \mathrm{Out}(\pi_1(S, b_0))$ $\simeq ((C_\infty \times C_\infty) \propto C_2) \propto GL(2, Z)$. The generator of $GL(1, Z)$ is represented by the automorphism χ_0 sending t to t^{-1} and fixing a and b, while $\mathrm{Hom}(\pi_1(S, b_0), Z)$ is generated by χ_1 and χ_2 defined by $\chi_1(t) = \chi_2(t) = t$, $\chi_1(a) = at$, $\chi_1(b) = b$, $\chi_2(a) = a$, and $\chi_2(b) = bt$. We have $\chi_0\chi_i\chi_0^{-1} = \chi_i^{-1}$ for $i = 1,2$. The elements of $\mathrm{Out}(\pi_1(S, b_0))$ are represented by automorphisms of the form $\Theta(t) = t$, $\Theta(a) = \gamma(a)$, and $\Theta(b) = \gamma(b)$ for $\gamma \in \mathrm{Out}(\pi_1(S, b_0))$. Now $\Theta\chi_0\Theta^{-1} = \chi_0$, while if

$$\gamma_*^{-1} = \begin{bmatrix} a & b \\ c & d \end{bmatrix},$$

then

$$\Theta\chi_1\Theta^{-1} = \chi_1^a\chi_2^b$$

and

$$\Theta\chi_2\Theta^{-k} = \chi_1^c\chi_2^d .$$

Putting together lemma 1.1.2 and the observation that all of these automorphisms can be realized by homeomorphisms, we have

PROPOSITION 3.1.3. $\Gamma(M(\phi_0)) \simeq (C_\infty \times C_\infty) \propto (C_2 \times GL(2, Z))$.

3.2. THE CASE OF $\det(\phi_*) = -1$ AND $\mathrm{tr}(\phi_*) = 0$

By proposition 2.1.2, ϕ_* is conjugate to exactly one of

$$\begin{bmatrix} 0 & 1 \\ 1 & 0 \end{bmatrix} \quad \text{or} \quad \begin{bmatrix} 1 & 0 \\ 0 & -1 \end{bmatrix},$$

so $M(\phi)$ is either $M(J)$ or $M(K)$ where $J(a) = b$, $J(b) = a$, $K(a) = bab^{-1}$, and $K(b) = b^{-1}$.

From propositions 2.3.1 and 2.2.3(h), and lemma 1.2.6(b), we have $FP(M(J)) \simeq C_2 \times C_2$ generated by $H(\delta, 1)$, and $H(1, -1)$. Define an automorphism χ of $\pi_1(M(J))$ by $\chi(t) = t$, $\chi(a) = at$, and $\chi(b) = bt$, and choose a self-homotopy-equivalence f inducing χ. Since $\chi^k(a) = at^k$, χ generates an infinite cyclic subgroup of $E(M(J))$. A calculation shows that up to inner automorphisms, $H(\delta, 1)_{\#}\chi H(\delta, 1)_{\#} = \chi^{-1}$ and $H(1, -1)_{\#}\chi H(1, -1)_{\#} = \chi^{-1}$. So $\langle f\rangle$, $\langle H(\delta, 1)\rangle$, and $\langle H(1, -1)\rangle$ generate a subgroup of $E(M(J))$ isomorphic to $C_\infty \propto (C_2 \times C_2)$.

PROPOSITION 3.2.1. $E(M(J)) \simeq C_\infty \propto FP(M(J)) \simeq C_\infty \propto (C_2 \times C_2)$ with the infinite cyclic normal subgroup generated by $\langle f\rangle$, and $\Gamma(M(J)) \simeq C_\infty \propto FP(M(J)) \simeq C_\infty \propto (C_2 \times C_2)$ with the infinite cyclic normal subgroup generated by $\langle f^2\rangle$. Therefore, the image of $\Gamma(M(J))$ in $E(M(J))$ has index two.

Proof: First we show $\langle f\rangle$, $\langle H(\delta, 1)\rangle$, and $\langle H(1, -1)\rangle$ generate $E(M(J))$. Let g be a self-homotopy-equivalence. We have $H_1(M(\phi)) \simeq Z \oplus Z$ generated by $\bar{t}$ and $\bar{a}$. It is easy to check that t^2 generates the center of $\pi_1(M(\phi))$, hence $g_{\#}(t^2) = t^{\pm 2}$ and $g_*(2\bar{t}) = \pm 2\bar{t}$. Since g_* is an automorphism, we must have $g_*(\bar{a}) = \varepsilon\bar{a} + k\bar{t}$ for some k, and $\varepsilon = 1$ or -1. Therefore $(f^{-\varepsilon k}g)_*(\bar{a}) = \pm\bar{a}$. By lemma 1.3.2, $f^{-\varepsilon k}g$ is homotopic to a fiber-preserving homeomorphism. This proves the assertion about $E(M(J))$. Now $f_{\#}(c) = a^2b^{-2}$ so f cannot be homotopic to a homeomorphism. However, $f^2_{\#}(c) = c$ and $f^2_{\#}(t) = t$, so $f^2_{\#}$ preserves the peripheral structure and therefore f^2 is homotopic to a homeomorphism. The rest of the proposition now follows using lemma 1.1.2.

One can visualize the "exotic" homotopy equivalence of M(J) as follows. There is a 2-sided Möbius band in M(J) which arises from a properly imbedded arc α in S left invariant by the attaching map. Splitting M(J) along this band yields the product $K \times I$ of a Klein bottle and an interval, with a copy of the band in each boundary component. Let $t \mapsto \theta_t$ be an action $S^1 = \mathbb{R}/\mathbb{Z}$ on K which leaves the Möbius band invariant, and such that $\theta_{1/2}$ reflects the Möbius band across its center circle. Define F on $K \times I$ by $f(x, t) = (\theta_{t/2}(x), t)$. To form M(J) we glue the ends of the product $M \times I$ of a Möbius band and an interval to the copies of the Möbius band in $\partial(K \times I)$. To extend the homeomorphism f of $K \times I$ over $M \times I$, we must use a self-map of $M \times I$ which is not a homeomorphism. However, f^2 is the identity on $\partial(K \times I)$, and extends using the identity on $M \times I$.

The analysis of M(K) is quite similar, except that $H_1(M(K)) \simeq Z \oplus Z \oplus Z/2$ generated by $\bar{t}$, $\bar{a}$, and $\bar{b}$, and χ should be defined by $\chi(t) = t$, $\chi(a) = ab^{-1}t$, and $\chi(b) = b$. One obtains the same statement as in proposition 3.2.1 with J replaced by K.

4. SUMMARY OF THE MAPPING CLASS GROUP CALCULATIONS FOR PUNCTURED-TORUS BUNDLES

If ϕ_* has infinite order, let $n = n(\phi_*)$, the divisibility of ϕ_* as defined in definition 2.2.2.

Suppose first that $\det(\phi_*) = 1$. Recall from definition 2.2.2 that ϕ_* is <u>special</u> if $\phi_* = \pm T^n$ for some $T \in GL(2, Z)$ with $\det(T) = -1$. Define the <u>type</u> of ϕ_* as follows. Type$(\phi_*) = S$ if ϕ_* is conjugate to ϕ_*^{-1} in $SL(2, Z)$, type$(\phi_*) = G$ if ϕ_* is conjugate to ϕ_*^{-1} in $GL(2, Z)$ but not in $SL(2, Z)$, and type$(\phi_*) = N$ if ϕ_* is not conjugate to ϕ_*^{-1} in $GL(2, Z)$.

Combining propositions 1.2.5, 1.2.6, 1.3.3, 2.2.3, 2.3.1, 2.3.3, 3.1.2, and 3.1.3, we have

THEOREM 4.1. Suppose $\det(\phi_*) = 1$. Then $\Gamma(M(\phi)) \to E(M(\phi))$ is an isomorphism, and the groups $\Gamma(M(\phi))$ are given in Table 1.

Suppose now that $\det(\phi_*) = -1$. Combining propositions 1.2.5, 1.2.6, 2.2.3, 2.3.2, and section 3.2, we have

THEOREM 4.2. Suppose $\det(\phi_*) = -1$. If $tr(\phi_*) \neq 0$, then $\Gamma(M(\phi)) \to E(M(\phi))$ is an isomorphism, while if $tr(\phi_*) = 0$, then the image of this homomorphism has index two. The groups $\Gamma(M(\phi))$ are given in Table 2.

5. RESTRICTION OF MAPPING CLASSES TO THE BOUNDARY

Throughout the section, genus$(S) = 1$. Denote $\pi_1(S, b_0)$ by F. We have chosen standard generators a and b of F so that $[a, b] = c$.

Recall the groups $Aut^+(F)$ and $Aut^\pm(F)$ defined in section 1.1.

5.1. $Aut^+(F)$ AND $Aut^\pm(F)$

LEMMA 5.1.1. If $\phi \in Aut(F)$, then $\phi(c) = wc^\varepsilon w^{-1}$ where $w \in F$ and $\varepsilon = \det(\phi_*)$.

Proof: It suffices to check that the lemma is true for a set of generators of $Aut(F)$, which may be found in [23] or [20, pp. 163-164].

Let ℓ, $u \in Aut^\pm(F)$ and $j \in Aut^\pm(F)$ be the automorphisms defined by $\ell(a) = ab$, $\ell(b) = b$, $u(a) = a$, $u(b) = ba$, $j(a) = b$, and $j(b) = a$.

PROPOSITION 5.1.2. (a) $Aut^+(F)$ is the subgroup of $Aut(F)$ generated by $\{\ell, u\}$.

(b) $Aut^\pm(F)$ is the subgroup of $Aut(F)$ generated by $\{\ell, u, j\}$.

(c) If $\tau \in Aut^\pm(F)$ and $\tau_* = I$, then $\tau = \mu(c^k)$ for some $k \in Z$.

Proof: It is well-known and easy to check that $\{\ell_*, u_*\}$ generates $SL(2, Z)$ and $\{\ell_*, u_*, j_*\}$ generates $GL(2, Z)$. By a theorem of Nielsen [23],

TABLE 1 ($\det(\phi_*) = 1$)

$tr(\phi_*)$	$type(\phi_*)$	is ϕ_* special?	$\Gamma(M(\phi))$
0	S	no	C_2
1	G	no	C_2
-1	G	no	$C_2 \times C_2$
2 and $\phi_* = I$	S	yes	$(C_\infty \times C_\infty) \propto (C_2 \times GL(2, Z))$
-2 and $\phi_* = -I$	S	no	$C_2 \times PGL(2, Z)$
2 and $\phi_* \neq I$	G	no	$(C_\infty \propto C_2) \times D_{2n}$
-2 and $\phi_* \neq -I$	G	no	D_{4n}
≥ 3	N	yes or no	$C_2 \times C_n$
≥ 3	S	no	$C_n \propto C_4$
≥ 3	G	no	$C_2 \times D_{2n}$
≥ 3	S	yes	$(4, n \mid 2, 2)$
≤ -3	N	yes or no	C_{2n}
≤ -3	S	no	$\langle 2, 2, n \rangle$
≤ -3	G	no	D_{4n}
≤ -3	S	yes	$\langle n, 2 \mid 2; 2 \rangle$

(The groups $(4, n \mid 2, 2)$, $\langle 2, 2, n \rangle$, and $\langle n, 2 \mid 2; 2 \rangle$ are defined after proposition 2.3.2.)

TABLE 2 ($\det(\phi_*) = -1$)

$tr(\phi_*)$	$\Gamma(M(\phi))$
0	$C_\infty \propto (C_2 \times C_2)$
$\neq 0$	C_{2n}

the kernel of $\mathrm{Aut}(F) \to GL(2, Z)$ is the group of inner automorphisms. Therefore the kernel of $\mathrm{Aut}^+(F) \to SL(2, Z)$ and that of $\mathrm{Aut}^{\pm}(F) \to GL(2, Z)$ consist of exactly those inner automorphisms that take c to $c^{\pm 1}$. Since the centralizer of any nonzero element is infinite cyclic and equal to the normalizer, the kernels are infinite cyclic. Since c is not a proper power, the kernels are generated by $\mu(c)$. Part (c) follows. But $\mu(c) = (\ell u^{-1}\ell)^4$, so (a) and (b) are also proved.

Let $\delta \in \mathrm{Aut}^+(F)$ be the automorphism $(\ell u^{-1}\ell)^2$. Explicitly, $\delta(a) = ca^{-1}$ and $\delta(b) = b^{-1}c$. The important properties of δ are summarized in the next proposition.

PROPOSITION 5.1.3. (a) $\delta_* = \pm I$

(b) $\delta^2 = \mu(c)$

(c) If $\sigma \in \mathrm{Aut}^+(F)$, then $\sigma\delta = \delta\sigma$

(d) If $\sigma \in \mathrm{Aut}^{\pm}(F) - \mathrm{Aut}^+(F)$, then $\sigma\delta = \delta^{-1}\sigma$

(e) If $\sigma, \tau \in \mathrm{Aut}^{\pm}(F)$ and $\sigma_* = \pm\tau_*$, then $\sigma = \delta^k\tau$ for some $k \in Z$.

Proof: (a) and (b) are easily checked. If $\sigma \in \mathrm{Aut}^{\pm}(F) - \mathrm{Aut}^+(F)$, then by lemma 5.1.1 the expression for σ in terms of ℓ, u, and j contains an odd number of appearances of j. Therefore (c) and (d) follow from proposition 5.1.2 and the facts that $\ell\delta = \delta\ell$, $u\delta = \delta u$, and $j\delta = \delta^{-1}j$. For (e), if $\sigma_* = \varepsilon\tau_*$ with $\varepsilon = \pm 1$, put $m = (1 - \varepsilon)/2$ so that $(\delta^m\tau\sigma^{-1})_* = I$. By proposition 5.1.2(c), $\delta^m\tau\sigma^{-1} = \mu(c^n) = \delta^{2n}$ for some n, hence $\sigma = \delta^{m-2n}\tau$.

The next lemma will be quite useful in the next two sections.

LEMMA 5.1.4. Suppose $\phi, \eta \in \mathrm{Aut}^+(F)$ and $[\phi_*, \eta_*] = I$. Then $[\phi, \eta] = 1$.

Proof: If ϕ_* or η_* is $\pm I$, then one of ϕ or η is of the form δ^m, and the result follows from proposition 5.1.3(c). Otherwise, by proposition 2.2.3 there is a $T \in SL(2, Z)$ with $\phi_* = \pm T^m$ and $\eta_* = \pm T^n$ for some m and n. Choose $\tau \in \mathrm{Aut}^+(F)$ with $\tau_* = T$. By proposition 5.1.3(e), we can write $\phi = \delta^{k_1}\tau^m$ and $\eta = \delta^{k_2}\tau^n$. By proposition 5.1.3, $\tau\delta = \delta\tau$ so

$$[\phi, \eta] = [\delta^{k_1}\tau^m, \delta^{k_2}\tau^n] = 1.$$

5.2. THE IMAGE OF THE RESTRICTION $\Gamma(M(\phi)) \to \Gamma(\partial M(\phi))$

PROPOSITION 5.2.1. The image of $\Gamma(M(\phi)) \to \Gamma(\partial M(\phi))$ equals the image of $FP(M(\phi)) \to \Gamma(\partial M(\phi))$.

Proof: Except for the exceptional cases of section 3, we know from proposition 1.3.3 that $FP(M(\phi)) = \Gamma(M(\phi))$. But by section 3, the additional generators needed in the exceptional cases may be chosen so that their restrictions to $\partial M(\phi)$ are the identity.

Assume for now that $\det(\phi_*) = 1$. If $\langle h \rangle \in FP(M(\phi))$, then $h_*(c) = \varepsilon_1 c$ for $\varepsilon_1 = 1$ or -1, and consequently $h_*(t) = \varepsilon_2 t + kc$ for $\varepsilon_2 = 1$ or -1 and some $k \in Z$. In the next section we will analyze the possible values of k. For now, define a homomorphism $\Phi: FP(M(\phi)) \to C_2 \times C_2$ by $\Phi(\langle h \rangle) = (\varepsilon_1, \varepsilon_2)$.

PROPOSITION 5.2.2. Let $\langle h \rangle \in FP(M(\phi))$. If $\Phi(\langle h \rangle) = (1, 1)$, then $h|_{\partial M(\phi)}$ is isotopic to $1_{\partial M(\phi)}$.

Proof: By lemma 1.2.4, we may assume h is of the form $H(\gamma, \varepsilon)$ for some $\gamma \in \mathrm{Aut}^{\pm}(F)$ and $\varepsilon = 1$ or -1. Since $h_*(c) = c$ and $h_*(t) = t + kc$, we have $\gamma \in \mathrm{Aut}^+(F)$ and $\varepsilon = 1$. By lemma 1.2.2, $[\phi_*, \gamma_*] = \mu(c^k)_* = I$ so lemma 5.1.4 implies $[\phi, \gamma] = 1$. Therefore $k = 0$ and the conclusion follows.

Putting together lemma 1.2.4 and propositions 5.2.1 and 5.2.2, we have

COROLLARY 5.2.3. Suppose $\det(\phi_*) = 1$. Let G_1 be C_2 if ϕ_* is special, and 0 otherwise. Let G_2 be C_2 if ϕ_* is conjugate to ϕ_*^{-1} and 0 otherwise. Then the image of $\Gamma(M(\phi)) \to \Gamma(\partial M(\phi))$ is isomorphic to $G_1 \times G_2$.

We now consider the case when $\det(\phi_*) = -1$, and hence $\partial M(\phi)$ is a Klein bottle. It is well-known that $\Gamma(\partial M(\phi)) \simeq \mathrm{Out}(\pi_1(\partial M(\phi))) \simeq C_2 \times C_2$, generated by the automorphisms α and β defined by $\alpha(t) = t^{-1}$, $\alpha(c) = c$, $\beta(t) = tc$, and $\beta(c) = c$.

PROPOSITION 5.2.4. Suppose $\det(\phi_*) = -1$. If $\mathrm{tr}(\phi_*) \neq 0$, then the image of $\Gamma(M(\phi)) \to \Gamma(\partial M(\phi))$ is isomorphic to C_2 and is generated by a mapping class inducing β. If $\mathrm{tr}(\phi_*) = 0$, then $\Gamma(M(\phi)) \to \Gamma(\partial M(\phi))$ is an isomorphism.

Proof: By proposition 5.1.3(d), $[\delta, \phi] = \delta^2 = \mu(c)$ so lemma 1.2.2 shows that β is in the image only when ϕ_* is conjugate to ϕ_*^{-1}; since $\det(\phi_*) = -1$, this occurs exactly when $\mathrm{tr}(\phi_*) = 0$. In section 2, we calculated that $\Gamma(M(\phi)) \simeq C_2 \times C_2$ when $\mathrm{tr}(\phi_*) = 0$, so the restriction will actually be an isomorphism in this case.

5.3. THE VALUES OF k

In section 5.2 we noted that if $\langle h \rangle \in \Gamma(M(\phi))$ then $h_\#(t) = t^{\pm 1} c^k$ for some $k \in Z$. In this section we describe the ways in which k can be nonzero. The case of $\det(\phi_*) = -1$ is described fully by proposition 5.2.4, so we consider only the case of $\det(\phi_*) = 1$.

PROPOSITION 5.3.1. Suppose $\det(\phi_*) = 1$. There exists a homeomorphism of $M(\phi)$ taking t to tc^k with $k \neq 0$ if and only if $\phi = \delta^{-k}\gamma^{2n}$ for some $\gamma \in \mathrm{Aut}^{\pm}(F) - \mathrm{Aut}^{+}(F)$. Such a homeomorphism can exist for only one nonzero value of k.

Proof: If $\phi = \delta^{-k}\gamma^{2n}$, then $[\phi, \gamma] = [\delta^{-k}\gamma^{2n}, \gamma] = [\delta^{-k}, \gamma] = \delta^{-2k} = \mu(c^{-k})$, so lemma 1.2.2 shows that the homeomorphism exists. Conversely, if the homeomorphism exists then we may assume by proposition 5.2.1 that it is fiber-preserving, and hence by lemma 1.2.4 that it is of the form $H(\tau, 1)$ for some $\tau \in \mathrm{Aut}^{\pm}(F)$ with $[\phi, \tau] = \mu(c^{-k})$. If $\det(\phi_*) = 1$, then lemma 5.1.4 would imply that $k = 0$, hence $\det(\tau_*) = -1$. Therefore the centralizer of ϕ_* contains a matrix of determinant -1 so ϕ_* is special and $\mathrm{Cent}(\phi_*)$ is generated by $-I$ and γ_* for some $\gamma \in \mathrm{Aut}^{\pm}(F) - \mathrm{Aut}^{+}(F)$. For some n, $\phi_* = \pm\gamma_*^{2n}$ so proposition 4.1.3(e) shows $\phi = \delta^m\gamma^{2n}$ for some m. By lemma 5.1.4, $1 = [\phi, \tau\gamma] = [\phi, \tau]\tau[\phi, \gamma]\tau^{-1} = \mu(c^{-k})\tau\mu(c^m)\tau^{-1} = \mu(c^{-k-m})$ so $m = -k$.

For the last statement, if we also have $[\phi,\gamma'] = \mu(c^{-\ell})$ with $\ell \neq 0$, then by lemma 5.1.4, $\det(\gamma') = -1$. Again by lemma 5.1.4, we have $1 = [\phi,\gamma\gamma'] = [\phi, \gamma]\gamma[\phi, \gamma']\gamma^{-1} = \mu(c^{-k+\ell})$ so $\ell = k$.

PROPOSITION 5.3.2. Suppose $\det(\phi_*) = 1$. There exists an homeomorphism of $M(\phi)$ taking t to $t^{-1}c^k$ with $k \neq 0$ if and only if $\mathrm{tr}(\phi_*) \neq 0$ and there is a $\phi_0 \in \mathrm{Aut}^{+}(F)$ such that

(a) $(\phi_0)_* = \varepsilon\phi_*$ where $\varepsilon = \mathrm{sign}(\mathrm{tr}(\phi_*))$

(b) ϕ_0 is conjugate to ϕ_0^{-1} in $\mathrm{Aut}^{+}(F)$

(c) $\phi = \delta^k\phi_0$.

Such a homeomorphism can exist for at most one nonzero value of k.

Proof: If ϕ is the form $\delta^k\phi_0$, and $\gamma\phi_0\gamma^{-1} = \phi_0^{-1}$ with $\gamma \in \mathrm{Aut}^{+}(F)$, then $\gamma\phi\gamma^{-1} = \gamma\delta^k\phi_0\gamma^{-1} = \delta^k\phi_0^{-1} = (\phi_0^{-1}\delta^{-k})\delta^{2k} = \phi^{-1}\mu(c^k)$, and lemma 1.2.2 shows that $M(\phi)$ has such a homeomorphism. Conversely, if the homeomorphism exists, then as in proposition 5.3.1 we may assume it is of the form $H(\tau, -1)$ with $\tau\phi\tau^{-1} = \phi^{-1}\mu(c^k)$. Since $[\phi_*, \tau_*^2] = I$, lemma 5.1.4 shows $1 = [\phi, \tau^2] = [\phi, \tau]\tau[\phi, \tau]\tau^{-1} = \phi^2\mu(c^{-k})\tau\phi^2\mu(c^{-k})\tau^{-1} = \mu(c^{k-\det(\tau_*)k})$. Since $k \neq 0$, this implies $\det(\tau_*) = 1$ and therefore ϕ_* is conjugate to ϕ_*^{-1} in $SL(2, Z)$.

Suppose first that $|\mathrm{tr}(\phi_*)| \leq 2$. Proposition 2.1.3 describes these matrices up to conjugacy, and one checks that ϕ_* is conjugate to ϕ_*^{-1} in $SL(2, Z)$ only when $\phi_* = \pm I$ so $\phi = \delta^m$. But $\delta^{-m}\mu(c^k) = \tau\delta^m\tau^{-1} = \delta^m$, so $m = k$. Taking $\phi_0 = 1_F$, we have the conclusion.

Suppose now that $|tr(\phi_*)| \geq 3$. We may change ϕ up to conjugacy in SL(2, Z) without affecting the truth of the conclusion. Up to conjugacy in SL(2, Z), we know from section 2.4 that

$$\phi_* = \varepsilon U^{a_1} L^{a_2} U^{a_3} \cdots U^{a_{2n-1}} L^{a_n}$$
$$= \varepsilon (u^{a_1} \ell^{a_2} u^{a_3} \cdots u^{a_{2n-1}} \ell^{a_{2n}})_*$$

where u and ℓ were defined in section 5.1. Let $\phi_0 = u^{a_1} \ell^{a_2} u^{a_3} \cdots \ell^{a_{2n}}$ so that conclusion (a) holds. Since ϕ_* is conjugate to ϕ_*^{-1} in SL(2, Z), proposition 2.4.5 implies there is a k so that $a_{i+2k} = a_{2n-i+1}$ for $1 \leq i \leq 2n$. Let $\beta \in \mathrm{Aut}^+(F)$ be defined by $\beta(a) = b^{-1}$ and $\beta(b) = bab^{-1}$. One checks that $\beta u \beta^{-1} = \ell^{-1}$ and $\beta \ell \beta^{-1} = u^{-1}$. Therefore

$$(\beta \ell^{-a_{2k}} u^{-a_{2k-1}} \cdots \ell^{-a_2} u^{-a_1}) \phi_0 (u^{a_1} \ell^{a_2} \cdots u^{a_{2k-1}} \ell^{a_{2k}} \beta^{-1})$$
$$= \beta u^{a_{2k+1}} \ell^{a_{2k+2}} \cdots \ell^{a_{2n}} u^{a_1} \cdots \ell^{a_{2k}} \beta^{-1}$$
$$= \ell^{-a_{2k+1}} u^{-a_{2k+2}} \cdots u^{-a_{2n}} \ell^{-a_1} \cdots u^{-a_{2k}}$$
$$= \ell^{-a_{2n}} u^{-a_{2n-1}} \cdots u^{-a_{2k+1}} \ell^{-a_{2k}} \cdots u^{-a_1} = \phi_0^{-1}.$$

Then ϕ_0 satisfies conclusion (b). Using lemma 5.1.4 shows (c) and the last statement of the proposition.

6. SOME EXAMPLES OF SURGERY ON KNOTS

In this section we no longer assume genus(S) = 1.

6.1. THE FIRST CONSTRUCTION

Let A be the axis of symmetry for S shown in Fig. 1. Let

$$P = A \cap S = \{P_1, P_2, \cdots, P_{2g+1}\} .$$

Denote by θ a rotation about A through an angle of π; this is an involution of S with fixed-point set P. The quotient S/θ is the 2-disc and $S \to S/\theta$ is a 2-fold branched covering branched along P.

Let α_i, $1 \leq i \leq g$, and γ_j, $1 \leq j \leq g$ be θ-equivariant Dehn twists about A_i and C_j respectively. Now α_i interchanges P_{2i-1} and P_{2i}, and γ_j interchanges P_{2j} and P_{2j+1}. On S/θ, they induce the standard generators

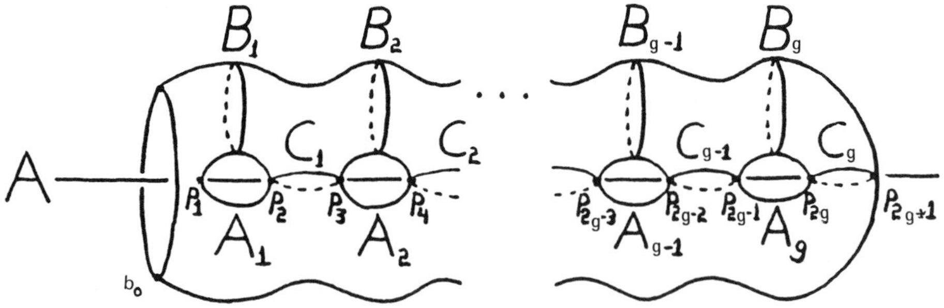

FIGURE 1

of the braid group B_{2g+1} [4]; specifically α_i induces σ_{2i-1} and γ_j induces σ_{2j}.

Let δ be a θ-equivariant homeomorphism which agrees with θ outside a small collar neighborhood of ∂S, but tapers off in the collar so that δ is the identity of ∂S. Clearly, $\delta_* = -I$ on $H_1(S)$, and δ^2 is a Dehn twist about ∂S, inducing $\mu(c^{\pm 1})$ on $\pi_1(S, b_0)$. We choose the direction of tapering so that δ^2 induces $\mu(c^{-1})$. Observe that $\delta\alpha_i\delta^{-1} = \alpha_i$ and $\delta\gamma_i\delta^{-1} = \gamma_i$ for $1 \leq i \leq g$. On S/θ, δ induces a Dehn twist about the boundary, equal to $(\sigma_1\sigma_2\sigma_3 \cdots \sigma_{2g})^{2g+1}$, the "full-twist" braid.

Finally, let ρ be a reflection on S through the plane that contains $A \cup \{b_0\}$. We assume ρ leaves $A_1, A_2, \cdots, A_g, C_1, C_2, \cdots, C_g$ individually invariant; thus $\rho\alpha_i\rho^{-1} = \alpha_i^{-1}$ and $\rho\gamma_i\rho^{-1} = \gamma_i^{-1}$ for $1 \leq i \leq g$.

The following easy geometric observation is crucial.

LEMMA 6.1.1. $[\delta,\rho] = \delta^2 = \mu(c^{-1})$.

We also have

LEMMA 6.1.2. Let γ be any product of the mapping classes $\{\alpha_i,\gamma_i \mid 1 \leq i \leq g\}$ and their inverses. Then $[\delta(\rho\gamma)^2,\rho\gamma] = \mu(c^{-1})$.

Proof: We have observed that δ commutes with γ. Therefore $[\delta(\rho\gamma)^2,\rho\gamma] = [\delta,\rho\gamma] = [\delta,\rho] = \mu(c^{-1})$.

Therefore lemma 1.2.2 immediately implies

PROPOSITION 6.1.3. Let γ be any product of the mapping classes $\{a_i,\gamma_i \mid 1 \leq i \leq g\}$ and their inverses. Then there is a homeomorphism of $M(\delta(\rho\gamma)^2)$ that carries t to tc. Consequently, +1 surgery on $N(\delta(\rho\gamma)^2)$ along K yields $N(\delta(\rho\gamma)^2)$.

6.2. THE HOMOLOGY OF N(ϕ) AND SOME EXAMPLES

Let ϕ_* denote the induced automorphism on $H_1(S) \simeq Z^{2g}$, with respect to the standard basis $\{A_1, B_1, A_2, \cdots, A_g, B_g\}$.

LEMMA 6.2.1 $N(\phi)$ is a Z-homology 3-sphere if and only if $\det(I - \phi_*) = \pm 1$.

Proof: Abelianizing $\pi_1(N(\phi)) \simeq \pi_1(M(\phi))/(t = 1) \simeq \langle a_1, b_1, a_2, \cdots, a_g, b_g \mid a_i = \phi(a_i), b_i = \phi(b_i), 1 \leq i \leq g\rangle$, we obtain the abelian presentation $H_1(N(\phi)) \simeq \langle A_1, B_1, A_2, \cdots, A_g, B_g \mid (I - \phi_*)(A_i) = 0, (I - \phi_*)(B_i) = 0\rangle$ so $I - \phi_*^t$ is a presentation matrix for $H_1(N(\phi))$. Therefore

$$|H_1(N(\phi))| = |Z^{2g}/\text{image}(I - \phi_*)| = |\det(\phi_*)| \quad \text{if } \det(\phi_*) \neq 0,$$

and the lemma follows.

Let $\rho\gamma$ be any automorphism as in proposition 6.1.3, and let $p(x)$ be the characteristic polynomial of $(\rho\gamma)_*$.

LEMMA 6.2.2. $N(\delta(\rho\gamma)^2)$ is a Z-homology 3-sphere is and only if $p(i)p(-i) = \pm 1$.

Proof: We have $\det(I - (\delta(\rho\gamma)^2)_*) = \det(I - \delta_*(\rho\gamma)_*^2) = \det(I + (\rho\gamma)_*^2) = \det(iI + (\rho\gamma)_*)\det(-iI + (\rho\gamma)_*) = p(-i)p(i)$. Now apply lemma 6.2.1.

LEMMA 6.2.3. $p(x) = (-1)^g x^{2g} p(-\frac{1}{x})$

Proof: Let J be the $2g \times 2g$ matrix

$$\begin{bmatrix} 0 & 1 & & & & & \\ -1 & 0 & & & & & \\ & & 0 & 1 & & & \\ & & -1 & 0 & & & \\ & & & & \ddots & & \\ & & & & & 0 & 1 \\ & & & & & -1 & 0 \end{bmatrix}$$

i.e., the intersection matrix of S. Since $\rho\gamma$ is induced by an orientation-reversing homeomorphism, we have $(\rho\gamma)_* J (\rho\gamma)_*^{-1} = -J$, and the result follows by elementary linear algebra.

For $g = 1$, $p(x)$ must be of the form $x^2 - ax - 1$, so $p(i)p(-i) = 4 + a^2$ and $N(\delta(\rho\gamma)^2)$ cannot be a Z-homology 3-sphere.

For $g = 2$, $p(x)$ must be of the form $x^4 + ax^3 + bx^2 - ax + 1$, and $p(i)p(-i) = 4a^2 + (2 - b)^2$, so $N(\delta(\rho\gamma)^2)$ will be a Z-homology 3-sphere only when $p(x) = x^4 + x^2 + 1$ or $p(x) = x^4 + 3x^2 + 1$. I have not found such an example.

For $g = 3$, $p(x)$ must be of the form $x^6 + ax^5 + bx^4 + cx^3 - bx^2 + at - 1$, so $p(i)p(-i) = 4(b-1)^2 + (2a-c)^2$. For this case I have found several examples by using a computer to calculate $\det(I - \delta_*(\rho\gamma)_*^2)$:

CALCULATION 6.2.4. For $g = 3$ and any of the following choices of γ, $N(\delta(\rho\gamma)^2)$ is a Z-homology 3-sphere containing a genus 3 fibered knot K such that +1 surgery on $N(\delta(\rho\gamma)^2)$ along K yields $N(\delta(\rho\gamma)^2)$:

$$\alpha_1\gamma_1^{-3}\alpha_2\gamma_2^{-1}\alpha_3\gamma_3^{-1}$$

$$\alpha_1^{-1}\alpha_2\alpha_3^{-1}\gamma_1^{5}\gamma_2^{-1}\gamma_3^{-1}$$

$$\alpha_1\alpha_2^{-3}\alpha_3\gamma_1^{-1}\gamma_2\gamma_3$$

$$\gamma_1^{-1}\gamma_2\gamma_3^{-1}\alpha_1\alpha_2\alpha_3^{-5}$$

One expects many more such examples for higher genera.

REMARK 6.2.5. If $\gamma = \alpha_1\gamma_1^{-3}\alpha_2\gamma_2^{-1}\alpha_3\gamma_3^{-1}$ and $\phi = \delta(\rho\gamma)^2$, then $M(\phi)$ has a hyperbolic structure. For the characteristic polynomial of $(\rho\gamma)_*$ is $p(x) = x^6 - x^5 + x^4 - x^3 - x^2 - x - 1$, hence the characteristic polynomial of ϕ_* is $q(x) = \det(xI - \phi_*) = \det(xI + (\rho\gamma)_*^2) = p(i\sqrt{x})p(-i\sqrt{x}) = x^6 - x^5 - 3x^4 + 7x^3 - 3x^2 - x + 1$. It is easy to check that $q(x)$ has no quadratic factors over Z (although $q(x)$ does factor as $(x^3 + x^2 - 2x + 1)(x^3 - 2x^2 + x + 1)$). The roots of $q(x)$ are approximately -2.1479, -0.4656, 0.5739 ± 0.3690i, and 1.2328 ± 0.7926i, hence none of the roots of $q(x)$ lie on the unit circle. By proof of lemma 4.3 in A. Casson's notes on surface homeomorphisms [5], ϕ is induced by a pseudo-Anosov homeomorphism, so $M(\phi)$ is hyperbolic [26].

6.3. IDENTIFICATION OF THE EXAMPLES

Since any homeomorphism $\phi = \delta(\rho\gamma)^2$ as in proposition 6.1.3 will be θ-equivariant, we can apply θ to each fiber of $M(\phi)$ to obtain an involution on $M(\phi)$. This extends over $S^1 \times D^2 = \mathrm{cl}(N(\phi) - M(\phi))$ by rotation in the S^1 factor, and the quotient of this involution on $N(\phi)$ is S^3. The fixed-point set of the involution is $L' = \cup_{i=1}^{2g+1}[P_i \times I]$, a link in $N(\phi)$, and thus $N(\phi)$ is a 2-fold branched covering over S^3 branched along the image of L', which we denote by L.

Since we know the homeomorphisms of S/θ induced by α_i, γ_i, and δ, we can easily draw a picture of L. First, recall that if χ is $\alpha_i^{\pm 1}$ or $\gamma_i^{\pm 1}$, then $\rho\chi\rho^{-1} = \chi^{-1}$. Hence if $\gamma = \chi_1\chi_2, \cdots, \chi_n$ with each χ_j an element of $\{\alpha_i, \alpha_i^{-1}, \gamma_i, \gamma_i^{-1} \mid 1 \leq i \leq g\}$, then $(\rho\gamma)^2 = \rho\gamma\rho\gamma = \rho\gamma\rho^{-1}\gamma = \chi_1^{-1}\chi_2^{-1} \cdots \chi_n^{-1}\chi_1\chi_2 \cdots \chi_n$. Recalling that α_i induces σ_{2i-1}, γ_i induces σ_{2i}, and δ induces $(\sigma_1\sigma_2 \cdots \sigma_{2g})^{2g+1}$, we can draw L, as illustrated in Figure 2 for the example $\gamma = \alpha_1\gamma_1^{-3}\alpha_2\gamma_2^{-1}\alpha_3\gamma_3^{-1}$.

We can now prove

THEOREM 6.3.1. For $g \geq 1$, the construction in section 6.1 can never yield $N(\phi) = S^3$.

Proof: By the Smith Conjecture for involutions [28], $N(\phi) = S^3$ if and only if L is the trivial knot in S^3. But L is the closure of a braid of the form $(\sigma_1\sigma_2 \cdots \sigma_{2g})^{2g+1}\chi_1^{-1}\chi_2^{-1} \cdots \chi_n^{-1}\chi_1\chi_2 \cdots \chi_n$. The algebraic sum of the exponents in this braid is $c = 2g(2g + 1)$. By theorem 3 of [2], the maximum Euler characteristic for a Seifert surface of L is less than or equal to $2g + 1 - c < 0$, hence L is nontrivial.

REMARK 6.3.2. By a more recent result [21], L is not even amphicheiral.

To "see" a homeomorphism of $M(\phi)$ that carries t to tc, consider any closed braid on 2g + 1 strings of the form shown in Figure 3. Note that the image K_0 of K is double covered by K, and a regular neighborhood of K_0 is the solid torus complementary to the solid torus $M(\phi)/\theta$. First, reflect through the plane of the paper. This changes X_1 to X_{-1}, X_{-1} to X_1, and changes D to a full right-hand twist. Now cut along a meridinal disc of $M(\phi)/\theta$, give two left-handed twists, and glue back together. This changes the right-handed twist back into D. Since a full twist is central in the braid group, the image of L in $M(\phi)$ is now isotopic back to L. Thus we have a self-homeomorphism of the pair $(M(\phi)/\theta, L)$, which restricts on $\partial(M(\phi)/\theta)$ to two Dehn twists about the boundary of a meridinal disc. Lifting $M(\phi)$, we have a self-homeomorphism that takes t to tc.

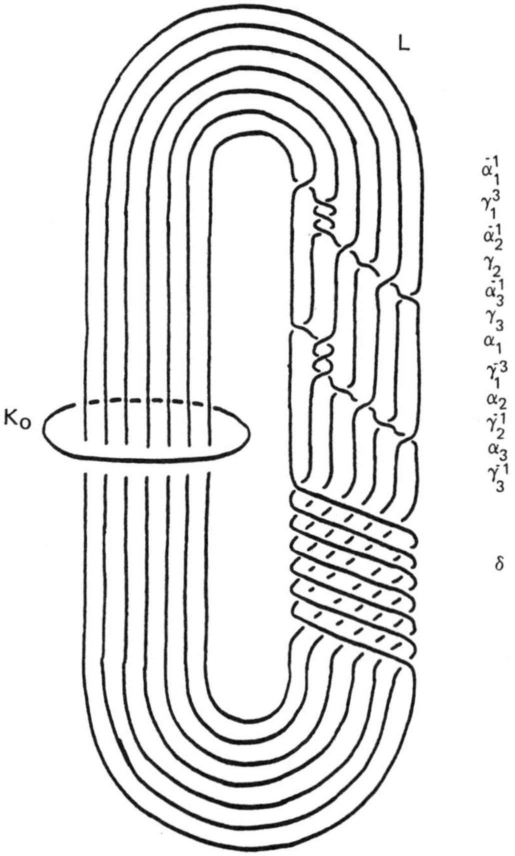
L
K_o
α_1^{-1}
γ_1^3
α_2^{-1}
γ_2
α_3^{-1}
γ_3
α_1
γ_1^{-3}
α_2
γ_2^{-1}
α_3
γ_3^{-1}
δ

FIGURE 2

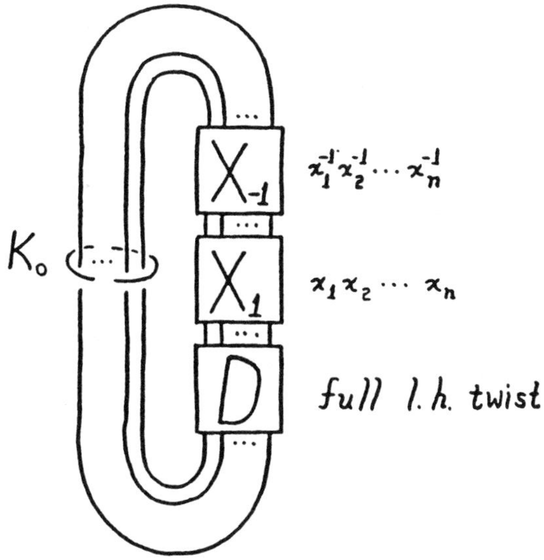

FIGURE 3

6.4. THE SECOND CONSTRUCTION

We will sketch a second construction, producing examples of $M(\phi)$ having homeomorphisms taking t to $t^{-1}c^{-1}$. Let β be the θ-equivariant homeomorphism of S which induces on S/θ a "half-twist" about the boundary (i.e., the braid called Δ in [4]). Note that $\beta^2 = \delta$. We have $\beta\alpha_i\beta^{-1} = \gamma_{g+1-i}$ and $\beta\gamma_i\beta^{-1} = \alpha_{g+1-i}$ for $1 \leq i \leq g$. In the braid group, this corresponds to $\Delta\sigma_i\Delta^{-1}$ and $\beta\gamma_i\beta^{-1} = \alpha_{2g+1-i}$.

Suppose τ is any product of the elements $\{\alpha_i, \alpha_i^{-1}, \gamma_i, \gamma_i^{-1} \mid 1 \leq i \leq g\}$. Then

$$\begin{aligned}\beta(\delta^{-1}\tau\beta\tau^{-1}\beta^{-1})\beta^{-1} &= \delta^{-1}\beta\tau\beta\tau^{-1}\delta^{-1}\\ &= \delta^{-1}\beta\tau\beta\delta^{-1}\tau^{-1}\\ &= \delta^{-1}\beta\tau\beta^{-1}\tau^{-1}\\ &= \beta\tau\beta^{-1}\tau^{-1}\delta^{-1}\\ &= (\delta^{-1}\tau\beta\tau^{-1}\beta^{-1})^{-1}\delta^{-2}\\ &= (\delta^{-1}\tau\beta\tau^{-1}\beta^{-1})^{-1}\mu(c^{-1}).\end{aligned}$$

By lemma 1.2.2, if $\phi = \delta^{-1}\tau\beta\tau^{-1}\beta^{-1}$, then there is a homeomorphism of $M(\phi)$ taking t to $t^{-1}c^{-1}$. Consequently +1 surgery on $N(\phi)$ along K yields $N(\phi)$.

As in the first construction, $N(\phi)$ is a 2-fold branched covering of S^3, branched along the closure of the braid corresponding to $\delta\tau\beta\tau^{-1}\beta^{-1}$. This braid will be of the form

$$(\sigma_1\sigma_2\cdots\sigma_{2g})^{2g+1}\sigma_{i_1}^{a_1}\sigma_{i_2}^{a_2}\cdots\sigma_{i_n}^{a_n}\sigma_{2g+1-i_n}^{-a_n}\sigma_{2g+1-i_{n-1}}^{-a_{n-1}}\cdots\sigma_{2g+1-i_1}^{-a_1}$$

so the proof of theorem 6.3.1 shows $N(\phi) = S^3$. Again, one can describe the homeomorphism on $M(\phi)/\theta$.

QUESTION. Is there any way other than these two constructions to produce a homeomorphism of $M(\phi)$ taking t to $(tc)^{\pm 1}$?

ACKNOWLEDGEMENT

The author was partially supported by an NSF grant.

REFERENCES

1. Applegate, H., and Onishi, H., Continued Fractions and the Conjugacy Problem in SL(2, Z), Comm. in Alg. 9, 11 (1981), 1121-1130.
2. Bennequin, D., Entrelacements et equations de Pfaff, Astérisque 107-108 (1983), 87-161.
3. Bing, R. H., and Martin, J., Cubes with Knotted Holes, Trans. Amer. Math. Soc. 155 (1971), 217-231.
4. Birman, J., Braids, Links, and Mapping Class Groups, Annals of Mathematics Study No. 82, Princeton University Press, 1975.
5. Casson, A., Automorphisms of Surfaces After Nielsen and Thurston, lecture notes by S. Bleiler, University of Texas, 1982-83.
6. Conner, P., and Raymond, F., Manifolds with few periodic homeomorphisms, Springer-Verlag Lecture Notes in Mathematics Vol 299 (1972), 1-75.
7. Conner, P., and Raymond, F., Deforming Homotopy Equivalences to Homeomorphisms in Aspherical Manifolds, Bull. Amer. Math. Soc. 83 (1977), 36-85.
8. Conner, P., Raymond, F., and Weinberger, P., Manifolds with no Periodic Maps, Springer-Verlag Lecture Notes in Mathematics Vol. 299 (1972), 81-108.
9. Coxeter, H.M.S., and Moser, W.O.J., Generators and Relations for Discrete Groups, Springer-Verlag, 1980.

10. Culler, M., Gordon, C. McA., Leucke, J., and Shalen, P., Dehn Surgery on Knots, MSRI Preprint 08712-85 (1985).

11. Culler, M., Jaco, W., and Rubinstein, H., Incompressible Surfaces in Once-Punctured Torus Bundles, Proc. London Math. Soc. (3), 45 (1982), 385-419.

12. Floyd, W., and Hatcher, A., Incompressible Surfaces in Punctured-Torus Bundles, Topology Appl. 13 (1982), 263-282.

13. Gabai, D., Detecting Fibred Links in S^3, MSRI Preprint 06912-85 (1985).

14. González-Acuña, F., 3-dimensional Open Books, Lectures, Univ. of Iowa Topology Seminar, 1974-75.

15. Heil, W., On $\mathbb{P}^2$-irreducible 3-manifolds, Bull. Amer. Math. Soc. 75 (1969), 772-775.

16. Hempel, J., 3-manifolds, Annals of Mathematics Study No. 86, Princeton University Press, 1976.

17. Jaco, W., and Shalen, P., Seifert Fibered Spaces in 3-manifolds, Mem. Amer. Math. Soc. 220 (1979).

18. Johannson, K., Homotopy Equivalences of 3-manifolds, Springer-Verlag Lecture Notes in Mathematics Vol. 761, (1979).

19. Laudenbach, F., Topologie de la dimension trois. Homotopie et isotopie, Astérisque 12 (1974), 1-152.

20. Magnus, W., Karrass, A., and Solitar, D., Combinatorial Group Theory, Interscience Publishers, 1966.

21. Morton, H. R., Closed Braid Representations for a Link, and its Jones-Conway Polyniminal, preprint.

22. Myers, R., Open Book Decompositions of 3-manifolds, Proc. Amer. Math. Soc. 72 (1978), 397-402.

23. Nielsen, J., Die Isomorphismen der allgemeinen unendlichen Gruppe mit zwei Erzeugenden, Math. Ann. 78 (1918), 385-397.

24. Przytycki, J., Nonorientable Imcompressible Surfaces in Punctured-Torus Bundles, preprint.

25. Riley, R., An elliptical path from parabolic representations to hyperbolic structures, Springer-Verlag Lecture Notes in Mathematics Vol. 722 (1979), 99-133.

26. Thurston, W., Hyperbolic Structures on 3-manifolds II: Surface Groups and 3-manifolds which Fiber over the Circle, preprint.

27. Waldhausen, F., On Irreducible 3-manifolds which are sufficiently large, Annals of Math. 87 (1968), 56-88.

28. F. Waldhausen, Über Involutionen der 3-Sphäre, Topology 8 (1969), 81-91.

Lattice Gauge Fields and Chern-Weil Theory

ANTHONY PHILLIPS / Mathematics Department, State University of New York, Stony Brook, New York

DAVID A. STONE / Brooklyn College, City University of New York, Brooklyn, New York

1. Our recent work [9] lies in an area that may be described as differential geometry without infinitesimals. This is an expository article about that work and some unsolved problems arising from it.

A first step towards discretization of a connection in a principal bundle over a smooth manifold is to triangulate the base; then one can keep the continuous geometry of the fibres while restricting attention to just the combinatorial topology of the base space. The two types of geometry, continuous and discrete, interact when one tries to do differential geometry by means of whatever is left of a connection in the original bundle. The resulting structure is essentially what is called a lattice gauge field. The concept was invented by the physicist K. Wilson as a tool for attacking fundamental problems of quantum field theory (see [14], which is reprinted in Rebbi's survey book [10]). We shall restrict our attention to some of the mathematical aspects of lattice gauge field theory.

Our particular interest is in the determination of global topological invariants from local geometric data in this context. The Chern-Weil theory solves this problem (under suitable hypotheses) when the data are those of a principal bundle with a connection. So we are led to ask how Chern-Weil theory is to be done in the absence of many of its key ingredients, such as differential forms on the base space or a connection 1-form on the total space. In case the structure group is U(1) or SU(2) we have some partial answers and some hints as to where more may be found; these results are the subject of this report.

1.1. DEFINITION: A gauge field (ξ,ω) consists of a principal bundle $\xi = (\pi:E \to M)$, whose structure group is a compact, connected Lie group G, and a connection 1-form ω on E. We shall assume that M is a smooth, compact, closed, oriented manifold, and that G is a subgroup of some unitary group. (For generalizations see [9].)

1.2. Now let Λ be a smooth triangulation of M. We denote its vertices by $\alpha, \beta, \cdots$ and its simplices by $\sigma = \langle\alpha\beta\gamma\rangle$ etc. Let C_α be the cell dual to vertex α. In (1.5) we shall make a construction from (ξ,ω) and Λ, whose result will form the basis of the definition (1.6) of a lattice gauge field. It will be convenient first to perform an intermediate construction (1.3), on the basis of which we shall define a coordinate bundle (1.4). In both constructions the idea is to use the cells C_α as (closed) charts for ξ and to construct transition functions from ω.

1.3. For each vertex α of Λ, pick an arbitrary identification of the fibre E_α of ξ over α with the group G. Now let $\langle\alpha\beta\rangle$ be any 1-simplex of Λ. Let $x \in C_\alpha \cap C_\beta$; then, since C_α and C_β are cones (in barycentric coordinates) from α and β respectively, there are corresponding generators $[\alpha,x]$ and $[\beta,x]$. Under parallel transport with respect to ω along $[\alpha,x] \cup [x,\beta]$, the group identity $I \in G = E_\alpha$ is carried into some element of $G = E_\beta$, which we denote by $v_{\alpha\beta}(x)$. Thus when G is abelian, $v_{\alpha\beta}(x) = \exp \int_{[\alpha,x] \cup [x,\beta]} \omega$. When G is non-abelian we must use the path ordered or "product" integral of Schlesinger [11] and Nijenhius [7]:

1.3.1. $v_{\alpha\beta}(x) = P \int_{[\alpha,x] \cup [x,\beta]} \omega$.

(See Figure 1.) We thus have a system of functions

1.3.2. $v_{\alpha\beta}: C_\alpha \cap C_\beta \to G$, one for each directed 1-simplex $\langle\alpha\beta\rangle$ of Λ.

Right multiplication by $v_{\alpha\beta}(x)$ is the transporter for $[\alpha,x] \cup [x,\beta]$.

They satisfy the "cocycle conditions":

1.3.3. $v_{\beta\alpha}(x) = (v_{\alpha\beta}(x))^{-1}$; and

1.3.4. $v_{\alpha\gamma}(y) = v_{\alpha\beta}(y) \cdot v_{\beta\gamma}(y)$, whenever $\langle\alpha\beta\gamma\rangle \in \Lambda$ and $y \in C_\alpha \cap C_\beta \cap C_\gamma$.

1.4. DEFINITION (Steenrod [13]). Given Λ, a coordinate bundle v on Λ is any system of continuous functions $v_{\alpha\beta}: C_\alpha \cap C_\beta \to G$, as in (1.3.2), that satisfy (1.3.3) and (1.3.4).

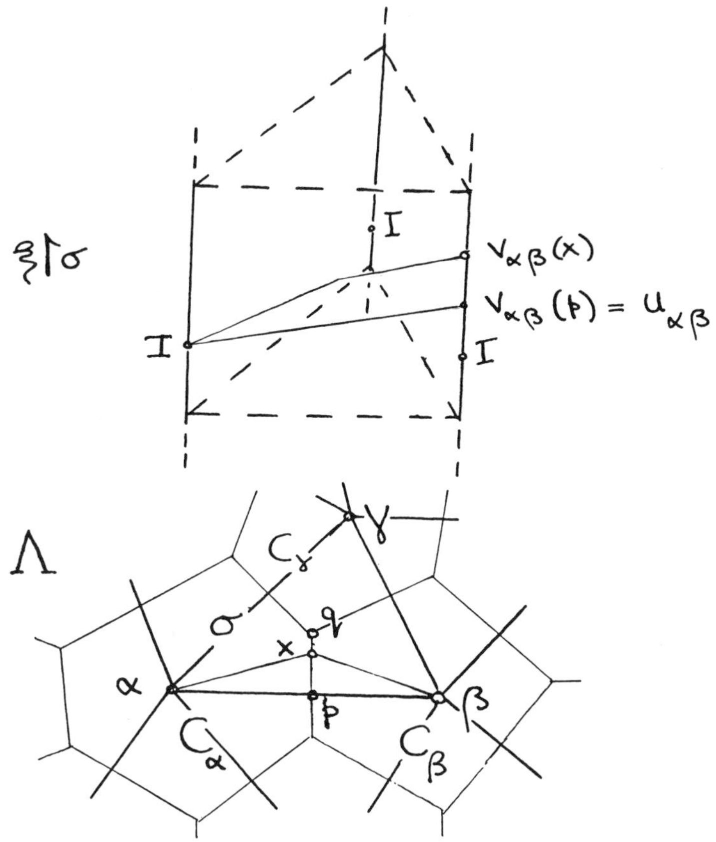

FIGURE 1. Construction of a coordinate bundle $\underline{v} = \{v_{\alpha\beta}\}$ and a lattice field $\underline{u} = \{u_{\alpha\beta}\}$ from a gauge field (ξ,ω) and a triangulation Λ of M.

Conversely, given $\underline{v}$ we can construct a bundle on M from the charts $\{C_\alpha \times G : \alpha \text{ is a vertex of } \Lambda\}$ by identifying, for every 1-simplex $\langle\alpha\beta\rangle$, and for every $x \in C_\alpha \cap C_\beta$, $(x,g) \in C_\alpha \times G$ with $(x, g \cdot v_{\alpha\beta}(x)) \in C_\beta \times G$. If $\underline{v}$ is defined by (1.3.1), then the resulting bundle is isomorphic to the original ξ.

1.5. To construct a lattice gauge field from (ξ,ω) and Λ we take the combinatorial point of view that the only paths in Λ are edge paths. So $v_{\alpha\beta}(x)$ can be defined only when x is the barycentre p of $\langle\alpha\beta\rangle$. Set

1.5.1. $u_{\alpha\beta} = v_{\alpha\beta}(p)$;

so that, from (1.3.1),

$$u_{\alpha\beta} = P\int_{\langle\alpha\beta\rangle} \omega .$$

The only constraint on the $u_{\alpha\beta}$'s comes from (1.3.3):

1.5.2. $u_{\beta\alpha} = (u_{\alpha\beta})^{-1}$.

1.6. DEFINITIONS. Given Λ, a <u>lattice gauge field</u> (henceforth abbreviated LGF) <u>u on</u> Λ assigns to each directed 1-simplex $\langle\alpha\beta\rangle$ of Λ an element $u_{\alpha\beta} \in G$, called a <u>transporter</u>, such that (1.5.2) holds.

The construction (1.5) of $\underline{u}$ from (ξ,ω) and Λ depends on choices in (1.3) of identifications of the fibres E_α with G. Any quantity constructed from $\underline{u}$ that is independent of these choices is called <u>gauge invariant</u>; these are the quantities that have geometric significance.

2. The work of [9] which we are to describe suggested the following diagram of constructions. The dashed arrows represent constructions which we know how to make in various special cases. Our problem is to understand how to carry them out with a fair degree of generality.

The purpose of this section is to explain the diagram and how our results and unsolved problems will be fit into it.

For the rest of this section we assume M, Λ and the structure group G to be fixed.

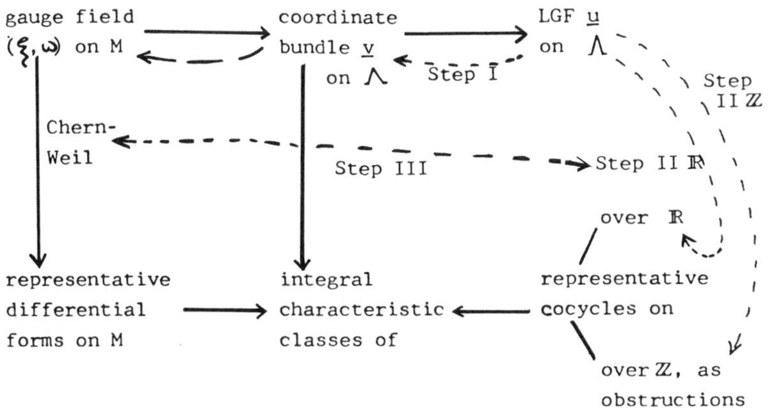

2.1. So far we have described (in (1.3) and (1.5)) the constructions from left to right in the top row. The other solid arrows need no comment as they are standard constructions [3, 13, 6]. We have also shown (in (1.4)) how to recover ξ from $\underline{v}$. It is not possible to recover ω exactly; but using a partition of unity whose domains are open neighborhoods of the cells C_α, it is a straightforward matter to use $\underline{v}$ to construct a connection ω' in ξ; the finer the triangulation Λ of M, the more closely ω' will approximate ω. We shall not discuss this construction further, but turn instead to the labelled dashed arrows.

2.2. STEP I: Find an algorithm that assigns to $\underline{u}$ a coordinate bundle $\underline{v}$ such that (1.5.1) holds. (Some restrictions must be placed on $\underline{u}$, and more compatibility conditions on $\underline{v}$; see Remarks (2.7) and (2.8).) Among the various possible ways of doing this we want one so explicit that we can carry out the next step.

2.3. STEP II $\mathbb{Z}$: Once $\underline{v}$ has been constructed it determines a unique principal bundle ξ, by (1.4). Each integral characteristic class c of ξ has various representatives as obstruction cocycles to a section of a bundle associated to ξ by c. We want a formula that will compute, directly from $\underline{u}$, one such representative, c_z, on Λ.

2.4. STEP II $\mathbb{R}$: When we allow real coefficients, more cocycle representatives of c become available, and we may hope to simplify c_z. The objective is to find such a cocycle, c_r, on Λ, in such a way that the last step can be performed.

2.5. STEP III: The Chern-Weil theory, given (ξ,ω) on M, assigns to c a representative differential form c_{cw} on M. We require a correspondence between the structures of the constructions $c \to c_r$ of Step II $\mathbb{R}$ and $c \to c_w$; that is, we wish to demonstrate that similar ingredients are used for similar purposes. Thus we shall be looking for analogues, in the semi-combinatorial context of LGF's, of the Lie algebra $\mathfrak{g}$ of G, G-invariant symmetric polynomials on $\mathfrak{g}$, and so on.

2.6. Our results in [9] cover the following cases. For G = U(1) and the first Chern class c_1, we can carry out the whole program. (This was first done, in essence, in [8].) The extension to Step II $\mathbb{Z}$ for $(c_1)^k$ is then immediate. For G = SU(2) and the second Chern class c_2 we can carry out Steps I and II $\mathbb{Z}$.

We shall also present some clues that may be useful in carrying out Steps II $\mathbb{R}$ and III in the cases G = SU(2) for c_2 and G = U(1) for $(c_1)^k$.

2.7. REMARKS. The set of LGF's over Λ can be identified with $G \times G \times \cdots$, one copy of G for each 1-simplex of Λ; this is a connected space. The set of coordinate bundles over Λ has distinct components, corresponding to different isomorphism classes of bundles. It is therefore in general impossible that there exist an algorithm $\underline{v} = A(\underline{u})$ which satisfies (1.5.1), is continuous in $\underline{u}$ and applies to all $\underline{u}$. So in our solution to Step I we shall impose a further condition on $\underline{u}$. This condition will be satisfied when $\underline{u}$ is constructed from (ξ,ω) and the triangulation Λ of M is sufficiently fine. (See Remarks (4.4) and (7.4).)

2.8. Without further conditions on $\underline{v}$ (besides (1.5.1)) regarding its compatibility with $\underline{u}$, the problem of constructing $\underline{v}$ from $\underline{u}$ would be trivial and its solution meaningless. For <u>any</u> coordinate bundle $\underline{v}$ on Λ can be easily modified so as to satisfy (1.5.1). Our extra condition says roughly that for each $\langle\alpha\beta\rangle$ and $x \in C_\alpha \cap C_\beta$, $v_{\alpha\beta}(x)$ is close to $u_{\alpha\beta}$ (for a precise statement, see Remark (7.5)).

2.9. Once Step I has been solved, Step II $\mathbb{R}$ can in some sense be carried out in terms of integrals of functions derived from the $v_{\alpha\beta}$, either by the method of Bott, Shulman and Stasheff [1] (see also Lüscher [5]), or by constructing ξ and ω' from $\underline{v}$ as in (2.1) and applying the Chern-Weil theory. We do not consider such a solution satisfactory until we have algorithms for evaluating any such integrals in terms of $\underline{u}$. Our proposed solution to Step II $\mathbb{R}$ for G = SU(2) and the class c_2 is in this spirit; see (7.6).

3. Before we can describe our results we need some notation.

3.1. For each simplex σ containing vertices α and β, set $C_\alpha^\sigma = C_\alpha \cap \sigma$, $C_{\alpha\beta}^\sigma = C_\alpha \cap C_\beta \cap \sigma$, and set $v_{\alpha\beta}^\sigma$ equal to the restriction $v_{\alpha\beta} \mid C_{\alpha\beta}^\sigma$. In our solution to Step I it is the functions $v_{\alpha\beta}^\sigma$ that we construct.

3.2. Let $\sigma = \langle\alpha\beta\gamma\rangle$ be a 2-simplex of Λ. Set $u[\alpha\beta\gamma] = u_{\alpha\beta} \cdot u_{\beta\gamma} \cdot u_{\gamma\alpha}$, and let $d[\sigma]$ denote the distance from I to $u[\sigma]$ (in the standard bi-invariant metric on G). Let $\underline{g}_{\alpha\beta}$ and $\underline{g}[\alpha\beta\gamma]$ denote the shortest geodesics in G from I to $u_{\alpha\beta}$ and to $u[\alpha\beta\gamma]$ respectively, provided that $u_{\alpha\beta}$ and $u[\alpha\beta\gamma]$ are not in the cut locus of I, so that these geodesics are unique. We use $\underline{g}_{\alpha\beta}^-$ and $\underline{g}^-[\alpha\beta\gamma]$ to denote the reverse geodesics.

3.3. Our algorithms depend on a preliminary choice of local ordering $\underline{o}$ of the vertices of Λ; that is, a partial ordering of the vertices such that those of every simplex are totally ordered. When we write $\sigma = \langle\alpha\beta\gamma\cdots\varepsilon\rangle$ it is to be understood that the vertices are listed in their $\underline{o}$ order; the same applies to notation such as $v_{\alpha\beta}$. Likewise, when $\sigma = \langle\alpha\beta\gamma\rangle$, we can now write $u[\sigma]$ and $\underline{g}[\sigma]$.

3.4. REMARK. In the context of our construction (1.5) of $\underline{u}$ from a gauge field (ξ,ω), $u[\sigma]$ is the result of parallel transport around $\partial\sigma$ in a particular direction, starting at a particular vertex. So $u[\sigma]$ is not quite gauge invariant; but $d[\sigma]$ is so.

We shall present our results as they arise in treating three examples of increasing complexity. For Example 1 (G = U(1), $c = c_1$, $\dim.\Lambda = 2$) we need only part of our solution to Step I, and only a small part of the Chern-Weil theory must be adapted to LGF's, since $\mathcal{U}(1)$ and $(c_1)_{cw}$ are so simple. Example 3 (G = SU(2), $c = c_2$, $\dim.\Lambda = 4$) requires the whole of our solution to Step I, and a new Lie algebra to replace $\mathfrak{su}(2)$, Finally Example 4 (G = U(1), $c = (c_1)^2$, $\dim.\Lambda = 4$) leads us to consider the structures on LGF's that correspond to G-invariant, symmetric polynomials on $\mathfrak{g}$. Example 2 is a useful auxiliary to the others.

4. EXAMPLE 1. $G = U(1)$, $c = c_1$, $\dim.\Lambda = 2$. (see Figure 2.)

4.1. STEP I. Our algorithm applies to those LGF's $\underline{u}$ on Λ that satisfy the following requirement.

4.2. <u>Continuity Hypothesis on u</u>: For every 2-simplex σ of Λ, $d[\sigma] < \pi$.

Let $\sigma = \langle\alpha\beta\gamma\rangle$. We specify that $v^{\sigma}_{\alpha\beta}$ and $v^{\sigma}_{\beta\gamma}$ must be constant functions. By (1.5.1) this forces $v^{\sigma}_{\alpha\beta} \equiv u_{\alpha\gamma}$, $v^{\sigma}_{\beta\gamma} \equiv u_{\beta\gamma}$. It remains to define $v^{\sigma}_{\alpha\gamma}$. Its domain $C^{\sigma}_{\alpha\gamma}$ is a 1-cell whose boundary points p and q are the barycentres of $\langle\alpha\gamma\rangle$ and σ respectively. We must set $v^{\sigma}_{\alpha\gamma}(p) = u_{\alpha\gamma}$ by (1.5.1), and

$$\begin{aligned} v^{\sigma}_{\alpha\gamma}(q) &= v^{\sigma}_{\alpha\beta}(q) \cdot v^{\sigma}_{\beta\gamma}(q), \text{ by } (1.3.4) \\ &= u_{\alpha\beta} \cdot u_{\beta\gamma} . \end{aligned}$$

The continuity hypothesis (4.2) implies that $u_{\alpha\gamma}$ and $u_{\alpha\beta} \cdot u_{\beta\gamma}$ are not antipodal in U(1). So we may define $v^{\sigma}_{\alpha\gamma}$ to map $C^{\sigma}_{\alpha\gamma}$ in the natural way to the unique shortest geodesic from $u_{\alpha\gamma}$ to $u_{\alpha\beta} \cdot u_{\beta\gamma}$. Using the notation of (3.2), we may write the image of $v^{\sigma}_{\alpha\gamma}$ as

4.3. $\text{im}.v^{\sigma}_{\alpha\gamma} = \underline{g}[\sigma] \cdot u_{\alpha\gamma}$.

We have thus solved Step I for this example. We shall return to Steps II$\mathbb{Z}$, II$\mathbb{R}$ and III after Example 2, in section (6).

4.4. REMARKS. In the context of our construction (1.5) of $\underline{u}$ from a gauge field (ξ,ω) we have (since G is abelian):

$$\begin{aligned} u[\sigma] &= \exp \int_{\partial\sigma}\omega, \text{ by } (1.5) \\ &= \exp \int_{\sigma}\Omega, \end{aligned}$$

where Ω is the curvature 2-form of ω, because on $\mathcal{U}(1)$, $\Omega = d\omega$. So for given (ξ,ω), (4.2) will hold for $\underline{u}$ provided that the triangulation Λ of M is fine enough that $|\int_{\sigma}\Omega| < \pi$ for every 2-simplex σ.

4.5. Let (ξ,ω) be given and let Λ satisfy the condition of the previous Remark. We can then construct $\underline{u}$ as in (1.5), hence $\underline{v}$ as in (4.1), and from $\underline{v}$ a principal bundle on M, by (1.4). It is shown in [8] that this bundle is isomorphic to the original ξ.

5.1. In general, our algorithm for $v^{\sigma}_{\lambda\mu}$ makes use of a process of reduction, which is illustrated in the next example. In the first place we use induction on $\dim.\sigma$. Next, let ρ be the face of σ whose vertices are

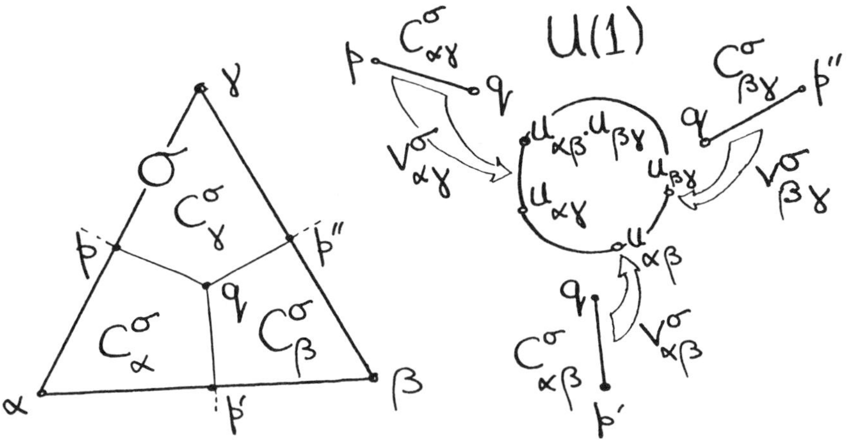

FIGURE 2. $G = U(1)$, $\dim.\Lambda = 2$. Construction of a coordinate bundle $\underline{v} = \{v^\sigma_{\lambda\mu}\}$ from a generic lattice gauge field $\underline{u} = \{u_{\lambda\mu}\}$.

λ, μ and all the vertices between these two in the local ordering $\underline{o}$. Under radial projection from $\mathrm{link}(\rho,\sigma)$, $C^\sigma_{\lambda\mu}$ is carried into $C^\rho_{\lambda\mu}$. If ρ is a proper face of σ, we define $v^\sigma_{\lambda\mu}$ to be the composition of this projection with $v^\rho_{\lambda\mu}$. So we are reduced to defining $v^\sigma_{\lambda\mu}$ in case λ is the first, and μ the last, vertex of σ. This agrees with the prescription used in (4.1).

5.2. EXAMPLE 2. $G = U(1)$, $c = c_1$, $\dim.\Lambda = 3$. We assume the continuity hypothesis (4.2) holds for $\underline{u}$, and concentrate on a simplex $\tau = \langle\alpha\beta\gamma\chi\rangle$ for which we suppose $u_{\alpha\chi} = u_{\beta\chi} = u_{\gamma\chi} = I$. (see Figure 3.) We shall investigate Step I in this situation.

5.3. Consider first $v^\tau_{\alpha\beta}$. Here $\rho = \langle\alpha\beta\rangle$ in (5.1), so we project the 2-cell $C^\tau_{\alpha\beta}$ into $C^\rho_{\alpha\beta}$, which is the barycentre p of $\langle\alpha\beta\rangle$. Since $v^\rho_{\alpha\beta}(p) = u_{\alpha\beta}$, $v^\tau_{\alpha\beta} \equiv u_{\alpha\beta}$ must be constant. Similarly $v^\tau_{\beta\gamma} \equiv u_{\beta\gamma}$ and $v^\tau_{\gamma\chi} \equiv u_{\gamma\chi} = I$.

Now consider $v^\tau_{\alpha\gamma}$. In this case $\rho = \langle\alpha\beta\gamma\rangle$. We project $C^\tau_{\alpha\gamma}$ into $C^\rho_{\alpha\gamma}$ from χ, and then apply $v^\rho_{\alpha\gamma}$ as constructed according to section (4). Thus, by (4.3),

5.3.1. $\mathrm{im}.v^\tau_{\alpha\gamma} = \underline{g}[\alpha\beta\gamma] \cdot u_{\alpha\gamma}$.

Similarly

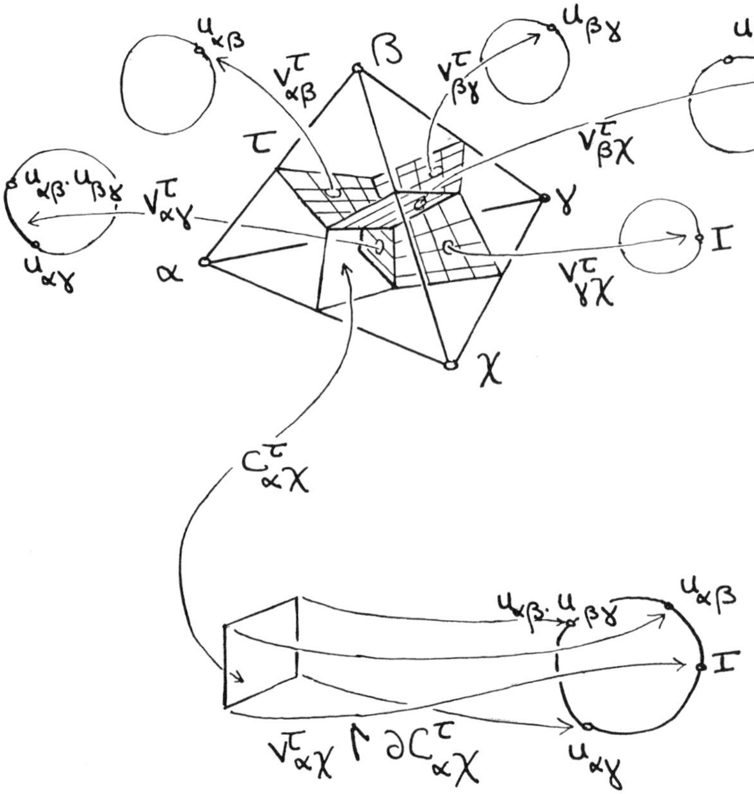

FIGURE 3. $G = U(1)$, $\dim.\Lambda = 3$. Obstruction to the construction of a coordinate bundle $\underline{v} = \{v^{\tau}_{\lambda\mu}\}$ from a lattice gauge field $\underline{u} = \{u_{\lambda\mu}\}$. In this example $u_{\alpha\chi} = u_{\beta\chi} = u_{\gamma\chi} = I$.

5.3.2. $$\text{im}.v^{\tau}_{\beta\chi} = \underline{g}[\beta\gamma\chi] \cdot u_{\beta\chi} = \underline{g}_{\beta\gamma}$$

in the notation of (3.2).

5.4. It remains to consider $v^{\tau}_{\alpha\chi}$. This map is completely determined on $\partial C^{\tau}_{\alpha\chi}$. For example, set $\sigma = \langle\alpha\beta\chi\rangle$; then on $C^{\tau}_{\alpha\chi} \cap \sigma$ (the front vertical edge of $C^{\tau}_{\alpha\chi}$ in Figure 3), we must have

$$v^{\tau}_{\alpha\chi} = v^{\sigma}_{\alpha\chi} = \underline{g}[\sigma] = \underline{g}_{\alpha\beta}.$$

On $C^{\tau}_{\alpha\chi} \cap C_{\beta}$ (the upper horizontal edge), we must have

$$v^{\tau}_{\alpha\chi} = v^{\tau}_{\alpha\beta} \cdot v^{\tau}_{\beta\chi} = u_{\alpha\beta} \cdot \underline{g}_{\beta\gamma}.$$

So we already have

5.4.1. $$v^{\tau}_{\alpha\chi} : \partial C^{\tau}_{\alpha\chi} \to U(1);$$

and there is, in general, an obstruction to extending $v^{\tau}_{\alpha\chi}$ over $C^{\tau}_{\alpha\chi}$, measured by the degree of this map. (We omit the details of sorting out orientations.)

5.5. REMARK. What has happened is that the method of Example 1 has enabled us to construct a bundle ξ' over $\partial\tau$. There is an obstruction to extending ξ' over τ, namely $c_1(\xi')$. Thus $c_1(\xi') = \deg.(v^{\tau}_{\alpha\chi} \mid \partial C^{\tau}_{\alpha\chi})$. This formula is the basis of our solution to Step II $\mathbb{Z}$ for Example 1.

6. We can now return to Example 1 and proceed with Steps II $\mathbb{Z}$, II $\mathbb{R}$ and III.

6.1. In this context $c_1(\xi)$ is the obstruction to trivializing ξ. This is the obstruction to extending ξ to a bundle over the cone on Λ, which we calculate (in view of (1.4)) as the obstruction to extending $\underline{v}$, constructed in section (4), to a coordinate bundle over this cone.

6.2. Let $\hat{\Lambda}$ be the cone on Λ from a new vertex χ. We extend $\underline{u}$ to a LGF $\hat{\underline{u}}$ on $\hat{\Lambda}$ by setting $\hat{u}_{\alpha\chi} = I$ for every vertex α of Λ, and $\hat{u}_{\alpha\beta} = u_{\alpha\beta}$ for every 1-simplex $\langle\alpha\beta\rangle$ of Λ.

To apply - or attempt to apply - our algorithm for Step I to $\hat{\underline{u}}$, we need a local ordering $\hat{\underline{o}}$ of the vertices of $\hat{\Lambda}$. We supplement $\underline{o}$ by the rule that χ follows every vertex of Λ. We also need to know that the continuity hypothesis (4.2) holds for $\underline{\hat{u}}$. This is considerably stronger than (4.2) for $\underline{u}$ alone; for example, we must now require that $\hat{u}[\alpha\beta\chi] = u_{\alpha\beta}$ be not equal to $-I$. The necessary continuity hypothesis is still generic (though no longer gauge invariant) in $\underline{u}$; so this is not a severe restriction.

6.3. Step II $\mathbb{Z}$. For each $\sigma = \langle\alpha\beta\gamma\rangle$ the method of section (5) gives an obstruction to the existence of a coordinate bundle function $v^{\tau}_{\alpha\chi}$, where $\tau = \langle\alpha\beta\gamma\chi\rangle$. In view of (6.1), we have thus found an integral cocycle $(c_1)_z$ representing $c_1(\xi)$, namely

6.3.1. $$(c_1)_z(\sigma) = \text{degree of } v^{\tau}_{\alpha\chi} \mid \partial C^{\tau}_{\alpha\chi}$$
$$= \deg.(\underline{g}_{\alpha\beta} \cup u_{\alpha\beta} \cdot \underline{g}_{\beta\gamma} \cup \underline{g}^{-}[\alpha\beta\gamma] \cdot u_{\alpha\gamma} \cup \underline{g}^{-}_{\alpha\gamma}).$$

This is our solution to Step II $\mathbb{Z}$.

6.4. Step II $\mathbb{R}$. When we turn to real coefficients, we may in the first place write

$$(c_1)_z(\sigma) = \frac{1}{2\pi} \text{ (sum of signed lengths of the arcs of (6.3.1))}$$

Three of these arcs are associated with edges of σ; so we can change $(c_1)_z$ by a coboundary to obtain a real-valued cocycle

$$(c_1)_r = \frac{1}{2\pi} \text{ (signed length of } \underline{g}^{-}[\alpha\beta\gamma] \cdot u_{\alpha\gamma}).$$

Our solution to Step II $\mathbb{R}$ is thus:

6.4.1. $$(c_1)_r = -\frac{1}{2\pi} \text{ (signed length of } \underline{g}[\sigma]).$$

6.5. REMARK. The right-hand side of (6.4.1) equals $\pm d[\sigma]/2\pi$, and is thus (allowing for orientations) gauge invariant. This is not the case with (6.3.1).

6.6. STEP III. Now assume that $\underline{u}$ was constructed from a gauge field (ξ,ω) according to (1.5), and that the condition of (4.4) holds. It then follows that

6.6.1. $$(c_1)_r = \int_\sigma -\frac{1}{2\pi i}\,\Omega\ .$$

Here Ω is the curvature of ω, with values in $\mathcal{U}(1) = i\,\mathbb{R}$, and $-\frac{1}{2\pi i}\Omega$ is the Chern-Weil form $(c_1)_{cw}$ representing c_1 (see Kobayashi and Nomizu [3]). We wish to establish a structural correspondence between the right-hand sides of (6.4.1) and (6.6.1). Now "$\int_\sigma$" means "evaluate on σ" and is implicit in (6.4.1). The factor "1/i" in (6.6.1) is the same as "logarithm," which measures signed lengths in U(1). What remains is to think of Ω as having the corresponding structure to $\underline{g}[\sigma]$. This leads us to propose:

6.6.2. There is a Lie algebra $\mathcal{U}(1)$ associated to U(1) which contains as a member any arc in U(1) that has I as one endpoint and has length less than π.

In this interpretation the arc $\underline{g}[\sigma]$ is itself the "curvature value" assigned to σ.

It is not clear at this stage how $\mathcal{U}(1)$ should be defined, partly because $\mathfrak{u}(1)$ has too few features to mimic. Our next example, in which G = SU(2), will provide further clues.

6.7. REMARK. The reason why the usual Lie algebra $\mathfrak{u}(1)$ is unsuited to u(1)-LGF's can be seen from the definition of a connection in a principal U(1)-bundle, for example. For there $\mathfrak{u}(1)$ is used to measure the effects of infinitesimal displacements in the base space, which are not available when we deal with LGF's. The problem is that a small rotation can be naturally represented by a unique element of $\mathfrak{u}(1)$, but this correspondence cannot be extended to rotations of arbitrary size.

7. EXAMPLE 3. $G = SU(2)$, $c = c_2$, $\dim.\Lambda = 4$.

7.1. STEP I. Our algorithm for the construction of $\underline{v}$ from a given $\underline{u}$ follows the reduction procedure of (4.1). This reduces us to the problem of constructing $v^\tau_{\alpha\chi}$, where $\tau = \langle\alpha\cdots\chi\rangle$. As in (5.4), $v^\tau_{\alpha\chi} \mid \partial C^\tau_{\alpha\chi}$ is already prescribed. Now $C^\tau_{\alpha\chi}$ is (in barycentric coordinates) a cone from the barycentre p of $\langle\alpha\chi\rangle$ on a subcomplex $\partial^1 C^\tau_{\alpha\chi} \subseteq \partial C^\tau_{\alpha\chi}$. For each $x \in \partial^1 C^\tau_{\alpha\chi}$, we should like to map the generator [p,x] into the unique shortest geodesic in SU(2) from $v^\tau_{\alpha\chi}(p) = u_{\alpha\chi}$ to $v^\tau_{\alpha\chi}(x)$. This can be done provided $v^\tau_{\alpha\chi}(x)$ is never antipodal to $u_{\alpha\chi}$; since $\dim.\tau \leq 4$, it follows that $\dim.\partial^1 C^\tau_{\alpha\chi} \leq 2$, and our proviso holds for generic $\underline{u}$. We can thus carry out Step I under a continuity hypothesis on $\underline{u}$ which, like those of (4.2) and (6.2), is generic in $\underline{u}$. In this form the hypothesis is rather tedious to state as an a priori condition on $\underline{u}$ (see [9]); so we shall content ourselves here with the unnecessarily strong continuity hypothesis (7.2). Our solution to Step I is given in theorem (7.3).

Figure 3 can be interpreted as an illustration of the construction of the last paragraph, in case dim.τ = 3. Assuming the continuity hypothesis holds for the LGF of that figure (the group now being SU(2)), im.$v^{\tau}_{\alpha\chi}$ will consist of two convex geodesic triangles, which we denote by

7.1.1. $\underline{h}[\alpha\beta\gamma] = \langle I, u_{\alpha\beta}, u_{\alpha\beta}\cdot u_{\beta\gamma}\rangle \cup \langle I, u_{\alpha\beta}\cdot u_{\beta\gamma}, u_{\alpha\gamma}\rangle$

In case τ is as in (7.1), with dim.τ = 4, we set

7.1.2. $\underline{k}[\;] = \text{im}.v^{\tau}_{\alpha\chi}$.

This image has the form of one of the 3-cubes illustrated on the left in Figure 4. The ruled surface shown in that figure arises as the image of the 2-face

7.1.3.
$$\begin{aligned} v^{\tau}_{\alpha\chi}(C^{\tau}_{\alpha\chi} \cap C_{\gamma}) &= \text{im}.v^{\langle\alpha\beta\gamma\rangle}_{\alpha\gamma} \cdot \text{im}.v^{\langle\gamma\delta\chi\rangle}_{\gamma\chi} \\ &= (\underline{g}[\alpha\beta\gamma] \cdot u_{\alpha\gamma}) \cdot (\underline{g}[\gamma\delta\chi] \cdot u_{\gamma\chi}). \end{aligned}$$

All the edges of $\underline{k}[\tau]$ are either $\underline{g}[\sigma]$'s, for σ a 2-face of τ, or else translates of $\underline{g}[\sigma]$'s under multiplication by transporters, as is the case with each factor of the last line of (7.1.3).

7.2. Continuity Hypothesis on $\underline{u}$. For every 2-simplex σ of Λ, $d[\sigma] < \pi/24$.

7.3. THEOREM ([9]). Let $\underline{u}$ be an SU(2)-LGF on a 4-dimensional Λ, such that (7.2) holds. Then there exists a unique principal SU(2)-bundle ξ with the following property: ξ can be trivialized over the 4-dimensional dual cells C_{α} of Λ in such a way that, for every 1-simplex $\langle\alpha\beta\rangle$, $v_{\alpha\beta} : C_{\alpha} \cap C_{\beta} \to SU(2)$ takes values in the ball of radius $\pi/8$ about $u_{\alpha\beta}$.

The proof of this theorem consists in part of showing that the algorithm of (7.1) can be carried out, thus solving Step I.

7.4. REMARKS. The continuity hypothesis (7.2) is in the spirit of Remarks (2.7) and (4.4).

7.5. Our algorithm provided a $\underline{v}$ with stronger properties than those claimed in theorem (7.3): first, (1.5.1) holds; and second, if (7.2) is known to hold for $\underline{u}$ with every $d[\sigma] <$ (some Δ) $\leq \pi/24$, then each $v_{\alpha\beta}$ takes values in the ball of radius 3Δ about $u_{\alpha\beta}$. This is the precise form of what we said in Remark (2.8), that $v_{\alpha\beta}(x)$ is close to $u_{\alpha\beta}$ for every $x \in C_{\alpha} \cap C_{\beta}$.

7.6. STEP II $\mathbb{Z}$. We shall present our solution to this step only in outline. The method is similar to that of (6-1)--(6.3). Let $\hat{\Lambda}$, $\hat{\underline{u}}$ and $\hat{\underline{o}}$ be as in (6.2). We regard $c_2(\xi)$ as the obstruction to extending $\underline{v}$, constructed in (7.1), to a coordinate bundle on $\hat{\Lambda}$. Let $\sigma = \langle\alpha\beta\gamma\delta\varepsilon\rangle$ be a 4-simplex of Λ, and set $\tau = \langle\alpha\beta\gamma\delta\varepsilon\chi\rangle$. Under a suitable continuity hypothesis on $\underline{u}$, which is generic though not gauge invariant, we obtain as our obstruction:

7.6.1. $(c_2)_z(\sigma)$ = degree of $v^\tau_{\alpha\chi} : \partial C^\tau_{\alpha\chi} \to SU(2)$.

To compute this degree we make use of the geometry of the map $v^\tau_{\alpha\chi} \mid \partial C^\tau_{\alpha\chi}$. Now $C^\tau_{\alpha\chi}$ is a combinatorial 4-cube; its boundary consists of eight 3-cubes, which are mapped into $SU(2) = S^3$ according to Figure 4. In S^3 the vertices of each region, simplex, pyramid or prism, are specified as products of transporters $u_{\lambda\mu}$. To evaluate the right-hand side of (7.6.1) we pick a point $y \in S^3$ in general position with respect to all the regions, and add the incidence numbers of y with each region (keeping track of orientations). Each incidence number can be calculated by solving quadratic equations and by working out the signs of 4×4 determinants specified by vertices of the region in question.

7.7. REMARK. For an application of this computation to quantum field theory, see [4].

7.8. STEP II $\mathbb{R}$. We proceed as in (6.4). With real coefficients, we may express $(c_2)_z(\sigma)$ in terms of the signed volumes of the regions in S^3 referred to in (7.6). Now σ has five 3-faces, to each of which is associated one of the eight 3-cubes of $\partial C^\tau_{\alpha\chi}$, in fact one of those on the left in Figure 4; compare (7.1). By changing $(c_2)_z$ by a coboundary we can eliminate the volumes of regions corresponding to these five 3-cubes. The three remaining ones can be described in terms of the notation of (7.1.1) and (7.1.2). The 3-cube of the type on the left in Figure 4 is $\underline{k}[\sigma]$. The other two 3-cubes are $\underline{h}[\alpha\beta\gamma] \cdot \underline{g}[\gamma\delta\varepsilon]$ and $\underline{g}[\alpha\beta\gamma] \cdot \underline{h}[\gamma\delta\varepsilon]$; thus each of them is divided into a pair of prisms. We are left with the formula:

7.8.1. $(c_2)_r(\sigma) = (1/2\pi^2)$ (sum of signed volumes of $\underline{k}$, $\underline{h} \cdot \underline{g}$ and $\underline{g} \cdot \underline{h}$)

However, we cannot call (7.8.1) a solution to Step II $\mathbb{R}$ until we can also carry out Step III.

7.9. For the group SU(2), the Chern-Weil differential form that represents c_2 is $(c_2)_{cw} = (1/4\pi^2)\mathrm{Det}(\Omega)$. (See [3].) We must therefore compare (7.8.1) to

$\partial C^{\tau}_{\alpha\chi}$ consists of

$\cup$

six 3-cubes,
each dissected into

two 3-cubes,
each dissected into

two
simplices,

$\cup$ pyramid $\cup$

two
simplices,

two prisms,

each of which

each of
which

each of which

geodesic
3-simplex

geodesic
3-simplex

cone on a
ruled surface

(geodesic simplex).

(geodesic simplex)

$v^{\tau}_{\alpha\chi}$

S^3

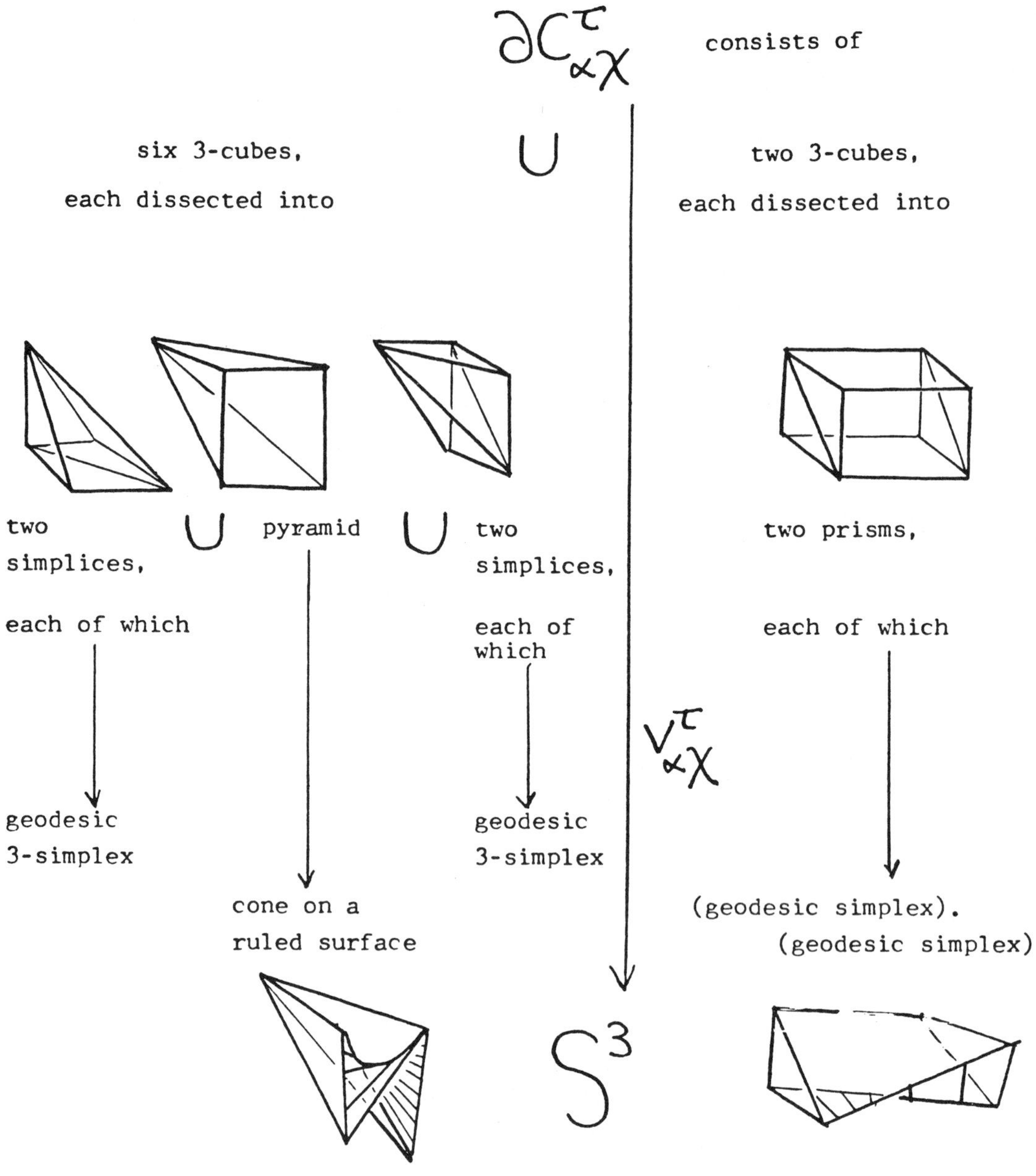

FIGURE 4. G = SU(2), dim.Λ = 4. Geometry of the map
$v^{\tau}_{\alpha\chi}$: $\partial C^{\tau}_{\alpha\chi}$ → SU(2) = S^3, where τ is the cone on a 4-simplex σ from an
extermal vertex χ.

7.9.1. $\int_\sigma (c_2)_{cw} = (1/2\pi^2) \int_\sigma \frac{1}{2} \mathrm{Det}(\Omega)$.

Now, as in (6.6), "$\int_\sigma$" means "evaluate on σ," which is implicit in the right-hand side of (7.8.1). Next, the determinant is the infinitesimal of volume on S^3, so "Det" in (7.9.1) and "signed volume" in (7.8.1) are in structural correspondence. Now we want the regions of SU(2) referred to in (7.8.1) to correspond to the values of Ω in (7.9.1). This suggests that we construct a Lie algebra $\mathscr{SU}(2)$ in the following way, amplifying (6.6.2).

7.10. DEFINITIONS. Let the Lie group G act on a space X. For any sets $R \subseteq G$ and $Y \subseteq X$ we define their set-product by

7.10.1. $R \cdot Y = \{g \cdot y : g \in R, y \in Y\}$.

Let $\mathcal{R}$ represent the set of all regions in SU(2) generated by its individual points and two operations:

i) taking the set-product of two regions; and
ii) taking the join of a region to a point by shortest geodesics, whenever that point is not antipodal to any point of the region.

7.10.2. We define $\mathscr{SU}(2)$ to be the chain complex over $\mathbb{R}$ generated by all those regions in $\mathcal{R}$ which have one vertex at I. The Lie product is determined by $[R_1,R_2] = R_1 \cdot R_2 - R_2 \cdot R_1$.

It remains to explain why the particular element $\underline{k} + \underline{h} \cdot \underline{g} + \underline{g} \cdot \underline{h}$ of $\mathscr{SU}(2)$ should appear in (7.8.1). We believe there may be an analogy between this element and the Chern-Simons form θ corresponding to $(c_2)_{cw}$. Let (ξ,ω) be given, with $\xi = (\pi : E \to M)$; then according to [2] the following 3-form on E satisfies $\pi^*(c_2)_{cw} = d\theta$:

7.10.3. $\theta = (1/2\pi^2)\mathrm{Tr}(\omega \wedge \Omega - (1/3)\ \omega \wedge [\omega,\omega])$.

Now let M* be a "partial section" of ξ; that is, an oriented cycle-with-boundary in E such that $\pi(\partial M^*)$ is a 0-dimensional set X in M, and $\pi : (M^*,\partial M^*) \to (M,X)$ has degree 1. Then $c_2(\xi) = \int_{\partial M^*} \theta$. Now when we interpret our construction of $(c_2)_z$ as the obstruction, not to an extension of ξ over $\hat{\Lambda}$, but to a section of ξ, then it turns out that we are in fact constructing a particular partial section M*; furthermore $(c_2)_z$--and hence also $(c_2)_r$--measures the obstruction to closing up M* in E. So we propose that $\underline{k} + \underline{h} \cdot \underline{g} + \underline{g} \cdot \underline{h}$ may be in structural correspondence with $\omega \wedge \Omega - (1/3)\ \omega \wedge [\omega,\omega]$. Perhaps ω corresponds to g -- a 1-form to a

1-dimensional chain -- and $\underline{h}$ to Ω, both being 2-dimensional. Then $\omega \wedge \Omega$ might correspond in structure to $\underline{g}.\underline{h}+\underline{h}.\underline{g}$, and (in view of (7.1)) $\underline{k}$ to $\omega \wedge [\omega,\omega]$. Such an interpretation of (7.8.1) appears, however, to conflict with the correspondence between $\underline{g}$ and Ω of (6.6); and we do not yet understand how to resolve this difficulty.

To complete Step III we need to know more generally what operations on $\mathcal{SU}(2)$ correspond to symmetric, SU(2)-invariant polynomials on $\mathfrak{su}(2)$. Our next example may provide a hint in that direction.

8. EXAMPLE 4. $G = U(1)$, $c = c_1^2$, $\dim.\Lambda = 4$.

8.1. STEPS I and II $\mathbb{Z}$. Under a continuity hypothesis like (7.2) on $\underline{u}$, the methods of (7.1) and (1.4) give a coordinate bundle $\underline{v}$ and hence a principal U(1)-bundle ξ. Note the necessity here of a strong (non-generic) continuity hypothesis like (7.2). There is a generic set of $\underline{u}$'s for which the bundle constructed by (1.4) cannot be extended over the 3-skeleton of Λ (see [8], section 4). We shall carry out what we can of Steps II $\mathbb{R}$ and III. There is a well-known solution to hand for Step II $\mathbb{Z}$, namely the Whitney-Alexander formula for the cochain product $(c_1)_z \cup (c_1)_z$ (see Spanier [12]).

8.2. STEP II $\mathbb{R}$. We propose to use the Whitney-Alexander formula again:

8.2.1. $(c_1^2)_r[\alpha\beta\gamma\delta\varepsilon] = (c_1)_r[\alpha\beta\gamma] \cdot (c_1)_r[\gamma\delta\varepsilon]$.

But to justify the use of this formula we must show how to derive it from a construction that corresponds in structure to the Chern-Weil construction; and this we can do only in part.

8.3. STEP III. First we identify $c_1^2(\xi)$ as the obstruction to a section of the U(1)*U(1)-bundle ξ^* associated to ξ. Here ξ^* denotes the fiberwise join of ξ with itself. Let ξ^+ be the complex line bundle associated to ξ, and let s_1 and s_2 be generic sections of ξ^+ with respective divisors D_1 and D_2; so $D_i = \{x : s_i(x)$ is in the 0-section of $\xi^+\}$. Then each D_i is Poincare dual to $c_1(\xi)$, so $D_1 \cap D_2$ is dual to $c_1^2(\xi)$. But $D_1 \cap D_2$ is the divisor of the section $s_1 \oplus s_2$ of $\xi^+ \oplus \xi^+$, namely the obstruction to pushing $s_1 \oplus s_2$ fibrewise into the complement of the 0-section of $\xi^+ \oplus \xi^+$; and this complement has ξ^* as fibrewise deformation retract.

8.3.1. We write $U_1 * U_2$ for $U(1) * U(1)$, and $X_1 * Y_2$ for the join of subsets X of the first factor and Y of the second. U(1) acts on $U_1 * U_2$ simultaneously on both factors.

8.4. Now $U(1) * U(1) = S^1 * S^1$ is homeomorphic to S^3, which is in turn homeomorphic to SU(2). So we can use the method of section (6), in which $G = SU(2)$. We must bear in mind, however, that the join metric on $U_1 * U_2$ is different from the bi-invariant one on SU(2), so that the geodesics used in the present construction will differ from those used in section 6.

8.5. Let $\hat{\Lambda}$, $\underline{\hat{u}}$ and $\underline{\hat{o}}$ be as in (6.2). In view of (6.1) and (8.1), $c_1^2(\xi)$ is the obstruction to using our algorithm for Step I to construct from $\underline{\hat{u}}$ a coordinate bundle $\underline{v}^*$ on $\hat{\Lambda}$ whose functions $v^*_{\lambda\mu}$ have values in $U_1 * U_2$.

We regard the LGF $\underline{\hat{u}}$ as having values in U_1. When the domain of some $v^{*\tau}_{\lambda\mu}$ reduces to a 1-dimensional cell $C^\rho_{\lambda\mu}$, according to (5.1), then $v^{*\tau}_{\lambda\mu}$ can be defined so as to have values in U_1 too. The same applies if $v^{*\rho}_{\lambda\mu}$ is required to map $C^\rho_{\lambda\mu}$ into a ruled surface, and also if the images of the 1-cells of $C^\rho_{\lambda\mu}$ are all congruent to $\underline{g}[\alpha\beta\gamma]$'s (where χ is excluded from these indices), because these arcs are so short, by the continuity hypothesis (7.2), that no obstruction can occur.

Let us now interpret Figure 3 as illustrating $v^{*\tau}_{\alpha\chi} : \partial C^\tau_{\alpha\chi} \to U_1$. For this situation our new rule is to map the centroid c of $C^\tau_{\alpha\chi}$ into $I_2 \in U_2$, and then, for each $x \in \partial C^\tau_{\alpha\chi}$, to map the segment [x,c] into $v^{*\tau}_{\alpha\chi}(x) * I_2$ in the obvious way.

8.6. Let $\sigma = \langle\alpha\beta\gamma\delta\varepsilon\rangle$ be a 4-simplex of Λ, and let $\tau = \langle\alpha\beta\gamma\delta\varepsilon\chi\rangle$. The method of (8.5) can be continued until, as in (7.6), we reach an obstruction, whose value on σ is the degree of the map $v^{*\tau}_{\alpha\chi} : \partial C^\tau_{\alpha\chi} \to U_1 * U_2$. As in (7.8), when we express this degree in terms of volumes, various coboundary-cancellations become possible. We end up with the volume of a certain region $T \subseteq U_1 * U_2$. In the notation of (7.10) and (8.3.1),

8.6.1. $T = \underline{g}[\alpha\beta\gamma] \cdot \{\underline{g}[\gamma\delta\varepsilon]_1 * I_2\}$.

Using (6.4.1) and (8.2.1) we find

8.6.2. $(c_1^2)_r(\sigma) = (1/4\pi^2)(\text{signed length of } \underline{g}[\alpha\beta\gamma])\cdot(\text{signed length of } \underline{g}[\gamma\delta\varepsilon])$
$= (1/4\pi^2)(\text{signed volume of } T)$.

8.7. For the group U(1) the Chern-Weil differential form that represents c_1^2 is $(c_1^2)_{cw} = (-1/4\pi^2)(\Omega \wedge \Omega)$; so (8.6.2) must be compared to

8.7.1. $\int_\sigma (c_1^2)_{cw} = (1/4\pi^2) \int_\sigma - (\Omega \wedge \Omega)$.

We could side-step the problem by rewriting the right-hand side of (8.7.1) as $(1/4\pi^2) \int_\sigma (\frac{1}{i}\,\Omega) \wedge (\frac{1}{i}\,\Omega)$. In view of (6.6) and the known structural correspondence between the Grassmann product of real-valued forms and the Whitney-Alexander product of real-valued cochains, (8.7.1) would then be in structural correspondence with the first equation of (8.6.2). For the sake of the general Chern-Weil homomorphism however, we want to concentrate on Lie-algebra-valued forms and cochains.

Now $(c_1^2)_{cw}$ is the image under the Weil homomorphism of the polynomial $P(ih,ik) = -hk$ on $u(1) \otimes u(1)$. As a structure to correspond to $u(1) \otimes u(1)$ one might define $\mathcal{U}(1)$ "$\otimes$" $\mathcal{U}(1)$ to be the chain complex over $\mathbb{R}$ generated by all regions of the forms $H \cdot (I_1 * K_2)$ and $H \cdot (K_1 * I_2)$, where H and K are arcs in $\mathcal{U}(1)$ as in (6.6.2), and where the notation is that of (7.10) and (8.3.1). Then (8.6.2) would be in $\mathcal{U}(1)$ "$\otimes$" $\mathcal{U}(1)$ by (8.6.1) and (6.6), and would correspond in structure to $\Omega \wedge \Omega$. One would still have to argue for a correspondence between the polynomial P defined above and the function "-signed volume" on $U_1 * U_2$.

ACKNOWLEDGEMENTS

The second author is grateful to the Science and Engineering Research Coucil (UK) for support under Award GR/C/57891, and to University College of London University for hospitality during the Summer of 1984.

REFERENCES

1. Bott, R., Shulman, H., and Stasheff, J., On the deRham Theory of Classifying Spaces, Adv. in Math. 20 (1976), 43-56.
2. Chern, S-S. and Simons, J., Characteristic Forms and Geometric Invariants, Ann. of Math. 99 (1974), 48-69.
3. Kobayashi, S., and Nomizu, K., Foundations of Differential Geometry, Wiley, 1969, New York.
4. Lasher, G., Phillips, A., and Stone, D., A Reliable Combinatorial Algorithm for the Topological Charge of SU(2) Lattice Gauge Fields, in Quark Confinement and Libesation: Numerical Results and Theory, Klinkhamer, F. R. and Halpern, M. B. (eds), World Scientific, 1985, Singapore.
5. M. Luescher, Topology of Lattice Gauge Fields, Commun. Math. Phys. 85 (1982), 39-48.
6. Milnor, J., and Stasheff, J., Characteristic Classes, Ann. of Math. Studies, 76, Princeton University Press 1974, Princeton.

7. Nijenhuis, A., On the Holonomy Groups of Linear Connections, I A B General Properties of Affine Connections, II Properties of General Linear Connections, Indag. Math. 15 (1953), 233-249 and 16 (1954), 17-25.

8. Phillips, A., Characteristic Numbers of U_1-valued Lattice Gauge Fields, Ann. Phys. 161 (1985), 399-422.

9. Phillips, A., and Stone, D., Lattice Gauge Fields, Principal Bundles and the Calculation of Topological Charge, Commun. Math. Phys., to 103 (1986), 599-636.

10. Rebbi, C., Lattice Gauge Fields and Monte Carlo Simulations, World Scientific 1983, Singapore.

11. Schlesinger, L., Parallelverschiebung und Krümmungs tensor, Math. Annalen 99 (1928), 413-434.

12. Spanier, E., Algebraic Topology, McGraw-Hill 1966, New York.

13. Steenrod, N., The Topology of Fibre Bundles, Princeton 1951, Princeton, N.J.

14. Wilson, K., Confinement of Quarks, Phys. Rev. D10 (1974), 2445-2459.

Intrinsic Skeleta and Intersection Homology of Weakly Stratified Sets

FRANK QUINN / Department of Mathematics, Virginia Polytechnic Institute and State University, Blacksburg, Virginia

The development of intrinsic skeleta in the PL category was the key to understanding how polyhedra are built up from manifolds. Topological analogs have been studied, most notably by Siebenmann [8] in the setting of locally conelike spaces. However technical difficulties have hindered effective development. In theorem 1 we give a useful characterization in the (more general) context of weakly stratified sets: if a space has a manifold weakly stratified filtration, then it has a unique minimal one which is essentially the topological intrinsic skeleton filtration. This makes it possible, for example, to understand the behavior of skeleta with respect to products.

Intersection homology is defined in terms of a filtration on a space ([1-4]), and has been very useful in investigating invariants of singular spaces. "Topological invariance" of this homology, with respect to a class of filtrations, means that the homology of a space does not depend on the choice of filtration in the specified class (presuming of course that one exists). Homology was shown to be independent of the choice of "stratified pseudomanifold" filtration by Goresky and MacPherson [2, part II]. More generally H. King [3] showed it to be independent of the choice of locally conelike filtration. In the corollary to theorem 3 we show that (more generally yet) it is independent of choice of manifold weakly stratified set filtration. This is also a philosophically more satisfactory setting for the result. Pseudomanifolds and locally conelike spaces are defined by local homeomorphism properties. Weakly stratified spaces are defined by local homotopy properties, which seem more appropriate for the study of a homology theory.

In the first section the results on intrinsic skeleta are derived, using the geometric material of [5]. The second section contains a basic technical fact about the local structure of weakly stratified sets, that they homotopically resemble cones on joins of "links." The intersection homology theorem is proved in the third section, using the results of the first two.

1. INTRINSIC SKELETA

Suppose X is a topological space. A subset $Y \subseteq X$ is homogeneous if any two points x, $y \in Y$ have homeomorphic neighborhoods (by a homeomorphism which carries x to y). Any point is contained in a maximal homogeneous subset. The (topological) intrinsic k-skeleton of X is defined to be the union of all maximal homogeneous subsets of dimension $\leq k$, and is denoted by X^k.

We note that the skeleta in the spaces we consider satisfy much stronger homogeneity properties. The definition given is intended as a simple way to identify the skeleta rather than an attempt to abstract a significant property. Also, in a generic topological space one expects no two points to have homeomorphic neighborhoods, so the whole space would be in the "intrinsic 0-skeleton."

A filtered space is a space together with closed subspaces $X \supseteq X^n \supseteq X^{n-1} \supseteq \cdots \supseteq X^0$, which are referred to as the sketeta of the filtration. If X is filtered we denote the underlying space by $|X|$. So for example the intrinsic k-skeleton of the underlying space is denoted $|X|^k$. If X and Y are filtered spaces with the same underlying space, then the identity map $X \to Y$ is a coarsening if strata in Y are unions of components of strata in X.

The strata of a filtration are the differences $X^i - X^{i-1}$. A manifold filtration is one whose strata are manifolds (here we assume the strata do not have boundary. Boundary is permitted in [5]). In particular a manifold weakly stratified set is a weakly stratified set (see [5], or the next section) with manifold strata. In this section we assume sets are filtered by dimension in the sense that components of $X^i - X^{i-1}$ have dimension i.

THEOREM 1.1. Suppose X has a manifold weakly stratified filtration. Then there is a weakly stratified filtration $X_{0,0}$ such that

1) for any manifold weakly stratified filtration Y of $|X|$, the identity $Y \to X_{0,0}$ is a coarsening,

2) if $k \geq 5$, or if X^4 is locally conelike, then $(X_{0,0})^k = |X|^k$, and the k-stratum of $X_{0,0}$ is a manifold, and

3) if $(X_{0,0})^0 = \phi$ then $(X \times \mathbb{R})^{k+1}_{0,0} = (X_{0,0})^k \times \mathbb{R}$.

The 0-skeleton problem in (3) arises when a point can be added to some higher stratum to get a homology manifold in which the point is not 1-LC embedded. The fundamental group in its complement identifies such a point as in the 0-skeleton of the original. But since the product of such a homology manifold with $\mathbb{R}$ is a manifold, the point lies in a higher stratum in the product. The point of (3) is that this is the only such problem which can occur.

By combining (2) and (3) we see that except possibly for points in the 0-skeleton, intrinsic skeleta behave in the expected way with respect to products. This resolves a problem implicit in Siebenmann's development of locally conelike sets [8]. We discuss the low dimensional hypotheses in (2) after the proof.

We recall that a filtered set is <u>locally conelike</u> [8,4] if a point $x \in X^k - X^{k-1}$ has a neighborhood U in X and a homeomorphism $\mathbb{R}^k \times cL \simeq U$ which carries $\{0\} \times$ (cone point) to x, and a filtration induced from L to the given one. Here cL denotes the cone on L, and the cone point is included as a skeleton of the cone. Locally conelike spaces are manifold weakly stratified sets, but the reverse is not true. For example if M is a manifold with an end which is tame and has stable fundamental group, then the 1-point compactification M^∞ is weakly stratified. But if the finiteness obstruction of the end is not trivial, M^∞ is not locally conelike at ∞.

PROOF OF THE THEOREM. Suppose X is a manifold weakly stratified set, filtered by dimension. We inductively construct coarser filtrations, by in the r,s step "promoting" as much of the s-stratum as possible into the r-stratum. More precisely we define filtrations $X_{r,s}$ for $r \geq s \geq 0$ so that if $(r,s) \geq (r',s')$ (lexicographic order) then $X_{r,s} \to X_{r',s'}$ is a coarsening.

If $r > \dim X$, let $X_{r,s} = X$. If $X_{r,0}$ is defined then let $X_{r-1,r-1} = X_{r,0}$. If $X_{r,s+1}$ is defined for $s \geq 0$ we describe how to construct $X_{r,s}$. Define $(X_{r,s})^k = (X_{r,s+1})^k$ for $k > r$, $(X_{r,s})^r = (X_{r,s+1})^r \cup K$, and $(X_{r,s})^k = (X_{r,s+1})^k - K$ if $k < r$, where K is a maximal collection of components of $(X_{r,s+1})^s - (X_{r,s+1})^{s-1}$ satisfying:

1) K is disjoint from the closure of $(X_{r,s+1})^{r-1} - (X_{r,s+1})^s$,

2) the fiber of $\mathrm{holink}(((X_{r,s+1})^4 - (X_{r,s+1})^{r-1}) \cup K, K) \to K$ is a homology sphere, and a homotopy sphere if $s = 0$, and

3) the filtration so defined is a weakly stratified set.

The first remark is that such a maximal collection exists. These conditions are all local, so if a collection of components of $(X_{r,s+1})^s - (X_{r,s+1})^{s-1}$ satisfies the conditions one at a time, the union also satisfies them. Also, by condition (3) the $X_{r,s}$ are automatically weakly stratified sets. Homotopy links (the holink in (2)) are defined in [5, §2].

The filtration of interest is the final one, $X_{0,0}$. It satisfies conclusion (3) of the theorem because the conditions above hold after $\times \mathbb{R}$ if and only if they hold before, except possibly if $s = 0$. The $s = 0$ case for X corresponds to the $s = 1$ case for $X \times \mathbb{R}$, so the homotopy link condition in (2) changes from homotopy to homology sphere.

The next step is to see that the k-stratum of $X_{0,0}$ is a manifold if $k \geq 5$, or if X^4 is locally conelike. Proceeding by induction, presume this is the case for $X_{r,s+1}$. In passing to $X_{r,s}$ only the r-strata are enlarged. If $k \geq 5$ then theorem 1.4 of [5] applies, by conditions (1) and (2) above. Next suppose $r \leq 4$ and $(X_{r,s+1})^4$ is locally conelike. The link L of a promoted stratum must be an ANR homology manifold which is a homology sphere. If $\dim L \leq 2$ then it is a genuine sphere, so the stratum has a manifold neighborhood. This covers all cases except the promotion of isolated points into the 4-stratum, in $X_{4,0}$. These points have neighborhood cones on homotopy spheres, so the end of the complement is proper homotopy equivalent to $S^3 \times [0,1)$. Freedman's characterization theorem asserts that this end is homeomorphic to $S^3 \times [0,1)$, so when the isolated point is replaced the result is a manifold.

Next recall from [5, corollary 1.2] that components of strata of manifold weakly stratified sets are homogeneous (in the sense above), provided adjacent strata have dimension at least 5. In fact they are isotopy homogeneous in the sense that there are ambient isotopies taking one point to another. Stratum components of locally conelike sets are also isotopically homogeneous, locally by direct construction in the cone neighborhood, and globally using this and connectnedness. If X^4 is locally conelike then the isotopy extension theorem [5, 1.1] gives extensions of such isotopies to all of X, so the low dimensional strata are also homogeneous in X.

Now (still presuming X^4 is locally conelike) compare these filtrations with the intrinsic skeletal filtration $|X|^*$. If K is a component of $X^i - X^{i-1}$ then the homogeneity observed above implies that if K intersects a stratum component of $|X|^*$ then it is contained in the component. In other words the identity map $X \to |X|^*$ is a coarsening. In particular $X_{0,0} \to |X|^*$ is a coarsening. To see the filtrations are the same, suppose they are not and let r be the highest dimension so that the r-strata are different. Since the identity is a coarsening, there must be a component K of some s-stratum of X, which is in the r-stratum of $|X|^*$ but is in a lower stratum of $X_{0,0}$. Assume that s is the largest such, then we get a contraction by showing that such a K would have been promoted in the construction of $X_{r,s}$.

K satisfies condition (1) in $|X|^*$ because the filtration is closed. Since a stratum of $X_{0,0}$ whose closure intersects K must have higher dimension, and since r,s are maximal, this holds in $X_{r,s+1}$ also.

Next note (again by the maximality of r,s) that $(X_{r,s+1})^r - (X_{r,s+1})^{r-1} \cup K$ is open in $|X|^r - |X|^{r-1}$, so is a manifold. K is a closed subset and also a manifold, so local Poincare duality implies that the fiber of the homotopy link is a homology sphere. If s = 0, K is an isolated point and therefore has homotopy link a homotopy sphere. K therefore satisfies condition (2).

Finally consider condition (3). Since $\dim\{|X|^r - (X_{r,s+1})^r\} < r$, in order for K to be in $|X|^r - |X|^{r-1}$ it is necessary for each point $y \in K$ to have a neighborhood homeomorphic to a neighborhood of some $x \in (X_{r,s+1})^r - (X_{r,s+1})^{r-1}$. Homeomorphisms preserve the intrinsic skeleton filtration. In a neighborhood of x this agrees with the given filtration on $X_{r,s+1}$. In a neighborhood of y it agrees with the filtration obtained from this one by promoting K into the r-stratum. Since the property of being a weakly stratified set is a local property and topologically invariant, it follows that the "promoted" filtration is weakly stratified.

We conclude that K satisfies the conditions (1) - (3) for promotion into the r-stratum, contradicting the initial assumption. Therefore $X_{0,0} = |X|^*$, under the assumption that the 4-skeketon is locally conelike. We can weaken this somewhat. The conelike assumption was used to show components of the lower strata are homogeneous, so get promoted as units into higher strata. However in the proof above if we are considering the intrinsic r-strata for $r \geq 5$ then we can avoid this. The first step was to show that the component K had no adjacent strata in $X_{r,s+1}$ of dimension $< r$, so the isotopy extension theorem [5, 1.1] applies to show it is homogeneous. Consequently $(X_{0,0})^k = |X|^k$ provided $k \geq 5$. This completes the proof of part 2 of the theorem.

It remains to prove statement (1) when X^4 is not locally conelike. Multiply by $\mathbb{R}^5$, then $(X \times \mathbb{R}^5)^4 = \emptyset$ so $|X \times \mathbb{R}^5|^* = (X \times \mathbb{R}^5)_{0,0}$. Using statement (3) of the theorem we see this is $(X_{0,0}) \times \mathbb{R}^5$, except for some points in the 0-skeleton. Precisely, there is a coarsening $X_{0,0} \to Y$ to a weakly stratified set with homology manifold strata, such that $|X \times \mathbb{R}^5|^* = Y \times \mathbb{R}^5$. $X_{0,0}$ can be recovered from Y by demoting back to the 0-skeleton the isolated points in strata of Y which have homotopy links which are not homotopy spheres. Now suppose X and Z are manifold weakly stratified sets with the same underlying space. By the above they both have as coarsenings the filtration which gives the intrinsic strata of $|X| \times \mathbb{R}^5$. Further both $X_{0,0}$ and $Z_{0,0}$ are identified as being obtained from this filtration by demoting certain points back to the 0-skeleton. The property used to identify these points (homotopy link not a homotopy sphere) is topologically invariant, so the same points get demoted in both. Therefore $X_{0,0} = Z_{0,0}$.

This completes the proof of the theorem.

REMARKS. We discuss the low dimensional problems. The first remark is that the isotopy extension theorem of [5] applies if the adjacent strata have dimension at least 5. Therefore the 4-stratum $(X_{0,0})^4 - (X_{0,0})^3$ is isotopy homogeneous if it is a manifold. It is assembled from strata of X so is a manifold weakly stratified set. It satisfies the criterion of [5, 1.4] to be a manifold, but the dimension is too low. However if the homotopy links of the strata in the whole space are homotopy spheres then the characterization of local flatness [6, part III] applies to show it is a manifold and the strata are locally flat. This local flatness necessarily happens if the set is locally conelike. More generally we ask: suppose $M \subseteq X$ is a manifold, tame and with homotopy link a fibration, with fiber a homology sphere, and X-M is a manifold. If the codimension of M is ≤ 3, must M be locally flat? It is sufficient to show that the fiber of the homotopy link is a homotopy sphere. A codimension 4 counterexample is obtained by taking the cone on a 3-dimensional manifold homology sphere.

For the 3-skeleton, we have already speculated that a 3-dimensional manifold weakly stratified set is a polyhedron. Note that to avoid the Poincaré conjecture the 3-strata should be assumed to have at most finitely many fake cells.

2. WEAKLY STRATIFIED SETS

In this section it is shown that weakly stratified sets satisfy a homotopy analog of the locally conelike condition. This is used in the proof in the next section. We require the spaces to be metric, but do not reqiure manifold strata. The proofs rely heavily on [5].

Suppose X is a filtered space, and $x \in X^i - X^{i-1}$. We describe the local structure of X "normal" to the stratum. If $k > i$, let L_k denote the fiber at x of the homotopy link [5, 2.1];

$$\text{holink}((X^k - X^{k-1}) \cup (X^i - X^{i-1}), X^i - X^{i-1}) \to X^i - X^{i-1} .$$

Define L_x to be the join $L_{i+1} * L_{i+2} * \cdots * L_n$ filtered by $(L_x)^k = L_{i+1} * \cdots * L_k$. Recall that the join $A * B$ is the homotopy pushout (double mapping cylinder) of the projections $A \leftarrow A \times B \to B$. Finally denote by cL_x the cone on L_x, and filter it by $(cL_x)^k = c(L_x)^k$ if $k > i$, and $(cL_x)^i$ is the cone point.

Now suppose U is a subset of filtered spaces X and Y, and $f : X \to Y$ is a stratum-preserving map which is the identity on U. We say f is a <u>stratum-preserving homotopy equivalence near U</u>, provided there is a neighborhood V of U in Y and a stratum preserving map $g : V \to X$ such that the compositions fg and gf (where defined) are stratum-preserving homotopic rel U to the inclusions.

We note that it is the stratum-preserving requirement which gives this definition force. If U is a neighborhood deformation retract in X, then the inclusion $U \subseteq X$ satisfies the definition except for preserving strata. This definition is a refinement of the "eventual homotopy equivalence of neighborhoods" of [6, part IV].

THEOREM 2. Suppose X is weakly stratified, and U is open in a stratum $X^i - X^{i-1}$ and contracts in the stratum to a point x. Then there is a map $U \times cL_x \to X$ which is a stratum-preserving homotopy equivalence near U.

There is a converse to this, in that such a homotopy locally conelike structure usually implies weakly stratified. But in fact it is one of the pleasant features of the theory that the structure of the full filtered space can be deduced from assumptions on how pairs of strata fit together. These simpler assumptions are often much easier to verify.

CONVERSE. Suppose X is filtered with locally contractible skeleta, and given $x \in X^i - X^{i-1}$ and $j > i$ then there is a neighborhood $U \subset X^i - X^{i-1}$, a space L, and a map $U \times cL \to (X^j - X^{j-1}) \cup U$ which is a stratum-preserving homotopy equivalence near U. Then X is weakly stratified.

Here both $U \times cL$ and $(X^j - X^{j-1}) \cup U$ have only two nonempty strata, and U is the smaller stratum in both.

PROOF OF THE CONVERSE. The definition of a weakly stratified set is a filtered space X so that for $k > i$, $(X^{i+1} - X^i) \subseteq (X^{k+1} - X^k) \cup (X^{i+1} - X^i)$ is tame and has homotopy link a fibration ([5, 3.1]). Both of these properties are local, so hold if each point in $(X^{i+1} - X^i)$ has a neighborhood U so that $U \subseteq (X^{k+1} - X^k) \cup U$ satisfies them. Tameness is preserved by stratum-preserving homotopy equivalences near U, and $U \subseteq U \times cL$ is certainly tame, so $U \subseteq U \times cL$ is certainly tame, so $U \subseteq (X^{k+1} - X^k) \cup U$ is tame. It also follows quickly from the definition that these equivalences induce fiber homotopy equivalences of homotopy links (see [5, 2.2, and the proof of 1.1 in §5]). Therefore it is sufficiently to show that holink $(U \times cL, U) \to U$ is a fibration. But the cone structure defines an inclusion $U \times L \to \text{holink}(U \times cL, U)$ which is a fiber homotopy equivalence over U. Since the product is a fibration, so is the holink.

PROOF OF THEOREM 2. We generalize the definition of the homotopy link [5,2.1]. If X is filtered and $Y \subseteq X$ then define the stratified homotopy link $\text{holink}_s(X,Y)$ to be the space of maps $\theta : [0,1] \to X$ such that $\theta^{-1}(Y) = \{0\}$, and $(0,1]$ is mapped into a single stratum. This is filtered by subsets for which $(0,1]$ is mapped into a particular skeleton of X. Evaluation at 0 defines a projection $p : \text{holink}_s(X,Y) \to Y$. Denote the mapping cylinder by Y_p, and define it precisely as $I \times \text{holink}_s(X,Y)/\simeq$, where the equivalence relation is given by $(0,\theta) \simeq \theta(0)$. Then Y_p is filtered by mapping cylinders of restrictions to skeleta of $\text{holink}_s(X,Y)$, with Y included as the lowest skeleton. Finally, evaluation defines a map $Y_p \to X$, by $(t,\theta) \to \theta(t)$. If Y is open in a stratum of X this map is stratum-preserving.

We recall ([5, §3]) that a neighborhood deformation retraction is "nearly" stratum-preserving if it preserves strata until it is required to map into the subspace. Precisely, it is a homotopy of a neighborhood $V \times [0,1] \to X$ which fixes Y, takes $V \times \{0\}$ into Y, is the inclusion on $V \times \{1\}$, and is stratum-preserving on $U \times (0,1]$. According to [5, prop. 3.2] if $Y \subseteq X^i - X^{i-1}$ is open then there is a nearly stratum-preserving deformation retraction of a neighborhood to Y. (The proposition is actually formulated for closed subsets. Applying it to $Y \subseteq (X - X^i) \cup Y$ gives the desired conclusion.) We use this to define a map from a neighborhood of Y into the mapping cylinder of the homotopy link: denote the deformation retraction by $r : V \times [0,1] \to X$, then $x \in V$ is mapped to $(d(x,Y), r : \{x\} \times I \to X)$. Here $d(x,Y)$ denotes the distance from x to Y, or 1 if this distance is greater than 1.

It is easily seen that evaluation $Y_p \to X$ and this map $V \to Y_p$ are stratum-preserving homotopy inverses near Y. The required homotopies can be written explicitly in terms of the data given (see [5, 2.2]).

The conclusion so far is that open sets in strata have neighborhoods with the stratified local homotopy type of mapping cylinders. The proof will be completed by analysing the map involved.

Suppose Y is a component of $X^i - X^{i-1}$, and U is open and contractible to x in Y. Suppose as an induction hypothesis that there is a stratum-preserving fiber homotopy equivalence $U \times L_x^k \to \mathrm{holink}_s(X^k, U)$, where L_x^k is the iterated join described before the statement of the theorem. The induction can start with $k = i$, and $L_x^k = \emptyset$. Define the stratified homotopy link of a triple, $\mathrm{holink}_s(X^{k+1}, X^k, Y)$, as in [5, 2.9]. Then there is a diagram

$$\begin{array}{ccc} \mathrm{holink}_s(X^{k+1}, X^k, Y) & \longrightarrow & \mathrm{holink}((X^{k+1} - X^k) \cup Y, Y) \\ \downarrow & & \downarrow \\ \mathrm{holink}_s(X^k, Y) & \longrightarrow & Y \end{array}$$

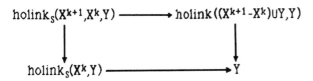

We now apply proposition 2.9 of [5]. The statement of this result is for unstratified homotopy links. However when the data is stratum preserving the proof produces stratum-preserving conclusions. The data required is a nearly stratum-preserving deformation retraction of a neighborhood in X to Y, covered by a fiber homotopy of homotopy links. The deformation retraction is provided by [5, prop. 3.2], and the fiber homotopy comes from [5, 3.3] and the fact [5, 3.2] that the homotopy links form stratified systems of fibrations. The conclusion of [5, 2.9(2)] is that the diagram is a pullback square over Y. By definition of weakly stratified, the right vertical is a fibration. By [5, 2.9 (3)], the homotopy pushout of the diagram is fiber homotopy equivalent to $\mathrm{holink}_s(X^{k+1},Y) \to Y$.

Now restrict to U. The restriction of $\mathrm{holink}_s(X,Y)$ to $p^{-1}(U)$ is exactly $\mathrm{holink}_s(X,U) \to U$. The restriction of a fibration to a contractible subset is fiber homotopy equivalent to a product, so the right vertical map is fiber homotopy equivalent to $U \times L_{k+1} \to U$. The lower map is $U \times L_x^k \to U$ by the induction hypothesis. The pullback of these is fiber homotopic to $U \times L_x^k \times L_{k+1}$, with the obvious product projections. The homotopy pushout of these projections gives $U \times L_x^k * L_{k+1}$, which is by definition $U \times L_x^{k+1}$. Since this pushout is fiber homotopy equivalent to $\mathrm{holink}_s(X^{k+1},U) \to U$, this completes the induction step.

3. INTERSECTION HOMOLOGY

In this section we recall the definition of (singular) intersection homology of a filtered space, and show it is invariant under certain coarsenings of the filtration. The "topological invariance" for manifold weakly stratified sets follows from this because it was shown in section 1 that any two manifold weakly stratified filtrations have a common coarsening.

A "perversity" p is a sequence of integers $p_0, p_1, \cdots$, which we assume satisfies $p_i \leq p_{i+1} \leq p_i + 1$, and $p_1 = 0$ ([1-4]).

Suppose X is a filtered space with dimension n. A p-singular i-simplex in X is a map $f : \Delta^i \to X$ so that $f^{-1}(X^k - X^{k-1})$ is contained in the $k + i - n + p_{n-k}$ skeleton of Δ^i. If G is a group then a p-singular i-chain (with G coefficients) is a G linear combination of p-singular i-simplices in X, whose boundary (in the chain complex sense) is a p-singular (i-1)-chain. Denote by $IC_*^p(X,G)$ the chain complex of p-singular chains and by $IH_*^p(X,G)$ the homology of this complex.

It is shown in [4] that this definition gives the compactly supported theory of [2, part I].

THEOREM 3. Suppose $X_1 \to X_2$ is a coarsening of ANR weakly stratified sets, so that if K is a component of the s-stratum of X_1 and lies in the r-stratum of X_2, then the fiber of $\mathrm{holink}((X_2)^r - (X_2)^{r-1},K) \to K$ is nonempty and has trivial reduced homology through dimension (r-s-2). Then the induced homomorphism of intersection homology groups is an isomorphism.

COROLLARY. If X is a manifold weakly stratified set (filtered by dimension), then the coarsening $X \to X_{0,0}$ induces isomorphisms of intersection homology. Consequently the intersection homology is independent of the choice of manifold weakly stratified filtration of $|X|$.

PROOF OF THE COROLLARY. The second statement follows from the first because theorem 1 states that if $|Y| = |X|$ then $Y_{0,0} = X_{0,0}$. The first statement follows from theorem 3 because the coarsening $X \to X_{0,0}$ satisfies the hypothesis of the theorem. It is a little easier to verify that the intermediate coarsenings $X_{r,s+1} \to X_{r,s}$ used in the definition of $X_{0,0}$ satisfy the hypothesis. In this case the only K which need checking are components of the s-stratum which are promoted into the r-stratum. One of the conditions for promotion is that $\text{holink}((X^{s+1}_{r,s+1} - X^{s}_{r,s+1}) \cup K, K) \to K$ have fiber a homology sphere. Comparison of duality in K and $(X^{s+1}_{r,s+1} - X^{s}_{r,s+1})$ shows that the fiber must be a homology (r-s-1)-sphere, therefore is (r-s-2) acyclic.

PROOF OF THE THEOREM. We begin with some generalities, because the explicit dependence of the groups on the indexing of the filtration requires more care in the description of how things are filtered. In sections 1 and 2 the collection of skeleta was important, and the indexing essentially a matter of convenience. For example a "stratum-preserving" map is required to take components of strata into components of strata but not necessarily of the same index.

In this section we include as part of the data of a filtered space a "dimension" such that $X^{\dim X} = X$. The "codimension k" skeleton is then $X^{\dim X-k}$. A map is "codimension-preserving" or "codimension-decreasing" if it maps strata into strata of equal codimension, or into the union of strata of less or equal codimension. For example a coarsening of a filtration is codimension decreasing. We caution that this formal dimension, codimension, etc. may not have any relation to geometric dimensions. For example in section 2 and below cone points are assigned a positive "dimension".

It follows directly from the definition that a codimension-decreasing map induces a chain map of p-singular chain complexes, hence of intersection homology groups. Further, if $Y \to X$ is codimension decreasing then we denote (somewhat imprecisely) the algebraic mapping cone of the induced chain map by $IC^{p}_{*}(X,Y;G)$. The homology of this complex is similarly denoted $IH_*^p(X,Y;G)$, and by definition fits into a long exact sequence with the homology of X and Y.

We now begin the details of the proof of the theorem. Proceed by induction on codimension: suppose the statement is true for coarsenings in which the codimension k skeleton of X_1 is empty, and verify it when the codimension k+1 skeleton is empty. Let dim X = n, and $K = X_1^{n-k}$. Compare the long exact sequences of the pairs $(X_1, X_1-K) \to (X_2, X_2-K)$. The map $X_1 - K \to X_2 - K$ satisfies the induction hypothesis, so is an isomorphism on intersection homology. By the 5-lemma, to conclude $X_1 \to X_2$ is an isomorphism it is sufficient to show the map of pairs induces an isomorphism.

The next reduction comes from the Mayer-Vietoris sequence. Suppose U, V are open in X. Then there is a natural exact sequence

$$\dashrightarrow IH_i^p(U\cap V, U\cap V-K) \longrightarrow IH_i^p(U,U-K) \oplus IH_i^p(V,V-K) \longrightarrow IH_i^p(U\cup V, U\cup V-K) \longrightarrow .$$

Comparing these sequences for the two filtrations, then again by the 5-lemma we conclude that $U \cup V$ satisfies the theorem provided the two filtrations on U, V, and $U \cap V$ do. By applying this observation repeatedly we see it is sufficient to cover K by open sets U_α so that they and all their intersections satisfy the theorem. Since we have assumed the strata are ANRs, which are locally contractible, the theorem will follow from the next lemma.

LEMMA. Suppose $U \subseteq X$ is open and $U \cap X_1^{n-k}$ is contractible in X_1^{n-k}. Then $(U_1, U_1 - X_1^{n-k}) \to (U_2, U_2 - X_1^{n-k})$ induces an isomorphism on intersection homology.

Here U_i denotes the space U with the filtration induced from X_i.

PROOF OF THE LEMMA. Denote $U \cap X_1^{n-k}$ by Y, and denote by x a point in the stratum to which Y contracts. Then according to theorem 2, there is a map $Y \times cL_x \to U_1$ which is a stratum-preserving homotopy equivalence near Y. Slightly more precisely, this equivalence is filtration-preserving. Since U_2 is a coarsening of U_1, the same maps and homotopies give an equivalence near Y to Y product with a coarsening of the filtration on cL_x. Denote this coarsening by $(cL_x)_2$.

Next observe that the filtration-preserving maps of pairs $(Y \times (cL_x)_i, Y \times (cL_x)_i - Y) \to (U_i, U_i - Y)$ for i = 1,2 induce isomorphisms of homology groups because they are homotopy equivalences near Y. Therefore to prove the lemma we show $(Y \times (cL_x)_1, Y \times (cL_x)_1 - Y) \to (Y \times (cL_x)_2, Y \times (cL_x)_2 - Y)$ induces isomorphisms. This map is the identity of Y product with $((cL_x)_1, (cL_x)_1 - c) \to ((cL_x)_2, (cL_x)_2 - c)$, and we claim it is sufficient to see this map of cones induces isomorphisms.

The claim follows from: if $(A_1,B_1) \to (A_2,B_2)$ is a filtration-increasing map which induces an isomorphism on intersection homology, and Y is a CW complex, then $(Y \times A_1, Y \times B_1) \to (Y \times A_2, Y \times B_2)$ also induces an isomorphism on intersection homology. Here Y is unfiltered, and the k-skeleton of a product with Y is the product of Y with the k-skeleton. In the application Y is an open set in an ANR, hence has the homotopy type of a CW complex. Changing Y by homotopy equivalence does not change the homology, so we may assume Y is CW.

This fact is proved by induction on dimension. Suppose it is known for k-complexes. A k+1 complex is covered by a neighborhood of the k-skeleton and the interiors of the k+1 cells. Applying the Mayer-Vietoris sequence to the corresponding product covers of $(Y \times A_i, Y \times B_i)$, we see that the 5-lemma gives isomorphism for the k+1 complex if it is known for the neighborhood of the k-skeleton, interiors of cells, and $S^k \times \mathbb{R}$. These are homotopy equivalent to the k-skeleton, points, and S^k respectively. As above, changing Y by homotopy equivalence does not change homology, so the induction hypothesis applies to all of these.

The proof is now reduced to showing that $((cL_x)_1, (L_x)_1) \to ((cL_x)_2, (L_x)_2)$ induces an isomorphism on intersection homology. The first step is to deduce from the induction hypothesis (that the theorem holds for codimension < k) that the coarsening $(L_x)_1 \to (L_x)_2$ induces isomorphism on homology.

Let $f : Y \times cL_x \to U$ be the filtration-preserving homotopy equivalence near Y, and let $g : V \to Y \times cL_x$ be its "inverse", where V is a neighborhood of Y in U. Define $f^\sim : Y \times cL_x \to V$ by restricting f to a subcone. By deleting Y we get maps

$$Y \times L_x \xrightarrow{f^\sim} V - Y \xrightarrow{g} Y \times L_x \xrightarrow{f} U - Y$$

which since they preserve both filtrations, induce homomorphisms on intersection homology

$$IH^p(Y \times (L_x)_i) \xrightarrow{f^\sim} IH^p(V_i - Y) \xrightarrow{g} IH^p((Y \times L_x)_i) \xrightarrow{f} IH^p(U_i - Y),$$

for i = 1,2. Since $g\tilde{f}$ is filtration-preserving homotopic to the identity, and since fg is homotopic to the inclusion, we conclude that $IH^p(Y \times (L_x)_i)$ is naturally isomorphic to the image of the inclusion $IH^p(V_i - Y) \to IH^p(U_i - Y)$.

Next consider

$$\begin{array}{ccc} V_1\text{-}Y & \longrightarrow & U_1\text{-}Y \\ \downarrow & & \downarrow \\ V_2\text{-}Y & \longrightarrow & U_2\text{-}Y \end{array}$$

By the induction hypothesis the vertical maps induce isomorphisms on intersection homology, so the horizontal maps have isomorphic images. According to the above this implies that $IH^p(Y \times (L_x)_1) \to IH^p(Y \times (L_2)_2)$ is an isomorphism. But this has $IH^p((L_x)_1) \to IH^p((L_x)_2)$ as a natural retract, so this is also an isomorphism.

Next consider the transformation between long exact sequences of pairs, induced by

$$\begin{array}{ccccc} (L_x)_1 & \longrightarrow & (cL_x)_1 & \longrightarrow & (cL_x)_1, (L_x)_1 \\ \downarrow & & \downarrow & & \downarrow \\ (L_x)_2 & \longrightarrow & (cL_x)_2 & \longrightarrow & (cL_x)_2, (L_x)_2 \end{array}$$

It is our goal to show that the right vertical induces isomorphisms. We have seen that the left vertical map induces isomorphisms, so it is sufficient to show that the center map (of cones) induces isomorphisms.

The intersection homology of cones is calcuated in [1-4]. If A is a filtered space with dimension n and $A^{n-k} = \emptyset$, and cA is filtered by cones on skeleta of A with the cone point as n-k skeleton, then the inclusion $A \to cA$ induces an isomorphism of IH_i^p for $i < n-p_{k+1}$, and $IH_i^p(cA) = 0$ for $i \geq n-p_{k+1}$. (The result in the references looks slightly different because the cone point is assumed to have filtration 0. This formulation follows formally by reindexing the filtration.) The filtration on $(cL_x)_1$ is obtained in this way.

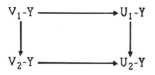

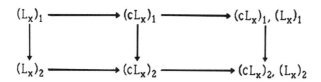

The proof now divides in two cases; either Y is contained in $(X_2)^{n-k}$, or in a higher dimensional stratum. In the first case the filtration $(cL_x)_2$ on the cone is also the cone on the filtration $(L_x)_2$, with the cone point included as the n-k skeleton. In the second case the cone point is included in a higher dimensional skeleton. In the first case since $(L_x)_1 \to (L_x)_2$ induces isomorphism, the calculations for cones implies $(cL_x)_1 \to (cL_x)_2$ does also. The proof is therefore complete in this case.

It is sufficient in the second case to consider $X_1 \to X_2$ so that the two filtrations agree on $X_1 - (X_1)^{n-k}$. In general define an intermediate coarsening, $X_1 \to X_3 \to X_2$, so that the skeleta of X_3 are the same as X_2 except that $(X_1)^{n-k}$ is included as the n-k skeleton of X_3. The "first case" discussed above applies to $X_1 \to X_3$, so we are reduced to considering $X_3 \to X_2$. This satisfies the conditions of the simplified second case. In this situation the two filtrations on the link L_x are the same, and the cones differ only in the filtration of the cone point.

According to the construction of L_x in theorem 2, $L_x = L_{n-r} * L^\sim$, where $L^\sim$ is the filtered space constructed from joins of links in higher strata, and n-r is the lowest non-empty stratum containing Y in its closure. Up to this point only the induction hypothesis and the general structure of weakly stratified sets have been used. Now we use the special hypothesis. Suppose Y lies in the n-q stratum of X_2. The fiber of the homotopy link at x of $(X_1)^{n-k}$ in $(X_2)^{n-q}$ is assumed to be non-empty and have trivial reduced homology up through dimension k-q-2. Since it is nonempty it must be that q = r, and L_{n-r} is this link.

We can now describe the filtration $(cL_x)_2$. There is a homeomorphism $c(L_{n-r} * L^\sim) \simeq c(L_{n-r}) \times c(L^\sim)$. Filter $c(L^\sim)$ by cones on the filtration of $L^\sim$ and with n-r skeleton the cone point, and filter the product of cones by products with this filtration. This agrees with the previous filtration except that the cone point has been promoted into the n-r skeleton, so is $(cL_x)_2$. Note that since $c(L_{n-r})$ is contractible, the projection $c(L_{n-r}) \times c(L^\sim) \to c(L^\sim)$ is a filtration-preserving homotopy equivalence. Therefore what we want to show is that $c(L_{n-r} * L^\sim) \to c(L^\sim)$ (with the filtrations specified above) induces isomorphism on intersection homology.

Let $y \in L_{n-r}$, then $c(L\ ) \simeq y * L^\sim \to L_{n-r} * L^\sim$ is a filtration-preserving map which when included in $c(L_{n-r} * L^\sim)$ gives a left inverse for the projection. It is sufficient to show that $y * L^\sim \to L_{n-r} * L^\sim$ is an isomorphism on IH_{p_i} for $i < n - P_{k+1}$, since the cone calculation shows $L_{n-r} * L^\sim \to c(L_{n-r} * L^\sim)$ is an isomorphism in this range, and the homology of both cones vanishes above this. Next consider the diagram

$$\begin{array}{ccccc} cy\times L^{\sim} & \longrightarrow & y*L^{\sim} & \longrightarrow & y*L^{\sim}, cy\times L^{\sim} \\ \downarrow & & \downarrow & & \downarrow \\ cL_{n-r}\times L^{\sim} & \longrightarrow & L_{n-r}*L^{\sim} & \longrightarrow & L_{n-r}*L^{\sim}, cL_{n-r}\times L^{\sim} \end{array}$$

where we have written $L_{n-r} * L^{\sim}$ as $(cL_{n-r} \times L^{\sim}) \cup (L_{n-r} \times cL^{\sim})$, and $y * L^{\sim}$ similarly. Applying homology gives long exact sequences of pairs horizontally. The top sequence is a retract of the bottom, so its homology is a direct summand. The left vertical induces isomorphism because cL_{n-r} is contractible. This shows the center vertical (which we want to know) is an injection with cokernal the same as the right vertical. By excision we can replace the right vertical by $y \times (cL^{\sim},L^{\sim}) \to L_{n-r} \times (cL^{\sim},L^{\sim})$. The homology cokernel of this is the homology of $(L_{n-r},y) \times (cL^{\sim},L^{\sim})$ (where this indicates the pair $(L_{n-r} \times cL^{\sim},\ L_{n-r} \times L^{\sim} \cup y \times cL^{\sim})$). Therefore we need to show that $IH_i^p(L_{n-r},y) \times (cL^{\sim},L^{\sim}) = 0$ if $i < n - p_{k+1}$.

Since the cone point in cL has filtration n-r, $IH_i^p(cL^{\sim},L^{\sim}) = 0$ if $i \leq n - p_{r+1}$. By hypothesis L_{n-r}, (which is unfiltered) satisfies $H_j(L_{n-r},y) = 0$ if $j \leq r - k - 2$. It follows from the Kunneth formula ([4], theorem 4) that $IH_i^p(L_{n-r},y) \times (cL^{\sim},L^{\sim}) = 0$ if $i \leq (n - p_{r+1}) + (r-k-2) + 1$. But it follows from the conditions on a perversity that $-p_{r+1} + r-k-1 \geq -p_{k+1} - 1$, so the homology does vanish for $i < n - p_{k+1}$ as required.

This completes the proof in the second case, and therefore the proof of the theorem. We remark on the use of the Kunneth formula in the last step. First, only the vanishing result is used, which is quite a bit weaker than the general formula. Second, the filtrations here and in [4] are different, because in [4] an effort is made to filter by dimension. If we start with a filtered space X and filter $M \times X$ by $(M \times X)^k = M \times (X^k)$ (rather than increasing the index to allow for the dimension of M) then the proof given in [4] works for M with the homotopy type of a CW complex.

ACKNOWLEDGEMENT

The author was partially supported by the National Science Foundation.

REFERENCES

1. Borel, A., et al, Intersection Cohomology, Progress in math. vol 50, Birkhauser 1984.

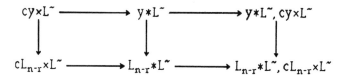

2. Goresky, M., and MacPherson, R., Intersection Homology Theory, Topology 19 (1980), 135-162, part II, Invent. Math 72 (1983), 77-130.
3. King, H., Intersection Homology and Homology Manifolds, Topology 21 (1982), 229-234.
4. King, H., Topological Invariance of Intersection Homology Without Sheaves, Topology and Applications, 20 (1985), 149-160.
5. Quinn, F., Weakly Stratified Sets, preprint 1985.
6. Quinn, F., Ends of Maps, I, Ann. Math. 110 (1979), 275-331. II, Invent. Math. 68 (1982), 353-424. III, J. Diff. Geom. 17 (1982), 503-521. IV, Am. J. Math, (to appear).
7. Quinn, Resolutions of Homology Manifolds, and the Topological Characterization of Manifolds, Invent. Math. 72 (1983), 267-284, ____, Erratum, ibid, (to appear.)
8. Siebenmann, L., Deformation of Homeomorphisms on Stratified Sets, Comment. Math. Helv. 47 (1972), 123-163.

Isolated Critical Points of Maps from R^4 to R^2 and a Natural Splitting of the Milnor Number of a Classical Fibred Link, Part II

LEE RUDOLPH / Adamsville, Rhode Island

0. INTRODUCTION; STATEMENT OF RESULTS

Let K be a fibred link in a 3-sphere in C^2, (K,ϕ_K,ψ_K) an open-book structure on the sphere. There is a field S_K of (not everywhere tangent) oriented real 2-planes on the sphere, built from (K,ϕ_K,ψ_K); homotopically S_K gives integers $(\lambda(K),\rho(K))$ which depend only on K. Here are some facts about $\lambda(K)$ and $\rho(K)$, proved in Part I of this paper [12]: the Milnor number $\mu(K)$ of K (rank of the first homology of any fibre surface $F_t = \psi^{-1}(\exp it)$) equals $\lambda(K) + \rho(K)$; if (K,ϕ_K,ψ_K) is a braided open-book structure on the (smoothed) bidisk boundary, then the set pos(K) (resp., neg(K)) of points Q in E(K), at which the oriented tangent 2-plane to the fibre surface is a complex line with its complex (resp., conjugate-complex) orientation, is an oriented 1-cycle in E(K); and $\lambda(K) = \ell k(K,\text{neg}(K)) + \ell k(\text{pos}(K),\text{neg}(K))$, $\rho(K) = \lambda(\text{Rev } K)$ (where Rev K is the mirror image of K). (See §1.)

In this Part, I show how $\lambda(K)$ and $\rho(K)$ can be computed for a totally braided open-book structure in terms of its derived closed braid (§2). At least when the derived braid is trivial, this simplifies their calculation. The closed strictly homogeneous braids of [13] (and a mild generalization thereof) are easily seen to have totally braided open-book structures with trivial derived closed braids (§3), and in this case the splitting of $\mu(K)$ has an interesting interpretation--$\lambda(K)$ and $\rho(K)$ turn out to be ranks of geometrically natural complements in $H_1(F_t;Z)$, corresponding to "negative" (or left-handed) and "positive" (or right-handed) parts of the fibre surface.

The paper ends (§4) with a proof of the curious fact that for some closed homogeneous braids (positive, negative, or 3-stringed), λ and ρ can be read off from the generalized Jones polynomial.

Some of this work was done at M.S.R.I., Berkeley, California.

1. REVIEW OF PART I (DEFINITIONS AND BASIC FACTS)

1.1. SPACES. The 3-sphere S^3 is the unit sphere of C^2. The 1-sphere S^1 is the boundary of the unit disk D^2 of C. These spaces all have their conventional orientations. The <u>smoothed bidisk</u> is

$$D = \{(z,w) \in C^2 : |z| \leq 1, |w| \leq 1 + h(|z|^2)\}$$

(oriented like C^2), where $h : [0,1] \to [0,1]$ is a homeomorphism, smooth on $[0,1[$, with $h(0) = 1$, $h'(0) = 0$, $h''(0) \neq 0$, $h(1) = 0$, and h^{-1} smooth and infinitely flat at 0; ∂D is diffeomorphic to S^3, and is given as the union of solid tori $\partial_1 D = \{(z,w) : |z| = 1, |w| \leq 1\} = S^1 \times D^2$ and $\partial_2 D = \partial D \backslash (\text{int } \partial_1 D)$.

1.2. BRAIDS. An oriented link K in Int $\partial_1 D$ is a <u>closed braid on n strings</u> if $pr_1|K : K \to S^1$ is an orientation preserving covering projection of degree n; K is a <u>trivial closed braid</u> if it is isotopic through closed braids to $S^1 \times \{w_1, \cdots, w_n\}$. An <u>adapted tubular neighborhood</u> N(K) of a closed braid K is one of the form $\{(z,w) + \delta(0,\eta) : (z,w) \in K, \eta \in D^2\}$ (for $\delta > 0$ sufficiently small).

1.3. OPEN BOOKS. A <u>braided open-book structure</u> on ∂D is a triple (K,ϕ_K,ψ_K) where K is a closed braid with adapted tubular neighborhood N(K) and exterior $E(K) = \partial D/\text{Int } N(K)$, and $\phi : N(K) \to K \times D^2$ and $\psi : E(K) \to S^1$ are smooth maps such that:

(OB1') $\phi(z,w+\delta\eta) = ((z,w),\xi(z,w)\eta)$ for $(z,w) \in K$, with $\xi : K \to S^1$ smooth (we call $\phi(Q\times D^2)$ the <u>meridional disk of</u> K at $Q \in K$);

(OB2) ψ is a fibration of E(K) by surfaces-with-boundary $F_t = \psi^{-1}(\exp it)$ such that the restrictions $\psi|\partial E(K)$ and $pr_2 \circ \phi|\partial N(K)$ agree on $\partial E(K) = \partial N(K)$ (we call F_t the tth <u>page</u>); and

(OB3) the map $\partial D\backslash K \to S^1$, which equals ψ on E(K) and $(z,w+\delta\eta) \rightsquigarrow \xi(z,w)\eta/|\eta|$ on $N(K)\backslash K$, is smooth.

If (K,ϕ_K,ψ_K) is a braided open-book structure then K is a fibred link; (the ambient isotopy type of) K is the <u>type</u> of braided open-book structure, and it is easy to see that there are braided open-book structures of every fibred link type. A fundamental braided open-book structure is (O, ϕ_O, ψ_O), the unbook: here O is the horizontal unknot $S^1 \times \{0\}$, $\phi_O(z,w) = ((z/|z|,0),w)$ for $(z,w) \in \partial_1 D = N(O)$, and $\psi_O(z,w) = w/|w|$ for $(z,w) \in \partial_2 D = E(O)$. The pages of the unbook are disks.

1.4. BRAIDED SURFACES. A surface F in ∂D (smooth, oriented, and without closed components) is a braided surface of degree n if it satisfies:

(BrS1) ∂F is a closed braid on n strings;

(BrS2) $F \cap E(O)$ is the union of n pages of (O, ϕ_O, ψ_O) (with their correct orientations); and

(BrS3) the real-valued function $(1/2\pi i)\log(z/|z|)|(\text{Int } F \cap N(O))$ (which is locally well-defined) is Morse without local extrema.

If $(1/2\pi i)\log(z/|z|)|(\text{Int } F \cap N(O))$ has k critical points on F, then F can be given a handle decomposition with n 0-handles (enlarged pages of the unbook) and k 1-handles ("bands") connecting them.

1.5. BASIC DEFINITION. Let (K, ϕ_K, ψ_K) be a braided open-book structure. Then pos(K) (resp., neg(K)) is the set of points Q in E(K) at which the oriented tangent plane to the page through Q is a complex line with its complex (resp., conjugate-complex) orientation. (Of course, pos and neg really depend on all the data of the braided open-book structure, not just K; the abuse of notation is mild.)

1.6. BASIC FACTS ([12],§§1-2). Both pos(K) and neg(K) are naturally oriented absolute 1-cycles in E(K); in fact, it is always possible (after a slight perturbation of ψ_K) to assume that they are oriented links. The integers $\lambda(K) = \ell k(K + \text{pos}(K), \text{neg}(K))$ and $\rho(K) = 1 - \ell k(\text{pos}(K), K + \text{Neg}(K))$ depend only on K, and $\lambda(K) + \rho(K) = \mu(K)$ is the Milnor number of K (the rank of the first integral homology group of a page of K). Let $\text{Rev} : \partial D \to \partial D : (z,w) \rightsquigarrow (z,\bar{w})$, so Rev(K) (the mirror image of K) is a fibred link if K is; then $\rho(K) = \lambda(\text{Rev}(K))$.

2. TOTALLY BRAIDED OPEN-BOOK STRUCTURES

2.1. DEFINITION (cf. [3], [1], [6]). Let (K, ϕ_K, ψ_K) be a braided open-book structure on ∂D. A closed K-braid is an oriented link B in ∂D which lies in Int E(K), such that $\psi_K|B : B \to S^1$ is an orientation-preserving covering projection.

2.2. DEFINITION. A <u>totally braided open-book structure</u> is a braided open-book structure for which each page is a braided surface. If (K,ϕ_K,ψ_K) is totally braided, then $\psi_K|\partial_2 D$ is a fibration and each of its fibres is the union of n pages of the unbook; with no loss of generality <u>we assume that</u> for $(z,w) \in \partial_2 D$, $\psi_K(z,w) = (\psi_0(z,w))^n$.

2.3. PROPOSITION. Let (K,ϕ_K,ψ_K) be totally braided. Then: (1) the 1-cycles pos(K) and neg(K) are smooth 1-manifolds (i.e., they have multiplicity 1 at each of their points); (2) pos(K) $\cap\ \partial_2 D$ is the <u>vertical unknot</u> $O' = \{0\} \times 2S^1$ (oriented as the intersection of ∂D with the complex line $\{z = 0\}$); (3) neg(K) $\cap\ \partial_2 D$ is empty. Furthermore, with their natural orientations, (4) pos(K) $\cup$ neg(K) is a closed K-braid; (5) $\mathrm{pos}_0(K) = \mathrm{pos}(K)\setminus O'$ is a closed braid (or empty); and (6) neg(K) is a closed braid with orientation reversed (or empty).

Proof: A point Q of E(K) belongs to pos(K) $\cup$ neg(K) iff the tangent plane at Q to the page of (K,ϕ_K,ψ_K) through Q is a complex line (up to orientation); and, when the order of contact between this plane and the page is exactly 1, Q has multiplicity 1 in pos(K) $\cup$ neg(K).

In $\partial_1 D = N(O)$, the unique complex line in the tangent space to ∂D at any point Q is (everywhere) tangent to the meridional disk of O through Q (this disk itself lies in an affine complex line); so $Q \in N(O) \cap E(K)$ belongs to pos(K) $\cup$ neg(K)) iff the page F of (K,ϕ_K,ψ_K) through Q is tangent to the meridional disk of O through Q, iff $Q \in \mathrm{Int}\, F$ and the real-valued, locally well-defined smooth function $(z,w) \rightsquigarrow (1/2\pi i)\log(z/|z|)\,|\,(\mathrm{Int}\, F \cap N(O))$ has a critical point at Q. By (BrS3), this function is Morse, without local extrema, on each page F; so the critical point is a saddle and the tangency is exactly first order. This proves (1) in N(O). In $\partial_2 D = E(O)$, the pages of (K,ϕ_K,ψ_K) have the same tangent planes as the pages of the unbook. The tth page of the unbook is parameterized by $\eta \rightsquigarrow (\eta,[1+h(|\eta|^2)]\exp it)$, $|\eta| \leq 1$; it is easy to prove (2) and (3), and complete the proof of (1).

Saddlepoints are stable: thus, as one passes from one page of (K,ϕ_K,ψ_K) or of (O,ϕ_O,ψ_O) to the next, the points of pos(K) $\cup$ neg(K) on the page can wander around but they cannot be created or destroyed, nor can they merge with each other or wander off the edge (by (BrS1)); so, up to orientation, (4), (5), and (6) are true.

The natural orientation of pos(K) $\cup$ neg(K) is that which makes (4) true. Then the orientation assertions in (5) and (6) follow by noting that at $Q \in (\text{pos}(K) \cup \text{neg}(K)) \cap N(O)$ the normal directions to the meridional disk of O and the page of K are equal or opposite depending on whether Q is in pos(K) or neg(K).

2.4. DEFINITION. Let (K, ϕ_K, ψ_K) be totally braided. The <u>derived closed braid</u> of K is $\text{der}(K) = \text{pos}_0(K) \cup (-\text{neg}(K))$.

2.5. NOTATION. Let $O^* = O$, and let (O^*, ϕ_O, ψ_O) be the totally braided open-book structure with $N(O^*) = S^1 \times (1/2)D^2$, $\phi_{O^*}(z,w) = \phi_O(z,2w)$, $\psi_{O^*}(z,w) = w/|w|$. (This is just the unbook with enlarged pages.)

2.6. PROPOSITION. Let $P : \partial D \to \partial D$ be a simple branched cover of degree n such that: (0) $P(\partial_j D) = \partial_j D$, j = 1, 2; (1) $P|\partial_2 D : (z,w) \rightsquigarrow (z, w^n/|w|^{n-1})$; (2) $pr_1 \circ (P|\partial_1 D) = pr_1$; (3) $P^{-1}(N(O^*))$ is an adapted tubular neighborhood N(K) of $K = P^{-1}(O^*)$ (a closed braid by (2)); and (4) P(crit(P)) can be oriented as a closed braid and as a closed O^*-braid. Define ϕ_K and ψ_K by (5) $\phi_{O^*} \circ P \circ \phi_K^{-1} = P \times \text{id} : K \times D^2 \to O^* \times D^2$ and (6) $\psi_K = (\psi_{O^*}) \circ P : E(K) \to S^1$. Then (K, ϕ_K, ψ_K) is a totally-braided open-book structure, and the unoriented link underlying der(K) is crit(P).

Proof: Given (0) - (2), for (6) to define a fibration it suffices that every page of O^* meet P(crit(P)) transversely, that is, that P(crit(P)) have an orientation as a closed O^*-braid--half of (4). And given (3), (5) does define a trivialization of the adapted tubular neighborhood. So (0) - (4) certainly assure that (K, ϕ_K, ψ_K) is an open-book structure. But the other half of (4)--that P(crit(P)) has an orientation as a closed braid--ensures that each page of (K, ϕ_K, ψ_K) is a braided surface.

2.7. REMARKS. (1) Of course hypothesis (3) is purely technical; the really key hypothesis is (4). (2) The converse to 2.6 seems to be true (that is, any totally-braided open-book structure arises as a "pull-back" of the unbook with enlarged pages); however, I have not found a convincing proof. (I thank the referee for pointing out that the proof proposed in the previous draft was wrong.) (3) Totally braided open-book structures of a specified type seem rather hard to find. The generalization of Alexander's Theorem to surfaces [6] shows that any fibred link type can be realized by a braided open-book structure with ONE page braided. A theorem of [1] shows that every type of braided open-book structure is the pull-back of $(O^*, \phi_{O^*}, \psi_{O^*})$ through a simple branched cover $P : \partial D \to \partial D$ (even, if desired, of degree 3) for which P(crit(P)) is orientable as a closed O*-braid; but it is unclear how to impose on P the further condition that P(crit(P)) be orientable as a closed braid.

2.8. CALCULATION. For any braided (K,ϕ_K,ψ_K), we have (according to 1.6) that $\lambda(K) = \ell k(K,\text{neg}(K)) + \ell k(\text{pos}(K),\text{neg}(K))$; so (by 2.3) if (K,ϕ_K,ψ_K) is totally braided, then we have

$$\lambda(K) = \ell k(K,\text{neg}(K)) + \ell k(O',\text{neg}(K)) + \ell k(\text{pos}_0(K),\text{neg}(K)). \qquad (*)$$

The first two terms on the right side are readily available (each is a "one page computation"): $\ell k(K,\text{neg}(K))$ is the number of strings of the closed K-braid neg(K), and $-\ell k(O',\text{neg}(K))$ the number of strings of the closed braid -neg(K). On the other hand, $\ell k(\text{pos}_0(K),\text{neg}(K))$ is a global quantity --one needs "the whole (open) book" to compute it. Now let P be as in (2.6); by construction, the number of strings of the closed K-braid neg(K) is the number of strings of the closed O*-braid P(neg(K)), and the number of strings of the closed braid -neg(K) is the number of strings of the closed braid P(-neg(K)). We arrive at the formula

$$\lambda(K) = \ell k(O,P(\text{neg}(K))) - \ell k(O',P(\text{neg}(K))) + \ell k(\text{pos}_0(K),\text{neg}(K)). \qquad (**)$$

This still has the term $\ell k(\text{pos}_0(K),\text{neg}(K))$, now doubly obnoxious because it has to be computed in a different space (the covering ∂D). HOWEVER, if P(der(K)) is a closed trivial braid, then certainly der(K) is trivial, and $\ell k(\text{pos}_0(K),\text{neg}(K))$ is 0. This special, but interesting, case is the topic of the next section.

3. CALCULATION OF λ AND ρ FOR CLOSED STRICTLY T-HOMOGENEOUS BRAIDS

3.1. CONSTRUCTION. Let $p : D^2 \to D^2$ be a smooth simple branched covering of degree $n > 0$ which restricts to $w \rightsquigarrow w^n$ on a neighborhood of S^1; so p has n-1 pairwise distinct critical points w_j in Int D^2, at any one of which it locally looks like $w \rightsquigarrow (w-w_j)^2$, while the critical values $\eta_j = p(w_j)$ are also pairwise distinct. We may assume 0 is a regular value of p. For $j = 1, \cdots, n-1$, let I_j' be a simple arc with endpoints 0 and η_j which contains no critical value other than η_j. Then $p^{-1}(I_j')$ is the disjoint union of n-1 simple arcs, one of which--call it I_j--contains w_j as an interior point and has two points of $p^{-1}(0)$ as its endpoints. The arcs I_j' may be chosen disjoint except for their common endpoint 0; then the union of the I_j is a tree T. By an appropriate choice of arcs I_j', any combinatorial type of tree on the n vertices $p^{-1}(0)$ can be realized as T.

Now let $P : \partial D \to \partial D$ be the smooth simple branched covering of degree n which takes $(z,w) \in \partial_1 D$ to $(z,p(w))$ and $(z,w) \in \partial_2 D$ to $(z,w^n/|w|^{n-1})$. The set crit(P) of critical points of P is $S^1 \times \{w_1, \cdots, w_{n-1}\}$, and the set of critical values P(crit(P)) is $S^1 \times \{\eta_1, \cdots, \eta_{n-1}\}$; of course, each of these links, when appropriately oriented, is a trivial closed braid on n-1 strings.

3.2. DEFINITION ([11]). A closed prebraid (implicitly, with respect to the covering P) is a 1-string closed braid Γ in $\partial_1 D$ which is disjoint from crit(P). The closed braid $\hat{\beta}(\Gamma)$ of a closed prebraid Γ is its total inverse image $p^{-1}(\Gamma)$; it is, in fact, a closed braid (on n strings). A criticized closed prebraid is the union of crit(P) and a closed prebraid Γ, oriented so as to be a closed braid (on n strings).

3.3. REMARKS. (1) Of course a closed prebraid is just an unknot; it is its placement in the criticized closed prebraid that is of real interest. (2) It is shown in [9] and [11] that up to isotopy through n-string closed braids every n-string closed braid is the braid of some closed prebraid; we won't need this.

3.4. DEFINITION. If there is a totally braided open-book structure $(\Gamma, \phi_\Gamma, \psi_\Gamma)$ for the closed prebraid Γ such that P(crit(P)) has an orientation making it a closed Γ-braid, then Γ is a closed strictly T-homogeneous prebraid, and $\hat{\beta}(\Gamma)$ is a closed strictly T-homogeneous braid. A closed braid which is strictly T-homogeneous for some T is simply a generalized strictly homogeneous braid.

3.5. PROPOSITION. If $K = \hat{\beta}(\Gamma)$ is a closed generalized strictly homogeneous braid on n strings, then there is a totally braided open-book structure on ∂D of type K for which der(K) is a trivial closed braid on n-1 strings.

Proof: There is an ambient isotopy of ∂D, supported in $\partial_1 D$, and respecting its product structure, which carries the totally braided open-book structure $(\Gamma, \phi_\Gamma, \psi_\Gamma)$ (for which P(crit(P)) is orientable as a closed Γ-braid) to $(O^*, \phi_{O^*}, \psi_{O^*})$; this isotopy replaces P with, say, P_1, for which $P_1(\text{crit}(P_1))$ is orientable as a closed O*-braid, and of course still orientable also as a closed braid. Now P_1 can be altered near N(O*), to P_2, so as to carry an adapted tubular neighborhood of $K = P_1^{-1}(O^*)$ onto N(O*). Then Proposition 2.6 applies to P_2, yielding (K, ϕ_K, ψ_K) with $P_2(\text{der}(K))$--and therefore der(K) itself--trivial.

3.6. EXPLICATION. The role of T is somewhat obscure in this treatment, as, indeed, are the words "strictly" and "homogeneous." Here is a brief explanation (cf. the discussion of positive braids in [11]).

First, one identifies the n-string braid group B_n with the fundamental group of the *configuration space* of unordered n-tuples of distinct points of Int D^2, with basepoint $p^{-1}(0)$. The (set valued) inverse p^{-1} of p maps the set of regular values of p into the configuration space. It is not hard to see that the fundamental group of $D^2\backslash\{\eta_1, \cdots, \eta_{n-1}\}$, with basepoint 0, is mapped onto B_n by $(p^{-1})_*$; this free group of rank n-1 is the *prebraid group* of [11].

Then a closed prebraid is precisely the graph, in $\partial_1 D = S^1 \times D^2$, of a loop $S^1 \to D^2 \{\eta_1, \cdots, \eta_{n-1}\}$; a closed braid is the "graph" (in the sense of n-valued functions) of a loop in the configuration space. In particular it is clear what is meant by *the closure of a (pre)braid*. Free homotopy of loops corresponds, on the one hand, to conjugacy in the (pre)braid group, and, on the other hand, to ambient isotopy of closed (pre)braids through closed (pre)braids. (The proof of Remark 3.3(2) should now be clear.)

The choice of arcs I_j' provides, in the usual way, a set of "standard" (free) generators x_j of the prebraid group. Their images $(p^{-1})_*(x_j) = {}^T\sigma_j$ in B_n may be called *T-standard generators* of B_n; they are determined up to equivalence (i.e., order and the action of automorphisms of B_n) by the combinatorial type of T. For example, if T = I is an arc, then the I-standard generators ${}^I\sigma_j$ are the "standard" standard generators σ_j; if n = 4 and T = Y is a triod, then one can take ${}^Y\sigma_1 = \sigma_1$, ${}^Y\sigma_2 = \sigma_1\sigma_2\sigma_1^{-1}$, and ${}^Y\sigma_3 = \sigma_1\sigma_2\sigma_3\sigma_2^{-1}\sigma_1^{-1}$; these are not equivalent to the I-standard generators.

A *T-braid word in* B_n *of length* k is an ordered k-tuple $\mathbf{b} = (b(1), \cdots, b(k))$ with $b(i) = {}^T\sigma_{j(i)}^{\varepsilon(i)}$, $\varepsilon(i) = \pm 1$ and $1 \leq j(i) \leq n-1$ for $i = 1, \cdots, k$, of T-standard generators of B_n and their inverses. The braid of $\mathbf{b}$ is $\beta(\mathbf{b}) = b(1), \cdots, b(k) \in B_n$; the closed braid of $\mathbf{b}$ is the closure $\hat{\beta}(\mathbf{b})$ of $\beta(\mathbf{b})$.

A T-braid word is T-*homogeneous* if $\varepsilon(i)$ depends only on $j(i)$ (i.e., for no j do ${}^T\sigma_j$ and ${}^T\sigma_j^{-1}$ both appear in the word), and in particular T-*positive* (resp., T-*negative*) if $\varepsilon(i)$ is always 1 (resp., -1); the adverb "strictly" is piled on if $j : \{1, \cdots, k\} \to \{1, \cdots, n-1\}$ is onto (i.e., each T-standard generator is represented, either by itself or by its inverse). A braid is (*strictly*) T-*homogeneous* (resp., T-positive;

T-negative) if it is the braid of a (strictly) T-homogeneous (resp., T-positive; T-negative) braid word. If T = 1, it is typically not mentioned.

Finally, then, it turns out that closed strictly T-homogeneous braids in the sense of 3.4 are exactly the closures of strictly T-homogeneous braids as just defined.

3.7. THEOREM. Let $\mathbf{b} = ({}^{T}\sigma_{j(1)}^{\varepsilon(1)}, \cdots, {}^{T}\sigma_{j(k)}^{\varepsilon(k)})$ be a strictly T-homogeneous braid word in B_n. Then $\lambda(\hat{\beta}(\mathbf{b})) = \#\{i : 1 \leq i \leq k, \varepsilon(i) = -1\} - \#\{j(i) : 1 \leq i \leq k, \varepsilon(i) = -1\}$, and $\rho(\hat{\beta}(\mathbf{b})) = \#\{i : 1 \leq i \leq k, \varepsilon(i) = +1\} - \#\{j(i) : 1 \leq i \leq k, \varepsilon(i) = +1\}$.

Proof: The assertion about $\lambda(K)$ follows from 3.5., 3.6., and 2.8(**). Since $\mu(K)$ is k-(n-1), the assertion about $\rho(K)$ follows.

3.8. COROLLARY. If **b** is strictly positive then $\lambda(\hat{\beta}(\mathbf{b})) = 0$.

3.9. REMARK. One can read 3.7 as saying that, for **b** strictly homogeneous, the natural (braided) fibre surface for $\hat{\beta}(\mathbf{b})$ is the union of subsurfaces of H_1-ranks $\lambda(\hat{\beta}(\mathbf{b}))$ (resp., $\rho(\hat{\beta}(\mathbf{b}))$), made up of the negative (resp., positive) 1-handles and the 0-handles to which they are attached. This suggests an alternative proof by "plumbing"; one will be given in a later part (joint research with Walter Neumann).

4. THE BIDEGREE OF THE GENERALIZED JONES POLYNOMIAL OF SOME CLOSED HOMOGENEOUS BRAIDS.

4.1. NOTATION. If K is an oriented link in an oriented 3-sphere, say ∂D, let [K] denote its <u>generalized Jones polynomial</u>, a Laurent polynomial in two variables over **Z**. Here, it will be convenient to use variables (v,x) in which the fundamental functional equation takes the form

$$\text{(FFE)} \qquad [K_+](v,x) = v[K_0](v,x) + vx^{-1}[K_-](v,x)$$

when K_+, K_0, and K_- are related in the usual way, which for our purposes can be expressed as follows: there are braid words **a** and **c** (in some braid group B_n, $n \leq 1$) and an index i (between 1 and n-1) such that the closures of $\mathbf{a}^\frown(\sigma_i,\sigma_i)^\frown\mathbf{c}$, $\mathbf{a}^\frown\sigma_i{}^\frown\mathbf{c}$, and $\mathbf{a}^\frown\mathbf{c}$ are K_+, K_0, and K_- respectively. (In this section, **b** is a <u>braid word of length</u> k <u>in</u> B_n iff **b** is an ordered k-tuple (b(1), ⋯ ,b(k)) with each b(j) either a standard generator σ_i of B_n or the inverse of a standard generator; that is, **b** is an I-braid word in the sense of 3.6. Also, if k = 1, we confuse (b(1)) with b(1); if k = 0, then

b is the empty tuple $\mathbf{o}^{(n)}$; and if **b** and **c** are braid words then **b**^**c** denotes their concatenation.) If **b** is a braid word in B_n we write [**b**;n] or simply [**b**] for [$\hat{\beta}$(**b**)].

As shown in, e.g., [4], (FFE) and the initial condition $[\mathbf{o}^{(1)}](v,x) = 1$ determine [K] for every K; the generalized Jones polynomial of K and that of its mirror image R(K) are related by [Rev(K)](v,x) = [K](-x,-v); and if # denotes connected sum (along any component of each summand, if the summands are links), then [K#L] = [K][L].

4.2. THEOREM. Let **b** be a strictly homogeneous braid word on n strings. If either (A) n is arbitrary and **b** is either strictly positive or strictly negative, or (B) n = 3, then $\lambda(\hat{\beta}(\mathbf{b}))$ is the x-degree of [**b**](v,x), and $\rho(\hat{\beta}(\mathbf{b}))$ is its v-degree.

4.3. REMARKS. (1) Hand calculations suggest that the theorem remains true for arbitrary strictly homogeneous braids on any number of strings; but the proof remains elusive. (2) Unfortunately, about the simplest example of a closed generalized strictly homogeneous braid which is not (of the link type of) a closed strictly homogeneous braid--namely, the closure of the square of $\sigma_1(\sigma_1\sigma_2\sigma_1^{-1})(\sigma_1\sigma_2\sigma_3\sigma_2^{-1}\sigma_1^{-1})$ in B_4--fails to satisfy the conclusion of 4.2. Perhaps equality can be replaced with an inequality, and something salvaged? (This is suggested by recent results of [5] on Jones polynomials of alternating knots.) (3) The statement for strictly positive or negative braids can be deduced from theorems on "positivity" (appropriately defined) of the generalized Jones polynomial of a closed positive braid ([2], [6], [14],...); cf., also [5]. The appearance of λ and ρ is, of course, new.

Proof of 4.2: The proofs of both parts are induction on the length of **b**, using (FFE) (of course!) and the fact that the degree of a sum of two (Laurent) polynomials is the larger of the degrees of the summands, when these degrees are distinct.

(A) The mirror image of closed strictly negative braid is a closed strictly positive braid, so the rule [Rev(K)](v,x) = [K](-x,-v) means we need consider the positive case only.

Make the inductive hypothesis that 4.2 is true for any strictly positive braid word of length at most k-1. Let $\mathbf{b} = (\sigma_{i(1)}, \cdots, \sigma_{i(k)})$ be a strictly positive braid word of length k in B_n. Then $\rho(\hat{\beta}(\mathbf{b})) = \mu(\hat{\beta}(\mathbf{b})) = k-n+1$ and $\lambda(\hat{\beta}(\mathbf{b})) = 0$, by 3.7 - 3.8.

According to a lemma embedded in the proof of the second proposition in the appendix to [10] (rediscovered in [2] and [14]), after possibly replacing **b** by another strictly positive braid word with the same braid (and therefore the same length), we may assume that either (1) there is exactly one j, $1 \le j \le k$, with $i(j) = n-1$, or (2) there is some j, $1 \le j < k$, with $i(j) = i(j+1) = n-1$.

In case (1), omitting the unique instance of σ_{n-1} in **b** replaces **b** by a strictly positive braid word **c** of length k-1 in B_{n-1}, with $\hat{\beta}(\mathbf{c})$ of the same link type as $\hat{\beta}(\mathbf{b})$; then by the inductive hypothesis, $\rho(\hat{\beta}(\mathbf{b})) = \rho(\hat{\beta}(\mathbf{c})) = (k-1)-(n-1) + 1 = k-n+1$ is the v-degree of $[\mathbf{b};n] = [\mathbf{c};n-1]$ and 0 is its x-degree, so 4.2 is true for **b**.

In case (2), either (2.1) there is an index m not equal to j or j+1 such that $i(m) = n-1$, or (2.2) there is no such m.

In case (2.1), we may (without changing the link type of the closed braid) assume that $j = k-1$, so $\mathbf{b} = \mathbf{b}_-{}^\wedge(\sigma_{n-1},\sigma_{n-1})$, where $\mathbf{b}_+ = \mathbf{b}$, $\mathbf{b}_0$, and $\mathbf{b}_-$ (as in 4.1) are strictly positive words in B_n, of lengths k, k-1, and k-2 respectively. By (FFE),

$$[\mathbf{b};n](v,x) = v[\mathbf{b}_0;n](v,x) + vx^{-1}[\mathbf{b}_-;n](v,x); \qquad (*)$$

by the inductive hypothesis, the first term on the right side of (*) has v-degree $1 + \{(k-1)-n+1\} = k-n+1$ and x-degree $0 + 0 = 0$, and the second term has v-degree $1 + \{(k-2)-n+1\} = k-n < k-n+1$ and x=degree $-1+0 = -1 < 0$; so by the ultrametric property of degrees, the left side of (*) has v-degree k-n+1 and x-degree 0, and 4.2 is true for **b**.

In case (2.2), let **c** be the strictly positive braid word of length k-2 in B_{n-1}, obtained by omitting both instances of σ_{n-1} in **b**. The link $\hat{\beta}(\mathbf{b})$ is of the type of the connected sum of $\hat{\beta}(\mathbf{c})$ and the positive Hopf link $\hat{\beta}((\sigma_1,\sigma_1))$. Easily, $[\sigma_1^2;2] = v-vx^{-2}+x^{-1}$; so by multiplicativity of [] under connected sum, and the inductive hypothesis applied to **c**, $[\mathbf{b};n](x,v) = (v-vx^{-2}+x^{-1})[\mathbf{c};n-1]$ has v-degree $1 + \{(k-2) - (n-1) + 1\} = k-n+1$ and x-degree $0 + 0 = 0$, and 4.2 is true for **b**.

Finally, to start the induction, note that if $k = 1$ then (by strictness) $n = 2$ and the link in question is an unknot.

(B) By (A) and a few obvious reductions, it is sufficient to study words of the form $\mathbf{b} = (\sigma_1^{p(1)},\sigma_2^{-p(2)},\cdots,\sigma_1^{p(k-1)},\sigma_2^{-p(k)})$ with $p(i) > 0$. The proof is formally the same as that of (A)--induction on k and repeated use of (FFE); the lemma on finding adjacent repeated letters ("squares") in positive words is replaced by the observation that is some p(i) is at least 2 then there is a square, while if on the contrary every p(i) is 1 (so **b** = $(\sigma_1, \cdots, \sigma_2^{-1})$) then either k = 2 and the closed braid is an unknot, or $k \geq 4$ and we can apply (FFE) to $\beta_+ = \beta(\mathbf{b}) = \sigma_1\sigma_2^{-1}\sigma_1\sigma_2^{-1}\gamma$, $\beta_0 = \sigma_1\sigma_2^{-2}\gamma$, and $\beta_- = \sigma_1\sigma_2^{-1}\sigma_1^{-1}\sigma_2^{-1}\gamma = \sigma_2^{-1}\sigma_1^{-1}\gamma$, where the latter two braids are again strictly homogeneous, and of smaller Milnor number.

4.4. REMARK. In light of Remark 3.9, this result suggests that the behavior of the generalized Jones polynomial under plumbing should be investigated.

REFERENCES

1. Birman, Joan, "A representation of theory for fibred knots and their monodromy maps," Topology of Low-Dimensional Manifolds (Proceedings, Sussex, 1977), ed. Roger Fenn, LNM 722 (Springer), 1-8.
2. Franks, J., and Williams, R. F., "Braids and the Jones-Conway Polynomial", preprint (January 1985).
3. Goldsmith, Deborah, "Motions of links in the 3-sphere," Bull. Amer. Math. Soc. 80 (1974), 63-67.
4. Lickorish, W.B.R., and Millett, K. C., "A Polynomial Invariant of Oriented Links," M.S.R.I. Preprint (1985).
5. Murasugi, Kunio, "Jones Polynomials of Alternating links," preprint (1985).
6. Rudolph, Lee, "Braided Surfaces and Seifert Ribbons for Closed Braids," Comment. Math. Helvetici 58 (1983), 1-37.
7. Rudolph, Lee, "Some Knot Theory of complex plane curves," L´Ens. Math. (Monographie 31, "Noeuds, Tresses et Singularités," 1983), 99-122.
8. Rudolph, Lee, "Constructions of quasipositive knots and links, I", ibid., 233-245.
9. Rudolph, Lee, "Some topologically locally-flat surfaces in the comlex projective plane," Comment. Math. Helvetici 59 (1984), 592-599.
10. Rudolph, Lee, "Keeping up with the Generalized Joneses; or, the Power of Positive Thinking," preprint (January 1985).
11. Rudolph, Lee, "Special positions for surfaces bounded by closed braids", Rev. Mat. Iberoamericana No. 3 (1985); MSRI Preprint 02512-85.

12. Rudolph, Lee, "Isolated Critical Points of Maps from R^4 to R^2 and a Natural Splitting of the Milnor Number of a Classical Fibred Link, Part I: Basic Theory; examples," submitted for publication (1985).
13. Stallings, J. R., "Constructions of fibred knots and links," Proc. Symp. Pure Math. XXXII, part 2 (1978), 55-60.
14. van Buskirk, James M., "A Version of the Generalized Jones Polynomial which is Positive for Positive Knots," preprint (January 1985).

Covering Theorems for Open Surfaces

URI SREBRO AND BRONISLAW WAJNRYB / Department of Mathematics, Technion-Israel Institute of Technology, Technion, Haifa, Israel

1. COVERING THEOREMS

This paper is based mostly on earlier work of the authors [4] and [3], where finite-to-one holomorphic functions between Riemann surfaces were considered. Here we consider finite-to-one open mappings between surfaces, reformulate the corresponding results with some extensions (see 2.5 and 2.6) and sketch the proofs.

By a surface we mean a connected orientable 2-manifold without boundary. A surface is closed if it is compact, and it is open otherwise. An open surface X of a finite genus can be embedded in a closed surface $\bar{X}$ of the same genus. By the number of boundary components of such a surface we mean the number of connected components of the boundary of X in $\bar{X}$. If this number is finite we may assume that the boundary of X in $\bar{X}$ is a disjoint union of topological circles, and the number of boundary components can be determined by the Euler characteristic of the surface.

Let $f:X\to Y$ be a finite-to-one open continuous function where X and Y are surfaces, and suppose that f is p-valent, i.e.,

$$\max_{y\in Y} \operatorname{card} f^{-1}(y) = p < \infty .$$

Then locally f is topologically equivalent at every point x in X to a polynomial mapping $z\to z^d$, $z \in \mathbb{C}$, where the positive integer $d = d(x)$ is the local degree of f at z and $b(x) = d(x)-1$ is the branch order of f at x. If $b(x) > 0$, x is a branch point and the point $f(x)$ is a branch value of f.

If X is closed, so is Y and then f is a covering, possibly ramified. Under these assumptions on X, Y and f the total branch order

$$b_f = \sum_{x \in X} b(x)$$

of f is finite and satisfies the Riemann-Hurwitz relation

$$b_f = 2[(\alpha-1)-p(\beta-1)] \tag{1.1}$$

where α is the genus of X and β is the genus of Y. Furthermore, in these circumstances, the multiplicity function

$$n(y) = \sum_{x \in f^{-1}(y)} d(x)$$

assumes the constant value p at every point in Y.

Also, the relation (1.1) implies that

$$\alpha \geq p\beta-p+1$$

is a necessary condition for the existence of a p-valent covering from a closed surface of genus α onto a closed surface of genus β, where equality holds if the covering is unramified. It is not hard to show that this condition, including the case of equality, is sufficient too.

We want to generalize these results to the case where X is open and Y and f are as before. Then in this case (1.1) does not hold, in general, $n(y) < p$ for at least one point y in Y and there is no restriction for the existence of a p-valent open continuous function $f : X \to Y$.

Thus the following questions arise:

1. How evenly can X cover Y? In particular, how big is the set of points y in Y for which $n(y) < p$ and how big is the set of points y in Y for which $n(y) \leq p-2$?

2. What can be said about the total branch order b_f of f? Can (1.1) be generalized?

1.2 In order to answer the first question we consider the so called deficiency set

$$A = \{y \in Y : n(y) \leq p-2\}$$

of f and ask for a lower bound for card A. The points of A will be called deficiency points.

The following two theorems (cf. [3], [4], and [2]) answer the first question.

THEOREM 1.3. Let X and Y be n-manifolds, $n \geq 2$. If X is open, Y is closed and $f : X \to Y$ is p-valent open continuous, then $n(y) < p$ for every point y in the cluster set

$$C(f) = \cap\{\overline{f(X-E)} : E \text{ compact in } X\} .$$

THEOREM 1.4. Let X be an open surface of genus α, Y a closed surface of genus β, and $f : X \to Y$ a p-valent, $p \geq 2$, open continuous function with total branch order b and a deficiency set A.

1. Any of the following conditions implies card $A \geq 1$:
   a) $\alpha = \infty$;
   b) $b = \infty$;
   c) b is finite and odd;
   d) $\alpha > \infty$ and X has exactly $k < \infty$ boundary components, and

   $$b > 2k+2\min\{[(\alpha-1)-(p-1)(\beta-1)],2[(\alpha-1)-(p-1)(\beta-1)]\}.$$

2. Let m be a positive integer. If $\alpha < \infty$ and

   $$2[(\alpha-1)-p(\beta-1)]-m(p-1) \leq b < 2[(\alpha-1)-p(\beta-1)]-(m-1)(p-1)$$

then card $A \geq m$.

3. The inequalities for card A in (1) and (2) are best possible.

1.5. We now turn to question (2). Unless the deficiency set A is empty, the total branch order of a p-valent open continuous function from an open surface to a closed surface can be infinite or assume almost every value. This will be apparent from the existence theorems which will be formulated later. It turns out (see Theorem 1.7) that the Riemann-Hurwitz relation (1.1) can be generalized for functions with empty deficiency set. In view of Theorem 1.3, such functions cover Y as evenly as possible and will thus be called almost-coverings.

DEFINITION 1.6. Let X be an open surface and Y a surface. A p-valent open continuous function $f : X \to Y$ is an <u>a-covering</u> (almost-covering) if f has a void deficiency set, i.e.,

$$p-1 \leq n(y) \leq p$$

for every point y in Y.

THEOREM 1.7. Let X be an open surface of genus α, Y a closed surface of genus β and $f : X \to Y$ a p-valent, $p \geq 2$, a-covering with total branch order b. Then

1. α is finite
2. b is finite, even and satisfies

$$(\alpha-1)-p(\beta-1) \leq \frac{b}{2} \leq k+\min\{[(\alpha-1)-(p-1)(\beta-1)],2[(\alpha-1)-(p-1)(\beta-1)]\} \tag{1.8}$$

where k denotes the number of boundary components of X.

REMARK 1.9. We shall prove in Theorem 2.1 that the inequalities in (1.8) are best possible in the sense that for every b satisfying (1.8) there exists a function f satisfying the assumptions of Theorem 1.7.

1.10. Suppose now that both X and Y are open and that $f : X \to Y$ is a p-valent covering. Then the Riemann Hurwitz relation has the form

$$b_f = p\chi(Y) - \chi(X)$$

and in particular

$$\chi(X) \leq p\chi(Y) \tag{1.11}$$

where χ is Euler characteristic.

It can be shown that if $\chi(Y)$ is finite then (1.11) is also a sufficient condition for the existence of a p-valent covering $f : X \to Y$, where f is unramified if and only if equality holds in (1.11).

If $\chi(X)$ and $\chi(Y)$ are finite and if

$$\chi(X) > p\chi(Y)$$

then there is no p-valent covering from X to Y. However, there is always a p-valent a-covering with a preassigned branch order b provided

$$b \geq 2\alpha - 2p\beta + p - 1$$

where α is the genus of X and β is the genus of Y. (See Theorem 2.5 below.)

1.12. The statements (3) in Theorem 1.4 and in Remark 1.9, about the sharpness of the results, will be restated in detail in the next chapter.

2. EXISTENCE THEOREMS AND MAPPING PROBLEMS

The next three theorems explain in detail the sharpness of Theorems 1.4 and 1.7. Theorem 2.1 restates Remark 1.9. Theorem 2.2 shows that the inequality card $A \geq 1$ in 1.4(1) is sharp and Theorem 2.3 shows that the inequality card $A \geq m$ in 1.4(2) is sharp.

THEOREM 2.1. Given a closed surface Y of genus β and integers $\alpha \geq 0$, $k \geq 1$, $p \geq 2$ and a non-negative even integer b which satisfy

$$(\alpha-1)-p(\beta-1) \leq \frac{b}{2} \leq k+\min\{[(\alpha-1)-(p-1)(\beta-1)],2[(\alpha-1)-(p-1)(\beta-1)]\} ,$$

there exist an open surface X of genus α with exactly k boundary components and a p-valent a-covering $f : X \to Y$ with $b_f = b$.

THEOREM 2.2. Let Y be a closed surface of genus β and let $p \geq 2$ be an integer. If either

1. α and b are non-negative integers satisfying $b \geq 2(\alpha-1)-2p(\beta-1)$, or
2. α is a non-negative integer and $b = \infty$, or
3. $\alpha = \infty$ and $b = \infty$,

then there exist an open surface X of genus α with one boundary component and a p-valent open continuous function $f : X \to Y$ with $b_f = b$ and with one deficiency point.

We say that X has one boundary component if the point ∞ in the Alexander compactification of X does not separate any of its neighborhoods.

THEOREM 2.3. Let Y be a closed surface of genus β. Given integers $\alpha \geq 0$, $b \geq 0$, $m \geq 1$, and $p \geq 2$ satisfying

$$2[(\alpha-1)-p(\beta-1)]-m(p-1) \leq b < 2[(\alpha-1)-p(\beta-1)]-(m-1)(p-1)$$

there exist an open surface X of genus α with one boundary component and a p-valent open continuous function $f : X \to Y$ with $b_f = b$ and with exactly m deficiency points.

REMARK 2.4. Suppose that X and Y are open surfaces. As noted in (1.11)

$$\chi(X) \leq p\chi(Y)$$

is a necessary condition for the existence of a p-valent covering $f : X \to Y$. On the other hand, one can obtain the following theorem: (The proof will be given in 3.11).

THEOREM 2.5. Given any two open surfaces X and Y of finite Euler characteristic and an integer $p \geq 2$, there exists a p-valent a-covering $f : X \to Y$. Furthermore, the branch order of f can be chosen to be infinite or to assume any integer value

$$b \geq 2\alpha - 2p\beta + p - 1 ,$$

where α is the genus of X and β is the genus of Y.

2.6. It is obvious that there are no finite to one coverings from a disc onto a surface Y unless Y is a disc. However, Theorem 2.5 yields the following results:

COROLLARY. Given an open surface Y of finite Euler characteristic and an integer $p \geq 2$, then there exists a p-valent a-covering from an open disc to Y. Furthermore, if Y is of positive genus, then the branch order of f can be chosen arbitrarily. If Y is a disc the branch order of f can be chosen to be infinite or assume any integer value $b \geq p-1$.

2.7. Consider again the case where X is open and Y is closed. One may ask for necessary and sufficient conditions for the existence of a p-valent a-covering $f : X \to Y$ with a given branch order. Such a condition is given in the following theorem, which is a consequence of 2.1 and (1.7) in view of the fact that $b_f \geq 0$.

THEOREM 2.8. Let Y be a closed surface of genus β. Then there exist an open surface X of genus α with k boundary components and a p-valent a-covering $f : X \to Y$ if and only if

$$(p-1)(\beta-1) + 1 - \frac{k}{2} \leq \alpha < \infty .$$

Furthermore, if equality holds, then f is unramified.

In the next five corollaries Y is assumed to be a closed surface.

COROLLARY 2.9. Given integers $\alpha \geq 0$, $k \geq 1$, and $p \geq 2$, there exist an open surface X of genus α with k boundary components and a p-valent a-covering from X onto a sphere.

COROLLARY 2.10. There exists a p-valent, $p \geq 2$, a-covering f from a k-connected planar domain D (i.e., a planar domain with k boundary components) onto Y if and only if

$$k \geq 2[(p-1)(\beta-1) + 1] .$$

COROLLARY 2.11. There exists an a-covering of Y by a disc if and only if Y is a sphere.

COROLLARY 2.12. There exists a surjective 2-valent open continuous function from a k-connected planar domain onto Y if and only if $k \geq 2\beta$.

COROLLARY 2.13. 1. There exists an a-covering of Y by an annulus or by a twice punctured disc if and only if Y is either a torus or a sphere;

2. If Y is a torus, then for every $p \geq 2$ there exists a p-valent a-covering by an annulus.

3. SKETCH OF PROOFS

We sketch the proofs and describe the main ideas. The details of most of the proofs can be found in [4].

3.1. The proof of Theorem 1.2 is elementary. For the arguments in dimension two see [3]. The same arguments are valid in higher dimensions. See also [2, Theorem 1].

3.2. The following modified and extended version (see [3] and [4]) of the Embedding Theorem of Lyzzaik and Styer [1], is applied several times in the proofs of Theorem 1.7, 1-3 and Theorem 1.4, 1-2.

THE EMBEDDING THEOREM. Let X be an open surface, Y a closed surface and $f : X \to Y$ a p-valent, $p \geq 2$, open continuous function with total branch order b and a deficiency set $A = \{y \in Y : n(y) \leq p-2\}$. If

1. $A = \emptyset$, i.e., f is an a-covering, or if
2. A and b are finite,

then there exist a closed surface $\hat{X}$, an embedding $\phi : X \to \hat{X}$, a p-valent covering $\hat{f} : \hat{X} \to Y$, and a neighborhood U of

$$E = \hat{X} - \phi(X) - \hat{f}^{-1}(A)$$

such that $f = \hat{f} \cdot \phi$ and such that $f|\hat{U}$ is an embedding.

3.3. Consider first the assumptions of Theorem 1.7. Since f is an a-covering, i.e., $A = \emptyset$, the Embedding Theorem can be applied. It thus follows that α is finite and that $b = b_f$ is finite and even. Application of the Riemann Hurwitz relation to f and the relation for the genus

$$\alpha = g(X) \leq g(\hat{X})$$

give the left inequality in (1.8).

The right inequality in (1.8) is more tricky. Since X is of finite genus, we may assume by virtue of the Embedding Theorem that X has a boundary consisting of k disjoint circles, say $C_1, \cdots, C_k$. By the Embedding Theorem, $\phi(C_1), \cdots, \phi(C_k)$ are disjoint circles in $\hat{X}$ and $\hat{f}\cdot\phi(C_1), \cdots, \hat{f}\cdot\phi(C_k)$ are disjoint circles in Y. The desired inequality is then obtained by invoking the Riemann-Hurwitz relation and by a repeated application of the following elementary lemma.

LEMMA 3.4. Let Y be a closed surface and let $C_1, \cdots, C_k$ be disjoint circles in Y. Suppose that $Y - \cup C_i$ has s connected components $V_1, \cdots, V_s$. Then

$$g(Y)-1 = k + \sum_{i=1}^{s} (g(V_i)-1)$$

where g denotes the genus.

3.5. The first part of Theorem 1.4 follows from Theorem 1.7. For the second part of Theorem 1.4 we use again the Embedding Theorem, and apply the Riemann-Hurwitz relation to $\hat{f}$. By studying the relation between b_f, $b_{\hat{f}}$, and card A and by applying $g(\chi) \leq g(\hat{X})$ we get a lower bound for b_f in terms of α, β, and card A. This yields the right inequalities in Theorem 1.4(1).

3.6. As noted above, the statements (3) in Theorem 1.4 and in Theorem 1.7 about the sharpness of the results, are included in the Existence Theorems 2.1-2.3. We now comment on the proof of the theorems in Chapter 2.

3.7. The proof of the Existence Theorems is constructive. Every construction is done in three steps.

STEP 1. We start with a construction of a p-valent covering $h : Z \to Y$, where Z is a closed surface and h has a branch order which is suitably chosen for each case. In every case the branch order is at least b. Then the genus of Z is uniquely determined and $g(Z) \geq \alpha$. Such constructions are known. Nevertheless we choose here a particular construction which is easy to handle and which suits our purposes. This construction is obtained by a modification of a cyclic covering as described below.

3.8. The p-cyclic covering. Given integers $\beta \geq 1$ and $p \geq 2$, consider the closed surface Z which is the boundary of a handlebody $B(p,\beta)$, where $B(p,\beta)$ is obtained from a solid torus by attaching to it p congruent arms each of which is a handlebody of genus β-1. Note that Z has genus $P(\beta-1) + 1$. Furthermore Z has a rotational symmetry. Let G denote the cyclic group of order p of rotations acting on Z. Then the quotient space $Y = Y(\beta) = Z/G$ is a closed surface of genus β. Since G acts without fixed points, then the natural projection $h : Z \to Y$ is a p-valent unramified covering map. We call this map a p-cyclic covering of Y.

3.9. Modification of coverings. Let $h : Z \to Y$ be a p-valent, $p \geq 2$, covering (possibly ramified) where Y is a closed surface and Z is a closed manifold (possibly not connected). We modify Z and h and get a new manifold Z_1 and a p-valent covering $h_1 : Z_1 \to Y$, such that $b_{h_1} = b_h + 2$ as follows:

Let D be a disc in Y such that $D \cap h(B_h) = \emptyset$. Let D' and D'' be two different connected components of $h^{-1}(D)$. D is simply connected, hence h is injective in D' and in D''. Choose a simple compact arc L in D. Let $L' = D' \cap h^{-1}(L)$ and $L'' = D'' \cap h^{-1}(L)$. Cut Z along L' and L'' and glue back crosswise along these cuts in such a way that each edge of L' is glued to the opposite edge of L''. By this operation we obtain a closed surface Z_1 and a p-valent covering map $h_1 : Z_1 \to Y$ such that $h = h_1$ on $Z-(L' \cap L'')$. Note that h_1 has two additional simple branch points at the end points of L'. The part of Z_1 obtained from D' and D'' by this operation is homeomorphic to the tube $S_1 \times (0,1)$. The pair (Z_1,h_1) will be called the modification of (Z,h) along the pair of arcs (L',L'').

3.10. STEPS 2 and 3. Let $h : Z \to Y$ be as constructed in Step 1. In Step 2, which is the most delicate, we choose a compact set E in Z such that the genus $g(Z-E) = \alpha$, $h|Z-E$ has the desired number of deficiency points (possibly zero) and Z-E has the possible minimal number of boundary components. This number turns out to be equal to one in case of Theorems 2.2 and 2.3 and at most k in Theorem 2.1 and 2.5.

Finally, we delete the set E from Z and, if necessary, delete a finite set of point from Z-E so that the remaining surface X has the desired number of boundary components. Then X and $f = h|X$ have the required properties.

3.11. Proof of Theorem 2.5. The surfaces X and Y are determined up to a homeomorphism by their genera α and β and the number of their boundary components k and ℓ, respectively. Consider first the case where $b \geq 2\alpha - 2p\beta + 2p - 2$ and where $\beta > 0$. In the first steps of the proof of

Theorem 4.1 in [4] we constructed a closed surface Z_1, a closed surface Y of genus β and a p-valent branched covering $h_1 : Z_1 \to Y$ with $b_{h_1} = b$. We fixed a point y in Y and then removed from Z_1 a set E, which is a union of certain arcs γ'_{ij}, δ'_{ij}, and ε'_i such that $Z_1 - E$ has genus α and $h_1|Z_1 - E$ has the point y as a unique deficiency point in Y. The mapping f needed for the present theorem is obtained from this construction by a further modification. If E does not contain all of $f^{-1}(y)$ we remove from Z_1 the additional set $E_1 = \bigcup_{j=1}^{p-1} \gamma'_{j,1}$ (see proof of 4.10 in [4]). Then $Z_2 = Z_1 - E - E_1$ is of genus α and has one boundary component and h_1 maps Z_2 onto $Y - \{y\}$. We now remove from Z_2 and $Y - \{y\}$ certain sets in order to get the right number of boundary components. First choose $\ell - 1$ points $y_1, \cdots, y_{\ell-1}$ in $Y - \Gamma$, where Γ is the set of points in Y which are covered at most p - 1 times by $h_1|Z_2$. Connect each y_i to y by p simple arcs η_{kj}, $1 \leq j \leq p$, such that η_{ij}, $1 \leq i \leq p-1$, $1 \leq j \leq p$, meet only at their end points, none of the η_{ij} meet the branch values of h_1 and such that each η_{ij} meets Γ only at y. For each i we lift each η_{ij}, $1 \leq j \leq p$, in Z_1 to η_{ij} such that for different j's the arcs η_{ij} start at different pre-images of y_i.

We now remove all the lifts from Z_2. We thus obtain from Z_2 an open surface X_1 of genus α with one boundary component. X_1 does not cover any of the points y, $y_1, y_2, \cdots, y_{\ell-1}$ and covers every other point in Y at least p-1 times. Finally we remove k-1 points from X_1 to obtain the required surface X. The case $\beta = 0$ can be treated in a similar way. By a similar argument we can prove the assertion of the theorem for the case

$$2\alpha - 2p\beta + p - 1 \leq b < 2\alpha - 2p\beta + 2p - 2$$

starting from Theorem 2.3 with m = 1.

ACKNOWLEDGEMENT. The first author wishes to thank the Technion's Foundation for Promotion of Research for a research grant.

REFERENCES

1. Lyzzaik, A. and D. Styer, A Covering Surface Conjecture of Brannan and Kirwan, Bull. London Math. Soc. 14 (1982), 39-42.
2. Srebro, U., Deficiencies of Immersions, Pacific J. of Math. 113 (1984), 493-496.

3. Srebro, U., Covering Theorems for Meromorphic Functions, J. D'Anal. Math. 44, (1984/85), 235-250.

4. Srebro, U., and B. Wajnryb, Covering Theorems for Riemann Surfaces, J. D'Anal. Math. (to appear).

Equivariant Handles in Finite Group Actions

MARK STEINBERGER / Department of Mathematics, Northern Illinois University, Dekalb, Illinois

JAMES WEST / Department of Mathematics, Cornell University, Ithaca, New York

0. INTRODUCTION

Handle decompositions have played a dominant role in the development of the theory of manifolds by both geometric and algebraic techniques since they exhibit the manifolds as unions of basic units (balls) assembled in relatively nice ways that capture the essential combinatorial aspects of their homological structure. Typically, one performs a geometric-topological argument in the individual handles or manipulates them, often passing from the handle decomposition to the cellular chain complex associated with it or to the CW-complex spine determined from the cores of the handles by collapsing the cocores ([39; Essay III]) and back again. In this way one obtains, among other fundamental things, the homology, homotopy type and simple-homotopy type, and in dimension ≥ 6, the h- and s-cobordism theorems. (cf. [38]).

The theory of handles in manifolds with group actions is more complex than it is in the inequivariant case. Of course this is to be expected because the interrelationships of the fixed point sets of the various subgroups of the acting group have to be taken into account. In the consideration of the action of the isotropy subgroup of a point on small neighborhoods of it, the theory of linear representations of the subgroups of the acting group enters irrevocably. However, complications arise at a more fundamental level because the sweeping existence theorems for inequivariant handle decompositions fail dramatically, even for locally smoothable actions of finite groups, and when they exist, torsion invariants computed from them are usually not topological invariants.

In this paper we discuss these closely related phenomena. Our main purpose is to provide a background against which we can outline our proof [60] that for locally smoothable actions of finite groups G on manifolds M in which all components of the fixed point sets of all subgroups have dimension ≥ 6 and none has codimension 1 or 2 in another, equivariant handle decompositions exist if and only if for each $\varepsilon > 0$ M is equivariantly ε-homotopy equivalent to a finite G-CW-complex (i.e., a CW-complex on which G acts by permuting the cells). We also state the analogous theorem from [56] giving the analogous obstruction to the existence of equivariant handle decompositions in the case that the above codimension restriction does not hold. It has essentially the same proof but uses isovariant rather than equivariant obstruction groups and hypothesizes the effect of Carter's vanishing theorem [10] (which ensures that in the codimension $\neq 1$ or 2 case there are no further obstructions).

1. REMARKS ON INEQUIVARIANT HANDLE THEORY

Let us first briefly review some of the salient points of the inequivariant theory. A handle of index k (k-handle) in an n-manifold M^n is a embedded copy of $D^k \times D^{n-k}$. Its core is $D^k \times 0$; its cocore is $0 \times D^{n-k}$. A handle decomposition of M on the subset $K \subset M$ is a filtration $M_0 \subset M_1 \subset \cdots \subset M_n = M$ of M by closed codimension zero submanifolds (with bicollared boundaries) such that each M_k is obtained from M_{k-1} by attaching k-handles M^k to a discrete collection of disjoint sites on the boundary of M_{k-1} by embeddings of $\partial D^k \times D^{n-k}$. $M_k = M_{k-1} \cup (D^k \times D^{n-k}, \partial D^k \times D^{n-k})_1 \cup \cdots$

In view of the precise manner in which the handles fit together, it is one of the wonders of topology that handle decompositions exist so generally as they do, especially in view of the elaborate and deep theories erected to treat the twin questions of smoothability and/or triangulability of manifolds. Morse theory provides smooth handle decompositions for smooth manifolds [44], and for piecewise linear manifolds the stars of the barycenters of a given combinatorial triangulation with respect to its second barycentric subdivision form a PL-handle decomposition. These are both quite natural developments, and so long as one could hope that all manifolds were smoothable or piecewise linearly triangulable, then it was natural also to anticipate the a priori less restrictive topological handle decompositions.

Interestingly, Kirby and Siebenmann [39] produced examples of manifolds not admitting piecewise linear structures and, essentially simultaneously, an existence proof for handle decompositions of all n-manifolds, $n \geq 6$. Siebenmann also obtained a proof [54] that some manifolds of dimension 4 and/or 5 do not admit handle decompositions, a phenomenon later shown by Quinn to be restricted to dimension 4 [49]. Thus, although the existence question remains of great interest for 4-manifolds, it is very satisfactorily settled otherwise.

Having considered existence of handle decompositions, we may think of uniqueness questions. The commonly computed algebraic invariants mentioned above are topological invariants: the cellular homology because it is isomorphic with singular homology, the chain-homotopy type of the cellular chain complex by cellular approximation and the Whitehead and Hurewicz theorems, and the simple-homotopy type by stabilization arguments [39; Essay III], [11], and [21] leading to a topologically invariant extension of simple-homotopy theory to locally compact ANR's via cell-like maps [12], [21], and [65].

The existence of handle decompositions and the topological invariance of Whitehead torsion lead to other strong manifestations of uniqueness: proofs of the h- and s-cobordism theorems in the topological category and the establishment of topological surgery [39; Essay III, Appendix C to Essay V].

2. GENERALITIES OF GROUP ACTIONS AND G-HANDLES; MOTIVATION OF THE CATEGORY OF LOCALLY SMOOTHABLE (LOCALLY LINEAR) ACTIONS

Now let G be a finite group and consider the category of G-spaces, i.e., spaces equipped with actions of G. The action introduces additional structure on a G-space X. In particular, X is the union of the orbits G(x) of its points x. Each orbit G(x) is naturally isomorphic as a G-space with the quotient G/G_x of G by the isotropy subgroup of x (the subgroup of elements fixing x). By varying x within G(x), we pick up a full conjugacy class (H) of isotropy subgroups. The equivariant isomorphism classes (G/H) of the orbits are termed orbit types. For subgroups $H \subset K$ of G we have natural maps of G-sets $G/H \to G/K$ producing a partial order on the category of orbit types by setting $(G/H) \geq (G/K)$. This provides a convenient framework for induction arguments. If $x \in X$ and $G_x = H$, then we denote by X_H the set of all points with isotropy subgroup H and by $X_{(H)}$ the set of all points of orbit type (G/H), i.e., with isotropy subgroups conjugate to H. By X^H we mean the fixed point set of H, and by $X^{(H)}$ we mean the union of all orbits of type $\leq (G/H)$.

Since G is finite, each point x has arbitrarily small G_x-invariant neighborhoods U. By continuity at x, we see that if U is small enough, then $G_y \subseteq H$ for all $y \in U$, so that $g \notin H$ implies that $gU \cap U = \phi$ and that $G(U)$ is comprised of $|G/H|$ many copies of U. In this case, gU is invariant under $gHg^{-1} = G_{gx}$, so we see that $G(U) \cong G \times_H U$. (If X is a right G-space and Y is a left G-space then the balanced product $X \times_G Y$ is the orbit space of $X \times Y$ by the action $g(x,y) = (xg^{-1},gy)$.) Because $G(U)$ is determined by U, a triangulation or cell in U determines one in each gU by translation, and a k-cell e^k contained in $U \cap X^H$ determines $|G/H|$ translates $gH(e^k)$ in U. This motivates the notion of a G-cell of type H as an isovariantly embedded copy of $G/H \times e^k = G \times_H e^k$, e^k being an open k-cell with the trivial action. A GCW-complex is then built up of G-cells attached by equivariant maps of their boundaries $G \times_H S^{k-1}$.

An equivariant handle of type H and index k is a balanced product $G \times_H (D^k \times D(V))$, where the subgroup $H \subset G$ acts trivially on D^k and where $D(V)$ is the unit disc of an orthogonal representation V of H. (There is no generality to be gained by using arbitrary real representations, as each such is equivalent to an orthogonal one.) Such a handle is attached to a manifold M by an equivariant (G-) embedding of $G \times_H (S^{k-1} \times D(V))$. Thus, as in the inequivariant case, a handle attachment is simply a thickening of a cell attachment, where the cells in question are G-cells $G \times_H D^k = G/H \times D^k$. Hence, the spine of an equivariant handlebody is a G-CW complex in the usual sense.

Note that our choice that the cocores $D(V)$ be representation discs places us in the category of locally linear actions (i.e., each point has a neighborhood which is equivariantly homeomorphic to a representation of its isotropy subgroup). This holds automatically for smoothable actions, via the exponential map of the tangent space. Thus, the locally linear category is the appropriate one for studying smoothability or studying G-homeomorphisms between smooth G-manifolds.

In the study of PL actions, local linearity is not automatic, and must be imposed. Such imposition has been shown to be vitally important for a coherent theory by many workers in the field (e.g., see [52; section 5] for a discussion of regular neighborhoods).

While one could consider more general types of cocore, this could seriously complicate such handle manipulations as subdivisions, cancellations, or even slides. For a successful theory, the restrictions to be imposed upon the cocore would not allow much greater generality than that of the locally linear category. For instance, a well-behaved handle theory would be highly unlikely in actions as general as those in [50].

Thus, we shall henceforth assume that all actions are locally linear in the appropriate category. We have found [41], [52], [7], [29], [31-36], and [55] to be valuable introductions to the study of such actions.

As is readily seen, local linearity ensures that such component M_α^H of M^H is a locally flat submanifold of M. Moreover, the orbit space M/G is a locally cone-like stratified space in the sense of [55] so the homeomorphism theory developed there applies and lifts to an equivariant theory in M. In particular, the principal results of [22] hold equivariantly, including the local contractibility of various spaces of equivariant embeddings and the equivariant homeomorphism group (if M is compact) and the ambient extension of locally extendable isotopies of compact sets.

As in the inequivariant case, smooth or PL G-manifolds admit smooth or PL handlebody decompositions. The first one is given in [63] and [4], and the second follows from the usual argument.

3. EQUIVARIANT WHITEHEAD TORSIONS AND s-COBORDISM THEOREMS FOR SMOOTH AND LOCALLY LINEAR PL ACTIONS

Equivariant s-cobordism theorems were given in the smooth and locally linear PL categories for compact manifolds and finite groups G by Browder and Quinn and by Rothenberg. For these theorems, we define an equivariant (G-) h-cobordism $(W^{n+1};M_0^n,M_1^n)$ by requiring each inclusion $i : M_i \to W$ to be an isovariant homotopy equivalence (isovariant means that isotropy subgroups are unchanged by all maps and homotopies).

It is easy and instructive to consider the argument. First we observe that if for each component $W_\alpha^{(H)}$ of each $W^{(H)}$ we choose an equivariant tubular or regular neighborhood $T_\alpha^{(H)}$ then the posited isovariant deformation retraction may be presumed to restrict to ones of $T_\alpha^{(H)}$ and $\partial T_\alpha^{(H)}$. Then we may dissect W as

$$\bigcup_{(H),\alpha} V_\alpha^{(H)}=(T_\alpha^{(H)} - \bigcup_{H<K} \operatorname{int}(T_\beta^{(K)})).$$

By extending trivializations of the compact G-manifolds $W_\alpha^{(H)} \cap V_\alpha^{(H)}$ to $V_\alpha^{(H)}$ we can infer that W is a trivial G-h-cobordism if and only if each $W_\alpha^{(H)} \cap V_\alpha^{(H)}$ is one. Now we can note that $W_\alpha^{(H)} \cap V_\alpha^{(H)} = G \times_H (W_{H\alpha} \cap V_{H\alpha})$ and that $W_{H\alpha} \cap V_{H\alpha}$ is acted on freely by some subgroup $K_{H\alpha}$ of NH/H. Thus, $W_\alpha^{(H)} \cap V_\alpha^{(H)}$ is trivial as a G-h-cobordism if and only if $(W_{H\alpha} \cap V_{H\alpha})/K_{H\alpha}$ is a trivial (inequivariant) h-cobordism. In this manner, we get an obstruction $\tau_{H\alpha}$ in $Wh(\pi_1(W_{H\alpha} \cap V_{H\alpha})/K_{H\alpha})$. Note that $\pi_1((W_{H\alpha} \cap V_{H\alpha})/K_{H\alpha}) \cong \pi_1(W_{H\alpha}/K_{H\alpha})$. Then

$$\tau(w) = \Sigma\, \tau_{H\alpha} \in \bigoplus_{(H)} \bigoplus_{(\alpha)} Wh(\pi_1(W_{H\alpha}/K_{H\alpha}))$$

vanishes if and only if $(W;M_0,M_1)$ is trivial as a G-h-cobordism. (Provided that $W_\alpha^{(H)} = M_{0\alpha}^{(H)} \times I$ if $\dim(M_{0\alpha}^{(H)}) \leq 4$.) (Here (α) runs over the equivalence classes under the action of NH on components of W_H.) We denote this group by $Wh_G^{iso}(W)$. This s-cobordism theorem and proof also works in the smooth category in the case that G is a compact Lie group.

The group $Wh_G^{iso}(X)$ has the following geometric model. Consider isovariant isomorphism classes of compact relative GCW-pairs (Y,X) in which the G-cells are all attached by isovariant maps and in which X is an isovariant strong deformation retract. Then $Wh_G^{iso}(X)$ is obtained by factoring out the equivalence relation generated by finite isovariant expansions and collapses. (This may be seen by analogs of the arguments in [29], [1], [28], and [19].)

The isovariant GCW category is a poor one in which to consider homotopy properties. For example, the inclusion $\cup_{H\neq(e)} X^{(H)} \to X$ is never an isovariant co-fibration when there are orbits of type G. Therefore, it is convenient to consider the equivariant maps. Illman [29] introduced an equivariant geometric Whitehead group $Wh_G(X)$ following Cohen's geometric construction of the equivariant Whitehead group [19]. Its description is the same as that given in the previous paragraph except one allows equivariant cell attachment, deformation retraction, and expansion and collapse. In [29], [35], [36], [1], and [28] it is shown that

$$Wh_G(X) \cong \bigoplus_{(H)} \bigoplus_{(\alpha)} Wh(\pi_1(X_\alpha^H/K_{H\alpha}))$$

where again on the right we have the classical Whitehead groups.

If X_α^H has codimension ≥ 3 in X_β^K, whenever $X_\alpha^H \subset X_\beta^K$, then the natural homeomorphism $Wh_G^{iso}(X) \to Wh_G(X)$ carries $Wh_G^{iso}(X)$ isomorphically onto the summand $Wh_G(X)^\rho$ generated by those pairs (Y,X) in which $Y_\alpha^H \subset \bigcup_{K>H} Y^K$ if

$$X_\alpha^H (= Y_\alpha^H \cap X) \subset \bigcup_{K>H} X^K.$$

Under the codimension hypothesis, then, $Wh_G(M^n)^\rho$ classifies smooth or PL G-h-cobordisms $(W_i^{n+1}; M, M')$ on M such that $W_\alpha^H = M_\alpha^H \times I$ if $\dim M_\alpha^H \leq 4$.

Under these codimension hypotheses, duality allows relaxation of the definition of G-h-cobordism to simply requiring that the inclusions $M_i \to W$ be equivariant homotopy equivalences.

4. EQUIVARIANT CONTROLLED SIMPLE-HOMOTOPY THEORY AFTER CHAPMAN AND FERRY

Chapman [12-17] and Ferry [23-25] developed a geometric controlled simple-homotopy theory from Cohen's geometric Whitehead group [19]. This theory automatically generalizes to the equivariant setting to provide a controlled version of Illman's Wh_G. In fact, almost all of the theorems of [16] and [17] have analogs in the equivariant PL category. We here give the minimum of detail necessary to our discussion of the obstruction to the equivariant generalization of the Product Structure Theorem of [39] in Section 7. For more detail, see the announcement [57], the exposition [59], [61], [62], and [56]. (The reader interested in controlled simple-homotopy theory applied to group actions will also wish to consult [47], [48], [49], [50], [2], and [3].)

In a nutshell, the theory is as follows. (We sketch it for compact spaces.) Given is a control space X and an $\varepsilon > 0$. Let P be a finite G-complex (simplicial) with a control map $p : P \to X$. Objects of $Wh_G(P)_\varepsilon$ are equivalence classes of triples (Q,P,r) where (Q,P) is a finite pair of simplicial G-complexes and $r : Q \to P$ is an equivariant retraction with the property that there exists an (unspecified) equivariant strong $p^{-1}(\varepsilon)$-deformation retraction r_t of Q to P with $r_1 = r$, meaning that the diameters in X of the homotopy tracks $pr_t(x)$, $0 \leq t \leq 1$, are less than ε for each x in Q. The equivalence relation is essentially "$p^{-1}(\varepsilon)$ expansion and collapse" (cf. [17], [61]). For $\delta < \varepsilon$ there is a control-relaxation homomorphism $Wh_G(P) \to Wh_G(P)_\varepsilon$.

If the control map p is simplicial or a simplical p-NDR in the sense of [48] (a neighborhood deformation retraction of a simplicial map in the category of maps) the inverse system defined as $\varepsilon \to 0$ is stable, i.e., for sufficiently small ε the relaxation homomorphisms are essentially isomorphisms. We denote the inverse limit group by $Wh_G(P)_c$ and call it the controlled group. This may be extended to compact GANR's X by approximating them with finite G-complexes. The resulting inverse system is still stable if controlled by the identity of X.

Controlled $\tilde{K}_0$ groups $\tilde{K}_{0,G}(X)_\varepsilon$, $\tilde{K}_{0,G}(X)_c$ are defined as the transfer-invariant elements of $Wh_G(X \times S^1)_\varepsilon$, $Wh_G(X \times S^1)_c$. For compact GANR X and control by the identity map, these inverse limit groups are functors of the system of fundamental groups of the fixed point set components X^H_α.

The Chapman-Ferry mapping torus construction [24], [17] (cf. [40]) now automatically produces elements $\sigma_\varepsilon(X)$ of the $K_{0,G}(X)_\varepsilon$ groups that are natural with respect to control relaxation and generate a controlled equivariant finiteness obstruction $\sigma_c(X)$ in $\tilde{K}_{0,G}(X)_c$. Relaxation yields $\sigma_c(X) \to \sigma(X)$, the equivariant finiteness obstruction in $\tilde{K}_{0,G}(X)$, for any compact GANR X. A functorial direct sum formula like that for Wh_G holds with inequivariant controlled groups [61].

Let X be a locally compact G-ANR which is dominated by a finite-dimensional G-complex. The controlled Whitehead and K-groups of X with control by the identity map may be computed as follows. There are a filtration of $Wh_G(X)_c$, the zeroth group of which we denote F_0, and exact sequences

$$H_2^{G,lf}(X;\tilde{K}_{0,G}) \to H_0^{G,lf}(X;Wh_G) \to F_0 \to 0$$

$$H_3^{G,lf}(X;K_{-1,G}) \to H_1^{G,lf}(X;\tilde{K}_{0,G}) \to Wh_G(X)_c/F_0 \to H_2^{G,lf}(X;K_{1,G})$$

$$\to H_0^{G,lf}(X,\tilde{K}_{0,G}) \to \tilde{K}_{0,G}(X)_c \to H_1^{G,lf}(X;K_{-1,G}) \to 0,$$

where the homology is Bredon homology [6] with locally finite chains and the coefficient systems are given by application of $K_{-i,G}$ and Wh_G to orbits.

The above sequence is in fact a collapsed Atiyah-Hirzebruch spectral sequence. The collapse is due to Carter's Vanishing Theorem [10], which states that $K_{-i,G}(G/H) = 0$ for $i \geq 2$ and all $H \subset G$ (when G is finite).

5. EQUIVARIANT FAILURE OF TOPOLOGICAL INVARIANCE OF WHITEHEAD TORSION

Up to this point the equivariant handle theory has paralleled the inequivariant theory with minimal and predictable deviation. Now, however, when we start to consider the locally linear topological category, the entire framework becomes unglued. This was not unexpected. It has been known since Milnor's disproof of the Hauptvermutung for polyhedra [43] that Reidemeister torsions, combinatorial invariants computed from equivariant chain complexes, are not necessarily topological invariants.

As far as we are aware, Illman [30] was the first actually to verify that the equivariant Whitehead torsions $\tau(f)$ are not necessarily zero in Wh_G or its analogs. Browder and Hsiang [8] gave a beautiful proof that if G is finite and acts smoothly and semi-freely on a manifold M with $\text{Dim } M \geq \dim M^G + 3 \geq 7$ and $\pi_1 M = \pi_1 M^G = 0$, then each element of Wh(G) may be represented by a smooth G-h-cobordism that is equivariantly homeomorphic with $M \times I$. Of course, it uses an infinite construction. In [52], there is an analogous result.

What is actually being discovered in these papers is that those elements in $Wh_G(X)$ that "live on orbits", i.e., are induced by naturality from $Wh_G(*)$ or $Wh_G(G/H)$, are never topologically invariant in G-h-cobordisms of n-manifolds. (This may be derived from the Equivariant α-Approximation Theorem of [58] (cf. [23], [18]), since elements that live on orbits have arbitrarily fine control. This theorem states that in the category of locally linear G-manifolds satisfying the codimension ≥ 3 hypothesis of the section 3, there is for each M^n an open cover α such that an equivariant α-homotopy equivalence $f : N^n \to M^n$ that is a homeomorphism on $f^{-1}(\cup\{M_i^H \mid \dim M_i^H \leq 4\})$ is equivariantly homotopic to an equivariant homeomorphism. It has as an immediate corollary an Equivariant Thin h-Cobordism Theorem, in the same category: given M^n there exists an open cover α of M^n such that each equivariant α-h-cobordism $(W^{n+1};M,M')$ on M such that $W_i^H = M_i^H \times I$ whenever $\dim M_i^h \leq 4$ is equivariantly homeomorphic with $M \times I$. (Here M is being used as the control space and either $f : N \times M$ or some deformation retraction $r : W \to M$, as the control map for the definition of α-homotopy equivalence.)

Dovermann and Rothenberg used this fact in [20] to produce an equivariant h-cobordism theorem.

THEOREM 5.1. ([20]). Let the finite group G act smoothly on the manifold M such that $\dim M^H \geq \dim M^K + 3 \geq 8$ whenever $K \subset H$. If M and each M^H are simply connected, then all equivariant smooth h-cobordisms on M are equivariantly homeomorphic with $M \times I$.

We have an analysis of this phenomenon of topological non-invariance from a geometric-topological viewpoint [57], [59], [61], [62], [56].

In $Wh_G(X)$, decree torsions to be equivalent if they have representatives that become equivariantly homeomorphic upon stabilization by Cartesian product with a linear G-disc or even by a universal linear G-action on a Hilbert cube (the product of infinitely many copies of the unit disc of the regular real representation $\mathbb{R}[G]$). Let $Wh_G^{htop}(X)$ denote the quotient of $Wh_G(X)$ by the resulting relation. It deserves to be called the topological equivariant Whitehead group because it is the group that classifies equivariant h-cobordisms on X admitting equivariant handle decompositions up to equivariant homeomorphism, when all fixed point components of X are of dimension ≥ 5 and no fixed point component has codimension ≤ 2 in another. To get a clean realizability statement, we need to use $Wh_G^{htop}(X)^{\rho}$, the quotient of $Wh_G(X)^{\rho}$.

THEOREM 5.2 [61]. Let G be a finite group acting locally linearly on the manifold M. If M satisfies the dimension and codimension hypotheses above, then an equivariant h-cobordism (W,M) which admits a relative equivariant handle decomposition is equivariantly homeomorphic with $M \times I$, rel. $M = M \times 0$, if and only if its equivariant torsion $\tau(W,M) \in Wh_G(M)$ vanishes in $Wh_G^{htop}(M)^{\rho}$; all torsions in $Wh_G^{htop}(M)^{\rho}$ are represented by such G-h-cobordisms on M.

It turns out that analogously with the inequivariant case ([13], [21], [65]) the stable homeomorphism classes of polyhedral pairs form the same group as the equivalence classes under cell-like mappings. Inequivariantly it is the same as the Whitehead group; equivariantly, it is the group of the above equivariant topological s-cobordism theorem (which turns out also to be a functor of the system $\{\pi_1(M_\alpha^H)\}$ of fundamental groups of fixed point components).

THEOREM 5.3 [61]. If in $Wh_G(X)$ we enlarge the equivalence relation to include the equivariant cell-like maps, the quotient group is naturally isomorphic with $Wh_G^{htop}(X)$.

6. MANIFOLDS WITHOUT EQUIVARIANT HANDLE DECOMPOSITIONS

In [48], Quinn gave the first explicit examples of locally linear actions on manifolds that do not admit equivariant handle decompositions. The construction is by an infinite process converging to a point in the fixed point set of an action on a disc and is an adaptation of a construction Ferry used to prove that compacta dominated by finite complexes may not be homotopy equivalent to finite complexes [26]. The controlled end theory is used to prove local linearity, and Oliver's equivariant finiteness theory [46] is used to show that the Euler characteristic of the fixed point set is not compatible with the action being equivariantly homotopy equivalent with a G-CW-complex.

In the same paper, Quinn gives an obstruction that must vanish if there is to be an equivariant mapping cylinder neighborhood in the manifold M (with boundary a manifold) of the singular set M* (the union of the fixed point sets of the various subgroups of G). If there is an equivariant handle decomposition of M then the union of the handles of non-free types is just such a neighborhood of the singular set, so this obstruction must vanish if there is a handle decomposition of M.

As noted in Section 4, a compact G-manifold M admitting an equivariant handle decomposition must have finiteness obstruction $\sigma(M) = 0$. By subdividing a handle decomposition, we conclude that also $\sigma_c(M) = 0$, using the stability of the inverse system for $\tilde{K}_{0,G}(M)_c$. Thus, the controlled finiteness obstruction of M is an obstruction to giving M a handle decomposition.

Interestingly, the relaxation map $\phi : \tilde{K}_{0,G}(X)_c \to \tilde{K}_{0,G}(X)$ is not always surjective, even for finite G-complexes [57] and [61]. This fact has the interesting consequence that not all elements of the group $\tilde{K}_{0,G}(X)$ are the obstructions to finiteness of compact G-manifolds, since a G-manifold's equivariant finiteness obstruction is the image of its controlled equivariant finiteness obstruction. (A finite simplicial G-complex can always be expanded to a compact G-manifold with boundary by embedding appropriately in a representation space and taking a regular neighborhood.)

More interestingly yet, the relaxation ϕ is not necessarily one-to-one [64]. Thus, there are locally linear, compact G-manifolds with non-vanishing controlled equivariant finiteness obstruction that are equivariantly homotopy equivalent (without control) to finite G-complexes.

A realization argument [59], [61] producing compact G-manifolds with arbitrary obstruction to controlled equivariant finiteness now produces compact G-manifolds that do not admit equivariant handle decompositions but do have the equivariant homotopy type of finite G-complexes [64]. A somewhat different program with similar ideas [56] shows that there are often equivariant (locally linear) h-cobordisms (W,M) in which M has a smooth or PL action yet the h-cobordism fails to admit equivariant handle decompositions on M. This is a very generally occurring phenomenon, since the aforesaid infinite process construction may often be employed to realize obstructions to controlled equivariant finiteness.

7. EQUIVARIANT HANDLE DECOMPOSITIONS FOR G-MANIFOLDS WITH VANISHING CONTROLLED FINITENESS OBSTRUCTION

In this section we indicate how to adapt the Kirby-Siebenmann argument [39; Essay III] for handle decompositions of n-manifolds, $n > 5$, to the equivariant, locally linear context when all fixed point components are of dimension > 5 and there are no incident pairs of them with one of codimension less than three in another. (A complete proof appears in [60].) For the rest of this section all actions are assumed to satisfy these hypotheses.

First let us review the essentials of the argument of Kirby and Siebenmann. Let M be a compact n-manifold without boundary covered by charts $U_1, \cdots, U_n$ containing compact sets $A_1, \cdots, A_n$, respectively, which also cover M. Suppose for induction that H_i is a handle decomposition of a closed codimension 1 submanifold M_i, with bicollared boundary, containing $A_1 \cup \cdots \cup A_i$ in its interior. Let Σ be the PL structure on U_{i+1} imported from R^n. Use the Product Structure Theorem [39; Essay I] to isotope Σ to a new PL structure Θ in which $\partial M_i \cap U_{i+1}$ is a PL submanifold. Now triangulate Θ so that $\partial M_i \cap U_{i+1}$ is a subcomplex and by subdividing and starring, obtain a finite handle decomposition of part of $U_{i+1} - \mathrm{int}(M_i)$ containing $A_{i+1} - \mathrm{int}(M_i)$. Adding this to H_i produces a handle decomposition H_{i+1} and the induction continues.

All of the above steps may easily be carried out equivariantly in general if the Product Structure Theorem holds, so it fails.

This is essentially an equivariant, controlled end problem. To be precise, let M be a G-manifold (without boundary, for convenience) and let Σ be a locally linear PL structure on $M \times R$. We shall say that Σ admits a product structure if there is an equivariant topological isotopy $f_t : M \times R \to M \times R$ from the identity to a PL homeomorphism from Σ to a product structure $\Theta = \theta \times R$ where θ is a locally linear equivariant PL structure on M. Consider now the projection $p : M \times R \to M$. Using this as a control map, consider the equivariant controlled end problem posed by Σ. The problem is whether there is a completion of $M \times R$ to a compact locally linear PL G-manifold N and extension of p to $p' : N \to M$. If there is such a completion, then the ends of Σ are products since the boundary of N is equivariantly PL collared. Using the sliced Concordance Implies

Isotopy Theorem of [39; Essay II], which unlike the unsliced one is not obstructed, together with results of [59] and [58] produces the desired product structure and isotopy Λ. The image $\lambda_1(M \times 0)$ can now be adjusted by an ambient homeomorphism h with support bounded in the R-direction so that $h\lambda_1(M \times 0) = M \times 0$. Now the pull-back $\theta^* = \lambda_1^* h^* (\theta \times R)$ is a "chart" for the desired adjustment Σ' of Σ in which $M \times 0$ is a pl submanifold (see [60]).

One obtains a controlled obstruction $\sigma_c(\varepsilon)$ in $\tilde{K}_{0,G}(M)_c$ from considering, say, the positive end (R_+). This is entirely analogous to the end invariant of [53].

THEOREM 1 (Equivariant Product Structures [60]). Σ admits a product structure if and only if the positive and invariant $\sigma_c(\varepsilon)$ vanishes.

Let us return to the Kirby-Siebenmann argument above and its notation. It would be nice if we could assert that the hypotheses under which we are working automatically guarantee that in the inductive step the end obstruction of the restriction of Σ to a small equivariant topological collar of $M_i \cap U_{i+1}$ in U_{i+1} is always zero. This is not the case. What is true, however, is that the obstruction is carried on the "equivariant 1-skeleton" of $N_{i+1} = M_i \cap U_{i+1}$.

It takes some work to produce the "equivariant 1-skeleton", which is by definition an embedded G-CW-complex L in N_{i+1} of dimension one such that each equivariant map $f : (S,T) \to (N,L)$ where (S,T) is an equivariant relative 1-complex, deforms equivariantly into L, rel. T.

PROPOSITION 1. Under the hypotheses of Theorem 1, suppose that $L \subset M$ is an equivariant 1-skeleton. Let $\Sigma' = \Sigma|(M - L) \times R$. Then the positive end invariant $\sigma_c(\varepsilon')$ of Σ' is zero.

Proof: This is immediate from the exact sequence from Section 4.

What we really need is handlebody 1-skeleta, i.e., handlebodies comprised of handles of index ≤ 1, the G-CW-spines of which are equivariant 1-skeleta. These we obtain by some embedded surgery which is made possible by the following results which follow from the equivariant product structure theorem above (Theorem 1). Cf. [60].

COROLLARY 1. Let V be a G-representation with dim $V^G \geq 5$ and satisfying our standing hypotheses. Then the stabilization map

$$\pi_0(Top_G(V)/PL_G(V)) \longrightarrow \pi_0(Top_G(V \oplus R)/PL_G(V \oplus R))$$

is onto.

COROLLARY 2. Let V and W be G-representations satisfying the hypotheses of Corollary 1. Then $V \oplus R$ is equivariantly homeomorphic with $W \oplus R$ if and only if V is equivariantly homeomorphic with W.

THEOREM 2. Let M be a locally linear G-manifold satisfying our standing hypotheses. Then equivariant handlebody 2-skeleta exit in M; moreover, each equivariant 2-skeleton in M is equivariantly simple-homotopy equivalent to the spine of one.

Now returning to the Kirby-Siebenmann proof (and its notation), we have M_i and $N = \partial M_i \cap U_{i+1}$ and the equivariant locally linear PL structure Σ on U_{i+1}. We can now use Proposition 1 and Theorem 2 to find a equivariant 1-skeleton L for N and then Theorem 1 to alter Σ on a collar neighborhood of N - L so that N - L is a PL submanifold. Extending L to a handlebody 1-skeleton, L^*, we get an extension of the handlebody decomposition over all of $M_i \cup U_{i+1} - L^*$. Inductively, we get a handle decomposition of an open subset of M which is the complement of the successive L^*'s. The argument is now completed by an inductive application of the following.

PROPOSITION 2. Let M be a manifold satisfying the hypotheses of this section, and let L be a handlebody in M with all handles of index ≤ 2. Suppose that M - L admits an equivariant handlebody decomposition. Then M admits an equivariant handle decomposition if and only if $\sigma_c(M) = 0$.

8. OBSTRUCTIONS TO EQUIVARIANT HANDLE DECOMPOSITION WITHOUT THE CODIMENSION RESTRICTION

The argument of the previous section shows that the equivariant controlled finiteness obstruction is also the obstruction to equivariant handle decomposition for locally linear finite group actions on manifolds provided that all fixed point components M^H_α have dimension ≥ 6 and none has codimension ≤ 2 in another. This codimension hypothesis comes from the use of equivariant Whitehead and $\tilde{K}_0$ functors that are unable to detect the influence of the infinite cyclic subgroup introduced in the codimension two cases and of Carter's Vanishing Theorem [10] for $K_{-i}(\mathbb{Z}[H])$ when $i \geq 2$ and H is finite. By using isovariant obstruction groups and hypothesizing the effect of Carter's theorem, we can state a theorem without the "dimension gap" hypothesis.

Let $\tilde{K}_{0,G}(M)^{iso}_c = \oplus_{(H)} \oplus_{(\alpha)} \tilde{K}_0(M^H_\alpha - M^{>H}_\alpha)/W_\alpha)_c$ where the control map is in each summand $(M^H_\alpha - M^{>H}_\alpha) \to M/G$. Let

$$\sigma_c(M)^{iso} = \Sigma_{(H)\ (\alpha)}\sigma_c(M^H_\alpha - M^{>H}_\alpha) \in \tilde{K}_{0,G}(M)^{iso}_c .$$

Here $M^{>H}_\alpha$ denotes the subset of M^H_α of points of orbit type smaller than G/H.

By the local representations of the locally linear action of G on M we mean the topologically defined linear isotropy subgroup representations. Note that there is essentially one for each conjugacy class (H) of subgroups of G and NH-class (α) of components of M^H.

THEOREM (see [56]). Suppose that the finite group G acts locally linearly on the manifold M and that there is an integer k such that for each local representation V of $H \subset G$, $K_{-i,H}(V)^{iso} = 0$ for all $i \geq k$. If for each subgroup H of G and component $M_{H,\alpha}$ of M^H we have $\dim M_{H,\alpha} \geq \max(2k + 1, 6)$, then M admits an equivariant handle decomposition if and only if $\sigma_c(M)^{iso} = 0$ in $\tilde{K}_{0,G}(M)^{iso}_c$.

Here $K_{-i,H}(V)^{iso} = \bigoplus_{(K)} \bigoplus_{(\alpha)} K_{-i}(\mathbb{Z}[\pi_\alpha])$, where (K) varies over the H-conjugacy classes of subgroups of H, (α) runs through the equivalence classes of components of V_K under the action of the normalizer $N_H K$ of K in H, and π_α is the fundamental group of a component α representing (α).

There is a relative version of this in the case that we wish to find a handle decomposition of M on a codimension zero submanifold of its boundary.

We observe that the hypotheses of this theorem are satisfied for k = 2 if G is of odd order and abelian, a 2-primary cyclic group, or if G is arbitrary and there are no codimension 2 inclusions of fixed-point components in M.

ACKNOWLEDGEMENT

Second author's talk at 1985 Georgia Topology Conference, Athens, Ga. Research was partially supported by the NSF.

REFERENCES

1. Anderson, D. R., Torsion Invariants and Actions of Finite Groups, Mich. Math. J. 29 (1982), 27-42.
2. Anderson, D. R., and Munkholm, H. J., The Simple Homotopy Theory of Controlled Spaces, preprint.
3. Anderson, D. R., and Munkholm, H. J., The Algebraic Topology of Controlled Spaces, preprint.

4. Bierstone, E., Equivariant Gromov Theory, Topology 13 (1974), 327-345.

5. Bing, R. H., A Homeomorphism Between the 3-sphere and the Sum of Two Solid Horned Spheres, Ann. Math. 56 (1952), 354-362.

6. Bredon, G., Equivariant Cohomology Theories, Springer Lecture Notes in Math. v. 34 (1967).

7. Bredon, G., Introduction to Compact Transformation Groups, Academic Press, New York, 1972.

8. Browder, W., and Hsiang, W.-C., Some Problems on Homotopy Theory, Manifolds, and Transformation Groups, Proc. Symp. Pure Math. XXXII, part 2 (1978), 251-267.

9. Browder, W., and Quinn, F., A Surgery Theory for G-manifolds and Stratified Sets, Manifolds, Univ. of Tokyo Press, Tokyo, 1973.

10. Carter, D., Lower K-theory of Finite Groups, Comm. Algebra 8 (1980), 1927-1937.

11. Chapman, T. A., Topological Invariance of Whitehead Torsion, Amer. J. Math. 96 (1974), 488-497.

12. Chapman, T. A., Cell Like Mappings, Springer Lecture Notes in Math. v. 482, 230-240.

13. Chapman, T. A., Simple Homotopy Theory for ANR's, Gen. Top. and App. 7 (1977), 165-174.

14. Chapman, T. A., Homotopy Conditions Which Detect Simple Homotopy Equivalences, Pac. J. Math. 80 (1979), 13-46.

15. Chapman, T. A., Invariance of Torsion and the Borsuk Conjecture, Can. J. Math. XXXII (1980), 1333-1341.

16. Chapman, T. A., Controlled Boundary and h-cobordism Theorems, Trans. Amer. Math. Soc. 280 (1983), 73-95.

17. Chapman, T. A., Controlled Simple Homotopy Theory, Springer Lecture Notes in Math. v 1009 (1983).

18. Chapman, T. A., and Ferry, S., Approximating Homotopy Equivalences by Homeomorphisms, Amer. J. Math. 101 (1979), 583-607.

19. Cohen, M. M., A Course in Simple-Homotopy Theory, Springer Grad. Texts in Math. v. 10 (1970).

20. Dovermann, K. H., and Rothenberg, M., An Equivariant Surgery Sequence and Equivariant Diffeomorphism and Homeomorphism Classification, Springer Lecture Notes in Mathematics v. 741 (1979).

21. Edwards, R. D., On the Topological Invariance of Simple Homotopy Type for Polyhedra, Amer. J. Math. 100 (1978), 667-683.

22. Edwards, R. D., and Kirby, R. C., Deformations of Spaces of Embeddings, Ann. Math. 93 (1971), 63-88.

23. Ferry, S., The Homeomorphism Group of a Compact Hilbert Cube Manifold is an ANR, Ann. Math. 106 (1977), 101-119.

24. Ferry, S., A Simple-Homotopy Approach to the Finiteness Obstruction, Shape Theory and Geometric Topology, Springer Lecture Notes in Math. v. 870 (1981), 73-81.

25. Ferry, S., Lectures at University of Kentucky, Spring, 1980.

26. Ferry, S., Finitely Dominated Compacta Need not have Finite Type, Springer Lecture Notes in Math. v. 870 (1981), 1-5.

27. Giffen, C. H., The Generalized Smith Conecture, Amer. J. Math. 88 (1966), 187-198.

28. Hauschild, H., Aequivariante Whiteheadtorsion, Manuscr. Math. 26 (1970), 63-82.

29. Illman, S., Whitehead Torsion and Group Actions, Ann. Acad. Sci. Fenn. Ser. Al Math. No. 588 (1974).

30. Illman, S., personal communication, 1976.

31. Illman, S., Smooth Equivariant Triangulations of G-manifolds for G a Finite Group, Math. Ann. 262 (1978), 199-220.

32. Illman, S., Approximation of G Maps by Maps in Equivariant General Position and Imbeddings of G-complexes, Trans. Amer. Math. Soc. 262 (1980), 113-157.

33. Illman, S., The Equivariant Triangulation Theorem for Actions of Compact Lie Groups, Math. Ann. 262 (1982), 487-501.

34. Illman, S., Recognition of Linear Actions on Spheres, Trans. Amer. Math. Soc. 274 (1982), 445-478.

35. Illman, S., Actions of Compact Lie Groups and Equivariant Whitehead Torsion, preprint.

36. Illman, S., Equivariant Whitehead Torsion and Actions of Compact Lie Groups, preprint.

37. Kahn, P. J., and Steinberger, M., Equivariant Structure Spaces, in preparation.

38. Kervaire, M., Le Theoreme de Barden-Mazur-Stallings, Comm. Math. Helv. 40 (1965), 31-42.

39. Kirby, R. C., and Siebenmann, L. C., Foundational Essays on Topological Manifolds, Smoothings, and Triangulations, Ann. Math. Studies v. 88 (1977).

40. Kwasik, S., Wall's Obstructions and Whitehead Torsion, Comment. Math. Helv. 58 (1983), 503-508.

41. Lashof, R., and Rothenberg, M., G-smoothing Theory, Proc. Symp. Pure Math. XXXII Part 1 (1978), 211-266.

42. Madsen, I., and Rothenberg, M., On the Classification of G-spheres III: Top Automorphism Groups, preprint.

43. Milnor, J., Two complexes which are homeomorphic but combinatorially distinct, Ann. Math. 74 (1961), 575-590.

44. Milnor, J., Morse Theory, Ann. Math. Studies v. 51, Princeton Univ. Press.

45. Milnor, J., Whitehead Torsion, Bull. Amer. Math. Soc. 72 (1966), 358-426.

46. Oliver, R., Fixedpoint Sets of Group Actions on Finite Acyclic Complexes, Comm. Math. Helv. 59 (1975), 155-177.

47. Quinn, F., Ends of Maps I, Ann. Math. 110 (1978), 275-331.

48. Quinn, F., Ends of Maps II, Inv. Math. 68 (1982), 353-424.

49. Quinn, F., Ends of Maps III, J. Diff. Geo. 17 (1982), 503-521.

50. Quinn, F., Weakly Stratified Sets, preprint.

51. Ranicki, A., Algebraic and Geometric Splittings of the K- and L-groups of Polynomial Extensions, preprint.

52. Rothenberg, M., Torsion Invariants and finite transformation groups, Proc. Symp. Pure Math. XXXII, Part 1 (1978), 267-311.

53. Siebenmann, L. C., The Obstruction to Finding a Boundary for a Manifold of Dimension ≥ 5, Thesis, Princeton University, 1965.

54. Siebenmann, L. C., Disruption of Low-dimensional Handlebody Theory by Rohlin's Theorem, Topology of Manifolds, ed. J. C. Cantrell and C. H. Edwards, Markham, Chicago, 1970.

55. Siebenmann, L. C., Deformation of Homeomorphisms on Stratified Sets I, II, Comm. Math. Helv. 47 (1972), 123-163.

56. Steinberger, M., The Equivariant s-cobordism Theorem, (preprint).

57. Steinberger, M., and West, J., Equivariant h-cobordisms and finiteness obstructions, Bull. Amer. Math. Soc. 12 (1985), 217-220.

58. Steinberger, M., and West, J., Approximation by Equivariant Homeomorphisms, Trans. Amer. Math. Soc. (to appear).

59. Steinberger, J., and West, J., Equivariant Geometric Topology, Proc. 1984 Banach center semester on topology, Warsaw (to appear).

60. Steinberger, J., and West, J., Controlled Finiteness is the obstruction to Equivariant Handle Decomposition, preprint.

61. Steinberger, M., and West, J., Equivariant Controlled Simple Homotopy Theory (in preparation).

62. Steinberger, M., and West, J., Universal Linear Finite Group Actions on Hilbert Cubes, Equivariant ANR Fibrations, and Hilbert Cube Manifold Bundles (in preparation).

63. Wasserman, A. G., Equivariant Differential Topology, Topology 8 (1969), 127-150.

64. Webb, D., Equivariantly Finite Manifolds with no Handle Structure, preprint.

65. West, J., Mapping Hilbert Cube Manifolds to ANR's, Ann. Math. 106 (1977), 1-18.

The Role of Knot Theory in DNA Research

D. W. SUMNERS / Department of Mathematics, Florida State University, Tallahassee, Florida

On the cover of Science, 12 July 1985, is a picture of a third century A.D. Roman bas relief, a representation of the knot 6_2. In the research article related to the cover picture [20], the knot 6_2 figures prominently in the proof of a model for genetic recombination. DNA molecules are long and string-like, and often naturally occur in closed circular form. In this arena, it is not surprising that knot theory is useful. It has been brought to bear on the study of the geometric action of various naturally occurring enzymes (called topoisomerases) which alter the way the DNA is embedded in 3-space. Large amounts of DNA are wound up and packed into the average cell. In fact, there is enough DNA in a 2 m. human body to stretch from the earth to the sun and back 50 times [13]! This of course means that the embedding of the DNA in the cell is exceedingly complicated. One role of the topoisomerase is to facilitate the central genetic events of replication, transcription, and recombination via geometric manipulation of the DNA. This manipulation includes promoting writhing (coiling up) of the molecule, passing one strand of the molecule through another via a transient break in one of the strands, and breaking strands and rejoining to different ends (a move performed by recombinant DNA enzymes).

The purpose of this article is to discuss the crucial role of knot theory in two of the most recent results [3], [20]. A great deal of activity has occurred in this area very recently, and a few of the papers of interest are included in the bibliography [3], [16], [20], and [21].

Any time a specialist in one area attempts to read papers in another, one finds similar (if not identical) ideas, but quite different terminology

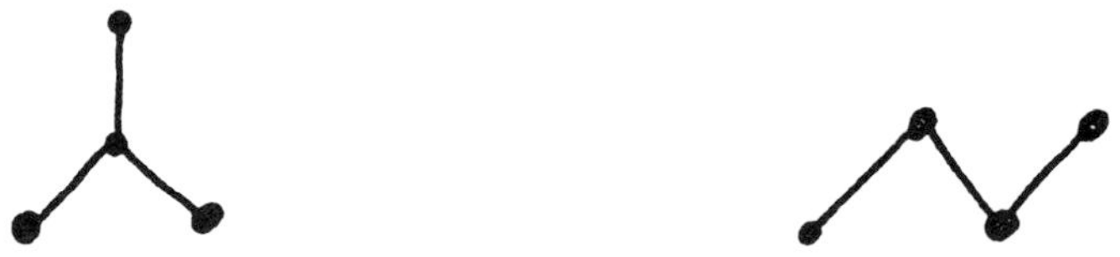

Isobutane **N-Butane**

FIGURE 1. Constitutional Isomers

for those ideas. I will begin by giving my version of the definitions of a few of the terms one sees when reading papers in chemistry and molecular biology. Any mistakes in the following interpretation of terminology and/or results are entirely my own.

MOLECULAR GRAPH. Vertices represent atoms, edges represent bonds (usually covalent bonds) between atoms. Often, peripheral structures (hydrogen atoms) are deleted from the picture. There is a great deal of interest in theoretical chemistry in the study of the intrinsic geometry (symmetries and enumeration) of molecular graphs [2], [6].

STEREOCHEMISTRY. The study of the extrinsic geometry of molecules -- the 3-dimensional configuration and symmetry of embedded molecular graphs. See the excellent survey article [16].

CONSTITUTIONAL ISOMER. Same constituent atoms, but non-isomorphic molecular graphs. See Figure 1.

TOPOLOGICAL ISOMER (Topisomer). Isomorphic molecular graphs, but embeddings are not isotopic. See Figure 2.

There is now and has been for some time a great deal of interest in the synthesis of exotically embedded molecules [16], [19], and [10]. At last report, no one has been able to synthesize (from scratch) a knotted molecule, but D. Walba of the University of Colorado is very close to the synthesis of the simplest knot, the trefoil. As we shall see, knots in circular DNA are produced in abundance <u>in vitro</u> (outside of the cell) by various enzymes (topoisomerases). Just recently, knotted circular DNA has been observed <u>in vivo</u> (in the cell) [18].

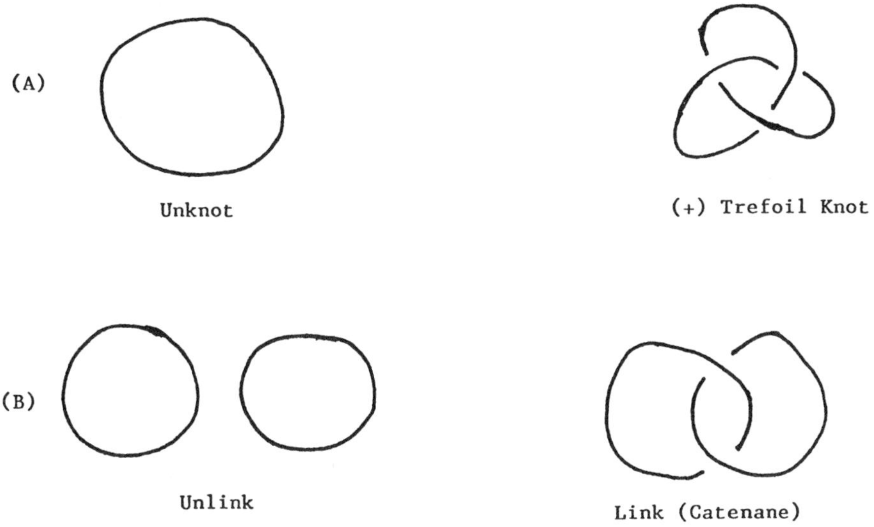

FIGURE 2. Topoisomers

CATENANE. A link of circular components. The name arises from the Latin "catena," or chain. A catenane is a molecule held together by topological (instead of chemical) bonds.

TOPOISOMERASE. Any chemical agent which forms topoisomers.

Now for some mathematical terms! If K is an unoriented knot (or link), then $\overline{K}$ denotes the mirror image of K, obtained by reversing the orientation of 3-space. As a practical matter, one obtains a projection of $\overline{K}$ from a projection of K by changing the sense of every crossover (node) in the diagram; i.e., the roles of over and under are reversed.

CHIRALITY. A knot K is chiral (or non-amphichiral) if K is not ambient isotopic to $\overline{K}$ ($K \neq \overline{K}$). One can make similar definitions for oriented knots and links. In chemistry, one calls a chiral pair ($K \neq \overline{K}$) a pair of enantiomers.

Chemists are very interested in chirality of molecules, both in their synthesis and detection. See [11] for an excellent example of the interaction of knot theory and synthetic chemistry in the arena of chirality.

One attempt to detect chirality of a knot (or oriented link) is the signed crossover number, first considered by Little [7].

CROSSOVER SIGN CONVENTION. (Known in chemistry as the "oriented skew lines convention" [16]). See Figure 3.

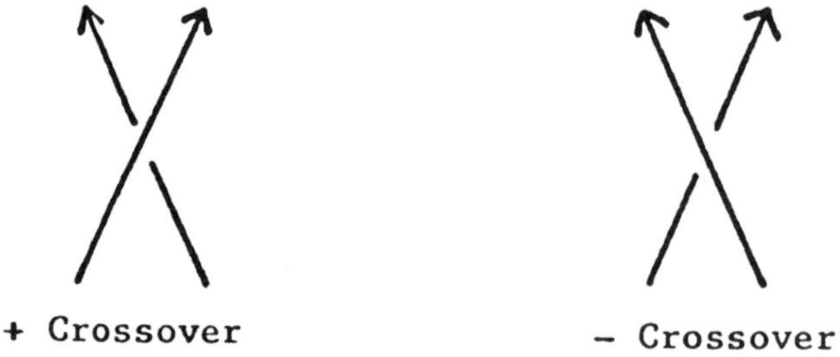

FIGURE 3. Crossover Sign Convention

SIGNED CROSSOVER NUMBER. Given a projection of a knot or oriented link K, assign either one of the two possible orientations to the knot, then add up the signed crossovers to obtain C(K). For a knot K, C(K) remains unchanged if the orientation of the knot itself is switched. In order to get a number which has a hope of being a topological invariant, one computes C(K) using minimal (minimum number of crossovers) projections.

As a chirality detector, if C(K) reveals an excess of right-hand (+) or left-hand (-) crossovers, then the knot or oriented link should be chiral. While this may be true for certain subclasses of knots and oriented links, in general, to my knowledge, this is unknown. I would like to thank S. Strogatz for clarifying conversations on C(K).

PROBLEM 1. Let C(K) denote the signed crossover number of a knot or oriented link K.

(a) For which knot families is C(K) a topological invariant? This is false in general. The infamous "Perko Pair" ($10_{161} = 10_{162}$) shown in Figure 4 is a pair of minimal 10-crossing projections of the <u>same</u> knot, with $C(10_{161}) = +8$ and $C(10_{162}) = +10$. The knot 10_{161} in Figure 4 is the mirror image of the knot labelled 10_{161} in the Rolfsen table [9].

The reason why they have different classification numbers in older knot tables is because knot theorists thought they were different knots! In 1974, K. Perko discovered the duplication in the tables. See Thistlethwaite's article [14] for an excellent discussion of this matter.

(b) For which families of knots does C(K) detect achirality (amphichirality)? That is, does $C(K) = 0 \Longrightarrow K = \overline{K}$? This is false in general, because $C(8_4) = 0$, but 8_4 is a chiral knot.

(c) For which families of knots does C(K) detect chirality? That is, does $C(K) \neq 0 \Longrightarrow K \neq \overline{K}$? This is also very likely false in general, but I am not aware at this writing of a specific counterexample for prime knots.

10_{161}

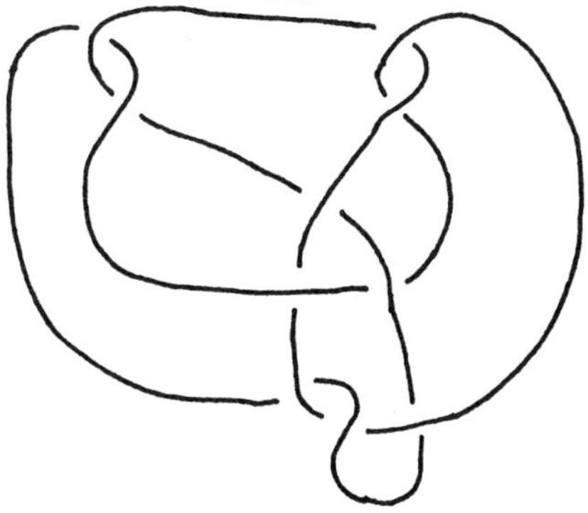

Twist(10_{161}) = +8

10_{162}

Twist(10_{162}) = +10

FIGURE 4. The Perko Pair (from Ref. 14)

One of the fundamental mathematical results in the modelling of the DNA double helix involves the differential geometry of a smoothly embedded ribbon in 3-space. The result described below was discovered by J. White [22] (see also [4]). A ribbon is a smooth embedding of $f : S^1 \times [-1,1] \to R^3$. The axis of the ribbon is $f(S^1 \times \{0\})$. We denote the ribbon by $R = f(S^1 \times [-1,1])$. Fix an orientation of the axis of R, and orient the two components of the boundary link ∂R the same way as the axis, as shown in Figure 5.

We wish to define geometric (but not topological) invariants of a ribbon R to correspond to "twist" (Tw) and "Writhe" (Wr). Obviously, the ribbons of Figure 6 (A and B) are topologically equivalent. We wish for Tw(R) to define a real number (geometrical, non-topological) invariant of the embedding of R that measures the amount the ribbon twists about its axis. We wish for Wr(R) to define a real number (geometrical, non-topological) invariant of the embedding of the axis of R that measures

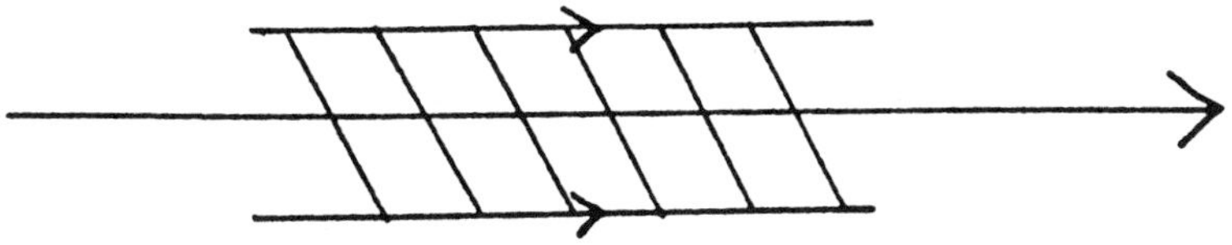

FIGURE 5. Orientation of ∂R

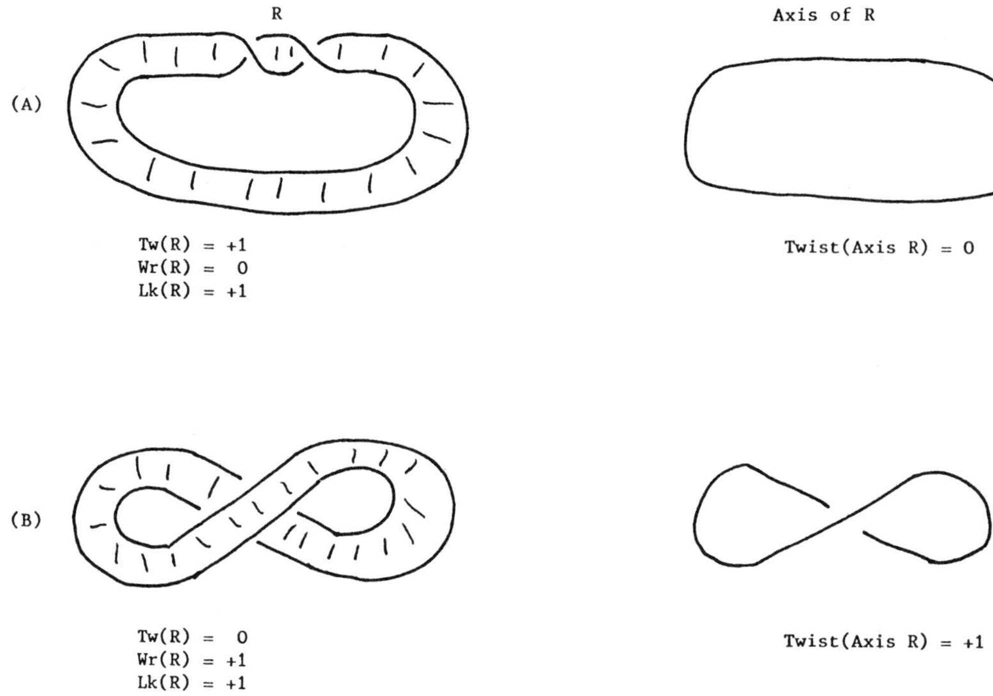

FIGURE 6. Twist and Writhe of R

the non-planarity of the axis. We define Lk(R) to be the homological linking number of the oriented 2-component link ∂R. Lk(R) is, of course, a topological invariant of R.

If the axis of R is planar (as in Figure 6A), then Wr(R) = 0 and Lk(R) = Tw(R) = +1. If the half-twists are taken out of the ribbon by an isotopy, and its axis is almost planar (as in Figure 6B), then Tw(R) = 0 and Lk(R) = Wr(R) = +1. In this special case, Wr(R) is claculated simply as the signed crossover number of Axis R, realizing of course that the projection of the axis is not minimal. The ribbon of Figure 6B is said to be "supercoiled," with one positive supercoil [1]. For precise definitions of Tw(R) and Wr(R) as integrals, the reader is referred to [22] and [4].

The result of fundamental importance to the mathematical modelling of DNA is the <u>Conservation Law</u> ([22], [4]).

$$Lk(R) = Tw(R) + Wr(R) \quad (1)$$

If (as is approximately true [12] for closed circular duplex DNA) $Tw(R) =$ constant, then we have the following geometric magnification effect due to the conservation law:

$$\Delta Lk = \Delta Wr \quad (2)$$

That is, a change of ± 1 in the linking number of ∂R (a small change in the geometry of R) is transformed to ± 1 supercoils of R itself (a large change in the geometry of R). As it turns out, changes in number of supercoils of a closed circular duplex (2-strand) DNA molecule are detectable in the laboratory.

The Crick-Watson DNA double helix consists of two strands (backbones) of alternating sugar and phosphate units. Each 5-carbon (pentagonal) sugar is attached to the phosphate unit on one side by the carbon atom designated 5', and to the phosphate unit on the other side by the carbon atom designated 3'. Therefore the backbone carries a natural chemical orientation 5' → 3'. Attached to each sugar is one of four bases: A = adenine, T = thymine, C = cytosine, G = guanine. The helical ribbon is formed by hydrogen bonding of the bases: A bounds only with its "dual" T (and vice versa), and C bonds only with its "dual" G (and vice versa). An A-T or C-G pair is called a "base pair." The backbones wind around each other to form a right-handed helical ribbon, the schematics of which are shown in Figure 7.

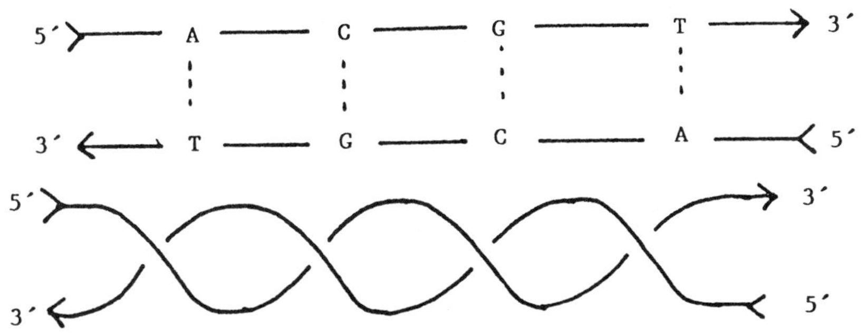

FIGURE 7. The Double Helix

As is evident in Figure 7, although each of the backbones of the double helix carries a natural chemical orientation, the axis of the ribbon is unoriented: the helix is right-handed from either direction. We adopt the same orientation convention for DNA as for the smoothly embedded ribbon. Orient the ribbon axis, and orient the backbones parallel to the axis (and forget the natural antiparallel orientation).

Since bonds can only be formed 3' to 5', closed circular duplex DNA forms a ribbon (not a Mobius band), and the dual backbones form a 2-braid link on the boundary of a solid torus neighborhood of the ribbon axis. The double helix has a natural preferred helical pitch = 10.5 base pairs per full twist in the ribbon. So, given a closed circular duplex DNA molecule whose axis is planar (Wr = 0), we say that such a molecule is "relaxed" (as opposed to "supercoiled"), and the relaxed linking number of a molecule with N base pairs is $Lk_0 = N/10.5$. Typically for closed circular duplex DNA, Lk_0 is large, because N is order 10^3 for a small virus, and order 10^6 for the DNA of a bacterium. Hence Lk_0 ranges from order 10^2 to 10^5. Given any starting configuration for a circular duplex DNA molecule, one can relax it by cutting one of the strands with the enzyme DNase, allowing the free ends to revolve until molecular strain is relieved, and then resealing the strand with the enzyme ligase.

In order to detect differences in the 3-dimensional configuration of DNA molecules, the main techniques are centrifugation and agarose gel electrophoresis. DNA has a net negative charge, so when placed in a gel across which exists an electrostatic potential difference, the DNA migrates through the gel towards the + electrode. In either centrifugation or electrophoresis, the speed at which a DNA molecule moves through the medium is controlled by its hydrodynamic radius (and by its charge, in the case of electrophoresis). In gel electrophoresis, given molecules of equal charge (topisomers), the more compact (supercoiled, knotted, etc.,) molecules migrate faster than the relaxed unknotted moleclues. Moreover, they migrate in discrete bands, which can be seen in gel electrophoresis by straining the gel with Ethidium Bromide (EB) and viewing under ultra-violet light.

When native (isolated from nature) circular duplex DNA's are viewed under the electron microscope, they appear very contorted (high writhing). They move faster under centrifugation and gel electrophoresis than the relaxed versions of the same molecule. Vinograd in 1965 proposed that this result is due to the fact that native DNA is underwound (relative to the relaxed state). That is, for the native DNA ribbon R, $Lk(R) - LK_0(R) < 0$. Or, native DNA has $(Lk(R) - Lk_0(R))$ negative supercoils, or, equivalently, has negative writhing number ($Wr < 0$). The size of a supercoil in native DNA is approximately 15 full twists of the band, or 158 base pairs. So, it is about 15 times as large as a full twist, and therefore much easier to detect in experiments.

The action of EB strain can be used in directed synthesis to change the linking and writhing number of circular duplex DNA molecules. First, one strand is cut with DNase, then the molecule is treated with EB. The EB untwists the ribbon, and the free ends of the broken strand swivel around. One then reseals the broken strand with ligase and washes out the EB. The result is a decrease in the linking number of the two strands, which is mirrored in the large-scale geometry of the molecule by an equal decrease in writhing (or increase in the number of negative supercoils).

I will now describe two exciting 1985 experimental results ([3], [20]) which were performed in the laboratories of N.R. Cozzarelli in Berkeley and A. Stasiak in Zurich. First a quotation from [3]: "In a topological approach to enzymology, the mechanism of an enzyme is deduced from topological changes that it introduces into its substrate. Because of the enormous variety in knot and catenane structure, much more information about a reaction can be garnered from the knot and catenane structure of products than from the product's primary structure or supercoiling. In addition, knot and catenane structure, unlike writhe, are topological invariants and thus cannot change during work-up and analysis of the DNA."

Topoisomerases are enzymes that abound in nature [17]. Type I topoisomerase (TOPO I) acts via the passage of single or double-stranded DNA through a transient single-strand break. Type II topoisomerase (TOPO II) acts via the passage of double-stranded DNA through a transient double-strand break. Figure 8 illustrates these processes.

It appears that the cells of all prokaryotic organisms (such as bacteria, whose genetic material is not enclosed in a nucleus) and all eukaryotic organisms (such as man, whose genetic material is enclosed in a nucleus) have both TOPO I and TOPO II [17]. The topoisomerases of prokaryotes and eukaryotes differ in mechanism. The elucidation of the differences among prokaryotic and eukaryotic topoisomerases (and the differences between TOPO I and TOPO II, for that matter) has used knot and link theory in non-trivial ways.

(A)

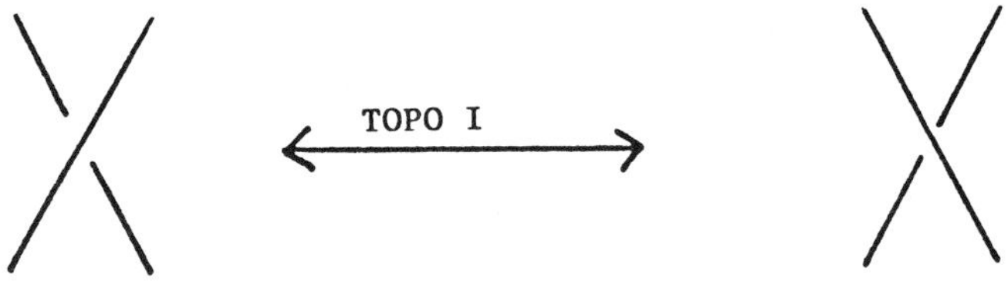

(B)

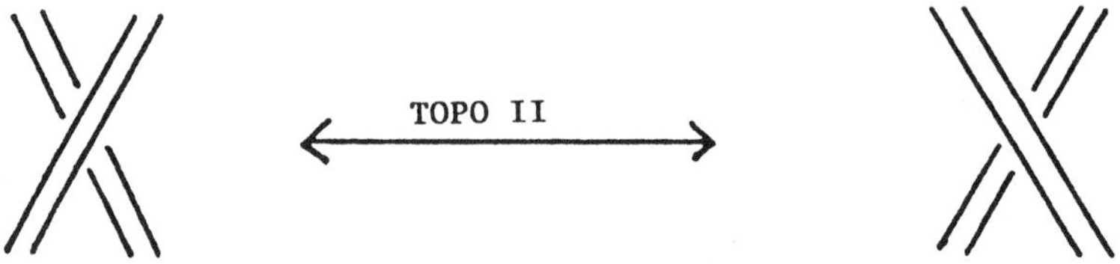

FIGURE 8. Topoisomerase Action

We begin with the study of the action of Escherichia coli TOPO I [3]. Prior to this research, knot theory had already proven to be very useful to molecular biologists. It was known that E. coli TOPO I relaxed negative supercoils in duplex circular DNA in increments of 1. Equivalently, it could change the linking number of the dual DNA backbone strands in increments of 1. The mechanism for this action is to nick one of the backbones, pass the other backbone through the break, then reseal the break. This enzyme was also known to form and remove knots from single stranded circular DNA, and to form links (catenanes) from unlinked single stranded circles. TOPO II, on the other hand, alters the linking number of circular duplex DNA in increments of two, and can make knots and links from unknotted and unlinked duplex DNA circles. TOPO I can change the knot or catenane structure of circular duplex DNA only if there is a pre-existing nick (break) in one of the strands. This paper studies the knotting action of TOPO I on uniquely nicked circular duplex DNA. The mechanism of TOPO I in this situation is to pass both intact strands through an enzyme-bridged transient break opposite the nick site, as shown in Figure 9.

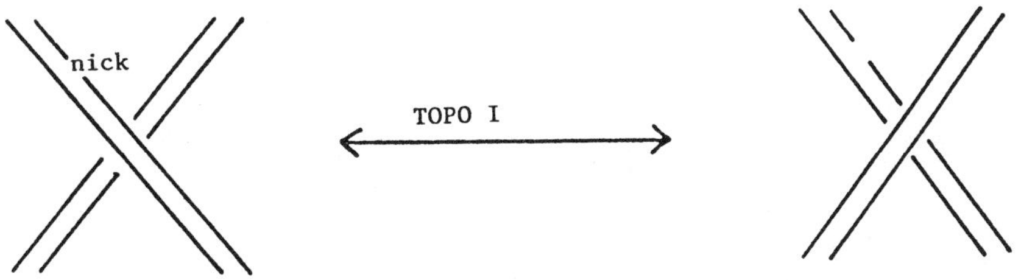

FIGURE 9. Mechanism of E. coli TOPO I

THE EXPERIMENT

The experiment proceeds by preparing unknotted circular duplex DNA with a single nick at a specific site. Because of the nick, the DNA is relaxed (no supercoiling). This substrate is then treated with E. coli TOPO I, and the reaction product is put through gel electrophoresis, forming a ladder of gel bands. The DNA is isolated from each of the rungs of the ladder, and then enhanced for electron microscopy (EM) viewing by coating with E. coli recA protein. This protein thickens the DNA from about 10 Angstroms to 100 Angstroms, allowing crossovers (nodes) in EM pictures to be unambiguously determined. The thickened molecules are stiffer than the unthickened ones, so some extraneous nodes are removed, and the knot projections end up closer to a minimal projection. Moreover, the right-hand helical structure of the DNA can be visualized in the thickened molecule, so that one can be certain that the true picture of the molecule (and not its mirror image) is being studied. The EM pictures are scored by a panel of two or three observers via isolating the crossovers one at a time with a mask which covers all of the molecule except the crossover under consideration.

THE RESULTS

Figure 10 shows in lane 1 a gel electrophoresis ladder of bands moving faster than the unknotted relaxed circle (top of band of lane 1).

In lane 1, the first knot band (second band from the top) corresponds to 3 crossing knots, the second knot band to 4 crossing knots, etc. That is, the knots in adjacent bands differ by one in crossover number (number of irreducible nodes) and the more complex knots migrate faster. Lane 2 shows a reference ladder of partially relaxed (un-nicked) supercoiled DNA generated by the action of TOPO I, with adjacent bands differing by one

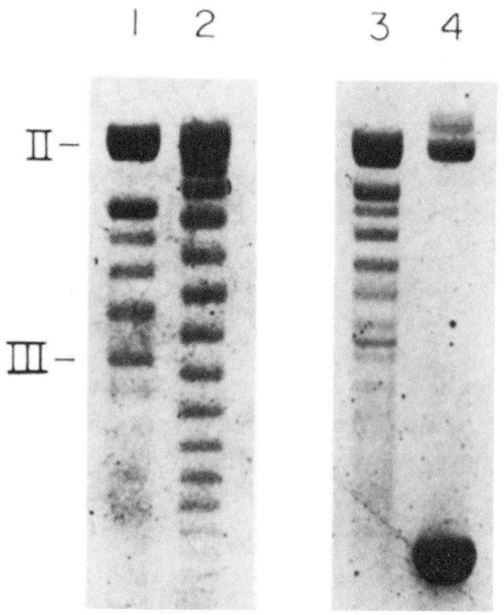

FIGURE 10. TOPO I Knot Ladder (from Ref. 3)

in the number of supercoils. The knot ladder is offset relative to the relaxation ladder, so partially relaxed unknotted molecules are not contaminating the knot ladder. Table I shows the correlation between number of irreducible nodes and band position. The authors draw the conclusion that the electrophoretic mobility is determined by the crossover number (n), despite the variety of knot types possible for an n-crossing knot.

Each knot in Table I was scored as to structure and chirality. The number of possible kinds of knots with a given number of nodes equals twice the number of chiral knots plus the number of amphichiral knots and composite knots.

TABLE I: Correlation between the number of irreducible knot nodes and electrophoretic mobility

Knot band	Number of knot nodes						
	3	4	5	6	7	8	9
	number observed						
1	73	1					
2		16	1				
3			8		1		
4			1	12			1
5				1	11		
6					2	4	2

Source: Ref. 3

Pictures of some of the DNA knots found are shown in Figure 11. The subscript "c" denotes composite (non-prime) knots. Knot 6_c is, for example, the square knot.

In terms of variety of structure, a total of 136 knots were observed. Counting mirror images (enantiomers) and composites, there are 15 distinct unoriented knots of six or fewer crossings. All were observed. There are 16 seven-crossing knots. Ten of these were observed. Of knot types with 2 enantiomers, both were observed in 8 of 9 cases in which more than one knot was observed. Table II shows the structural variety of DNA knots. As is evident, TOPO I produces all possible knots!

There are two 3-node knots, the (+) trefoil (C(K) = +3) and the (-) trefoil (C(K) = -3). The trefoils were examined for bias toward (+) and (-). The researchers found 44 (+) trefoils and 33 (-) trefoils, and concluded no bias, or that TOPO I does not prefer to manufacture one crossover type (+ or -) over the other.

TOPO I produces more (a yield of about 10%), and more complex knots than were expected in this experiment. Computer simulation [15], [8] of the knotting probability of a randomly embedded circular chain of the same length as the DNA predicts a 2% yield from a random strand-crossing reaction, and at that, all trefoils. The conclusion is drawn that TOPO I must promote deformation and folding up (writhing) of the molecule to create nodes, as well as inverting some nodes via strand passage. Moreover, TOPO I must stabilize folds in the molecule by binding to them, otherwise the more complex knots would become disfavored and untied by random crossing changes [8].

TOPO I must recognize the polarity (natural biological orientation) of the strand it breaks, because in this experiment it forms transient breaks only opposite the nick site. The polarity of the strand passed through, as well as the sign of the node inverted, is not recognized by TOPO I because it produces an even (racemic) distribution of enantiomers.

One other conclusion drawn is that TOPO I acts differently than the phage lambda integrative recombination system, where only (+) torus knots with odd node numbers are produced, and the recombinant enzyme Tn3 resolvase (to be discussed next), where only the knots 4_1 and 6_2 are produced, but no 3 or 5 node knots are produced.

We now move to a study of the mechanism of Tn3 resolvase, a recombinant DNA enzyme [20]. This is a site-specific enzyme that operates on circular duplex DNA. The resolution sites are relatively short, specific base pair sequences existing in the molecule. The base pair sequence of

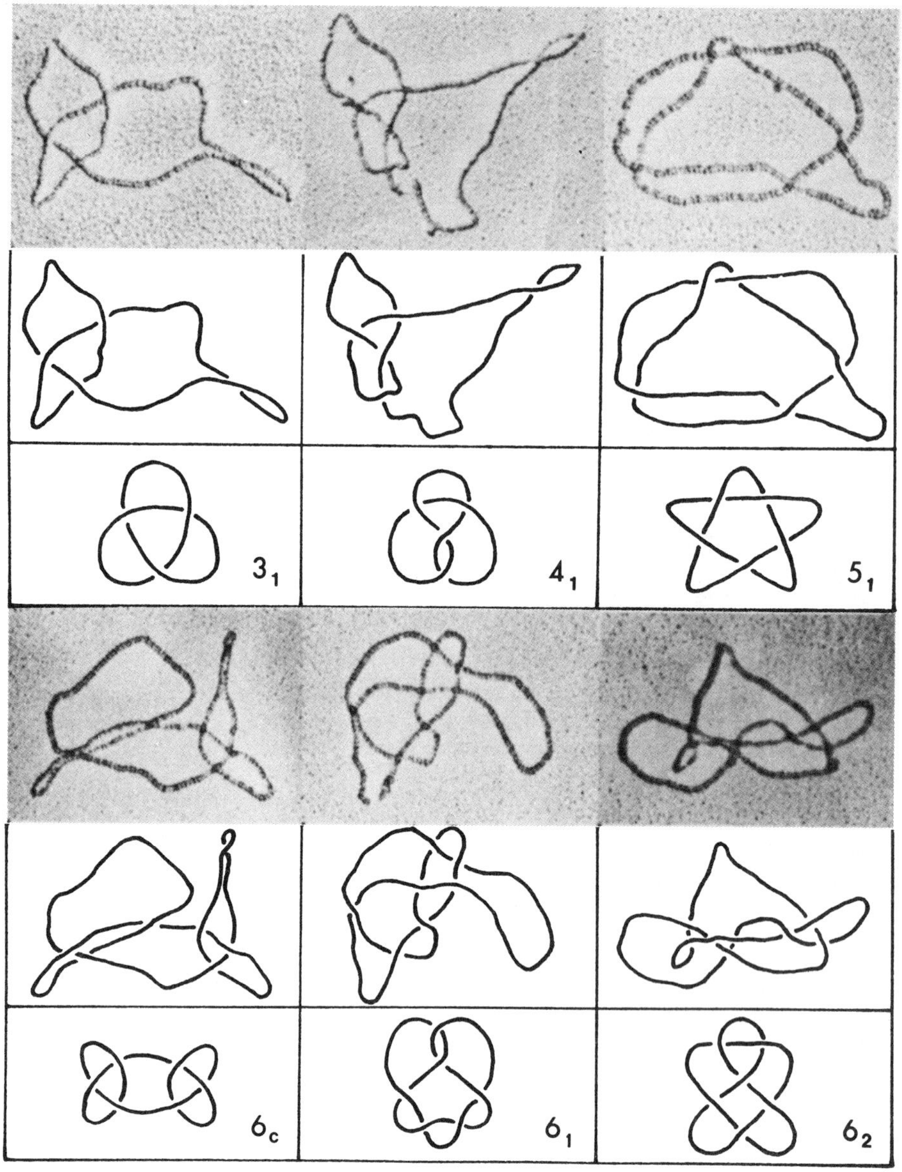

FIGURE 11. DNA Knots Produced by TOPO I.(from Ref. 3)

TABLE II: Variety of structure among topoisomerase I knots

Number of nodes in a knot	Number of observed knots	Observed kinds of knots	Possible kinds of knots
3	73	2	2
4	17	1	1
5	10	4	4
6	13	8	8
7	14	10	16
8	4	4	51
9	3	3	116

Source: Ref. 3

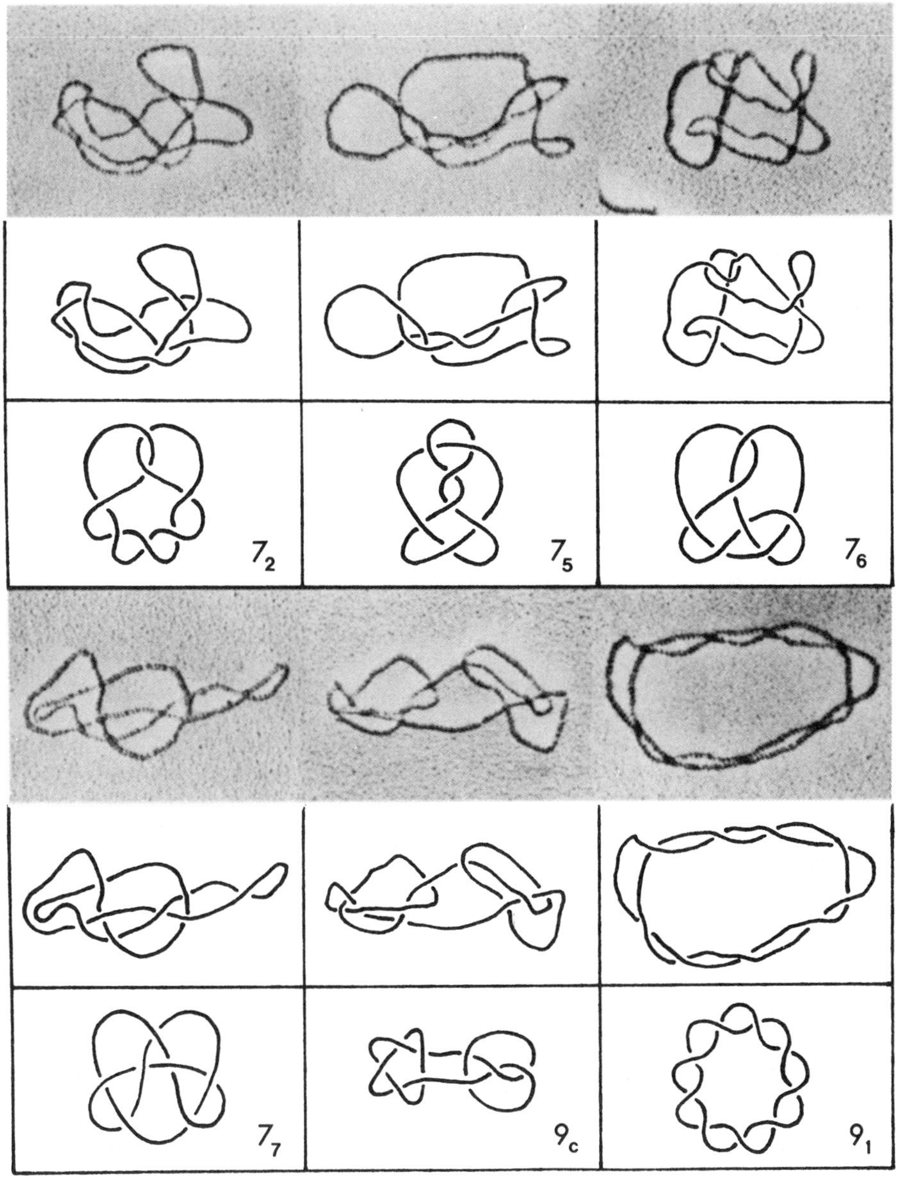

FIGURE 11, continued

the resolution site gives it a natural local biological orientation which Tn3 recognizes. The geometric mechanism of Tn3 is shown in Figure 12, where a single recombination (strand exchange) event is depicted.

If two resolution sites occur on the same circular duplex DNA molecule, they can be oriented in either parallel or anti-parallel fashion, giving rise respectively to links and knots upon recombination, as in Figures 13A and 13B. Tn3 resolvase operates only on parallel resolution sites.

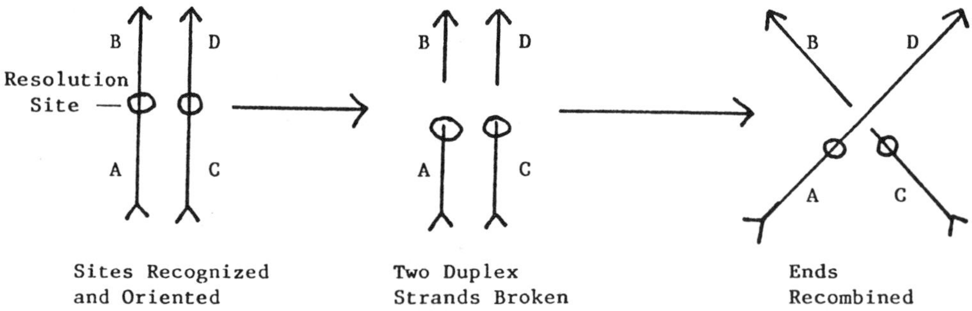

FIGURE 12. A Single Recombination Event

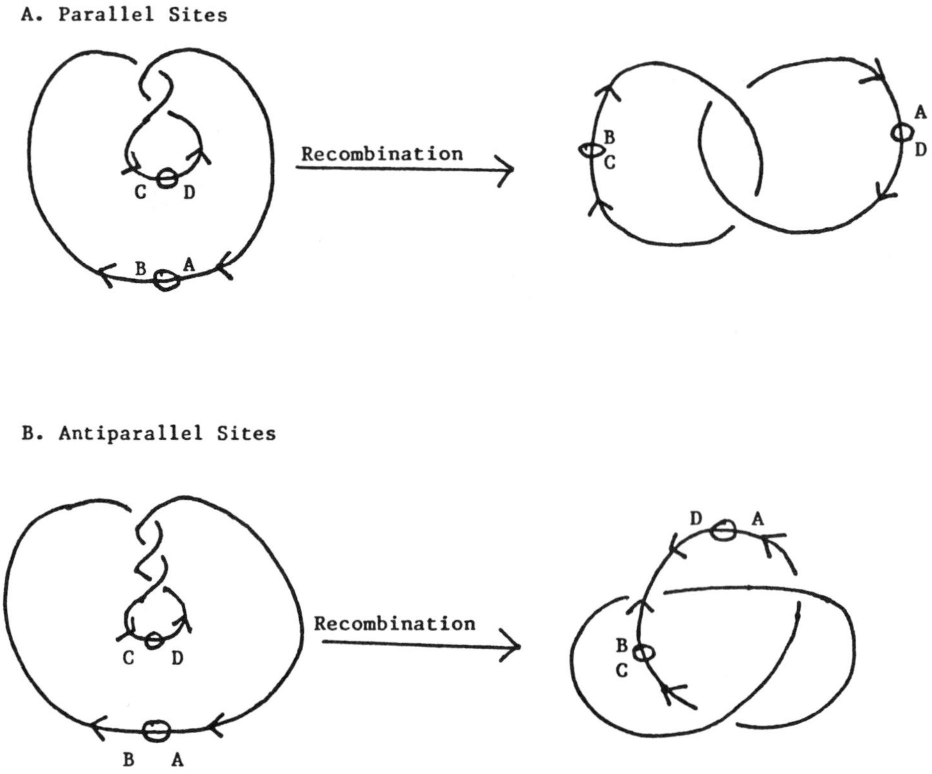

FIGURE 13. Possible Recombination Products (from Ref. 16)

Also, if a circular duplex DNA molecule contains two resolution sites, these sites divide the circle into two arcs (domains). As is evident in Figure 13, pre-existing interdomainal nodes (crossovers between domains)

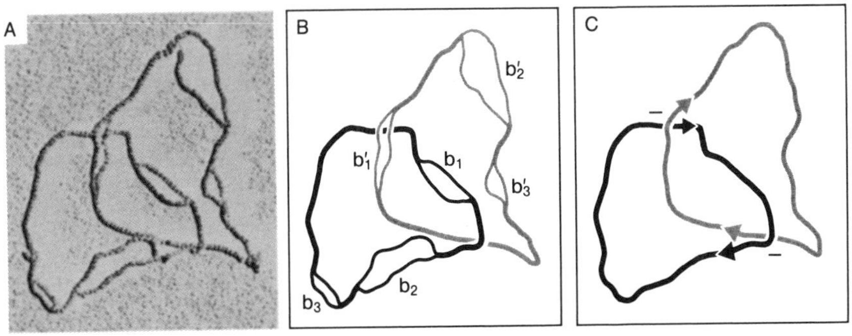

FIGURE 14. The (-1)-Link Produced by Tn3 (from Ref. 20)

can be trapped by recombination, and reflected in the topological complexity of the recombination products. Both intradomainal (self-crossovers of a domain) and interdomainal nodes contribute to the geometric complexity (supercoiling) of the recombination products.

This paper [20] gives experimental evidence to support a model for the action of Tn3 resolvase on circular duplex DNA with parallel resolution sites. One begins with unknotted DNA substrate, and then treats it with resolvase. The principal product was known to be the link of two unknotted components, exclusively of linking number (-1), and that a minor product was the figure 8 link (catenane), exclusively with a (+) crossover in the middle (see Figure 15). Since DNA carries no natural biological orientation, the proof that the mechanism always produces links with linking number = (-1) is most ingenious. This result was reported in [21], in which the DNA was engineered so as to carry a natural observable orientation. Regions rich in the A-T pair were inserted into the molecule. These regions can be more easily denatured (the H-bonds broken) than other regions on the molecule, so one partially denatures the duplex molecule, creating bubbles where the two backbone strands are spread apart. Given a circle with three distinct bubbles on it, an orientation can be unambiguously assigned. Figure 14 is the EM picture of the proof.

Tn3 was known to act dispersively in this situation -- to bind to the circular duplex DNA, to mediate a single recombination event (strand exchange), and then to release the catenated product. It was also known that Tn3 and free (unbound) catenanes do not react. However, once in 20 encounters, resolvase acts processively -- additional strand exchanges are promoted prior to the release of the product, forming more topologically

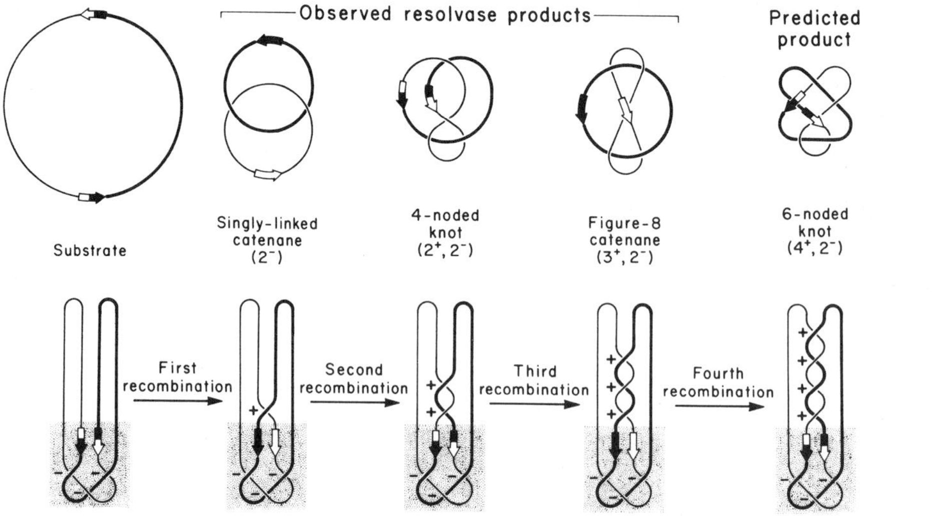

FIGURE 15. The Tn3 Model (from Ref. 19)

complex configurations. Of course, the more complex the configuration, the smaller the yield and the more difficult the detection of the reaction product.

The mechanism model being tested is that, in a single encounter between the substrate and Tn3, the enzyme binds to and fixes three negative (-) interdominal nodes, and that each processive single recombination event during that encounter introduces a single (+) interdomainal node. At the end of the encounter, the enzyme releases the molecule.

The hypothesized starting configuration for an encounter between unknotted DNA and Tn3 is the unknot with three negative interdomainal crossovers (3 negative supercoils) bound to Tn3. The first recombination event (strand exchange) produces the (-1) link (a chiral link since it is oriented), the second event produces the figure 8 knot (an achiral knot), the third event produces the (+) figure 8 catenane (a chiral link), and the fourth successive recombination event produces the chiral knot 6_2. The three negative supercoils remain bound during a single encounter (with possible multiple strand exchanges) with Tn3, and are released at the end of the encounter. This sequence is illustrated in Figure 15.

The model was tested by treating supercoiled unkontted circular DNA with parallel resolution sites with Tn3. One then singly nicks the reaction products, so that gel electrophoresis will separate the reaction products according to the number of crossovers in the minimal projection,

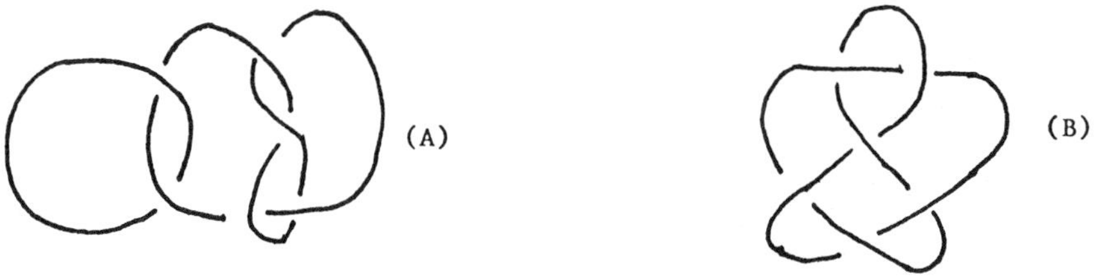

FIGURE 16. The Fourth Product Band

without contamination of the bands due to partially relaxed molecules. One then runs gel electrophoresis, forming a reaction product ladder opposite a reference T4 topoisomerase knot ladder. In the reference ladder, adjacent bands correspond to a difference of one in crossover number, and all crossover numbers (≥ 3) are represented. Sure enough, in the first product band (two crossovers) was found the (-1) link. In the second product band (four crossovers) was found the figure 8 knot. In the third product band (five crossovers) was found the (+) figure 8 catenane. In the fourth product band (six crossovers) was found the two six-crossing objects shown in Figure 16.

Figure 16A depicts the link of the figure 8 knot with an unknotted circle, and Figure 16B is the knot 6_2. (This knot is the mirror image of the knot labelled 6_2 in Rolfsen's table [9].) The link of 16A can be explained as two successive encounters with Tn3, the first encounter producing the figure 8 knot from the unknot via two successive strand exchanges, and the second encounter producing the link from the figure 8 knot via a single strand exchange. In order to separate the six-crossing single encounter knot products from the contaminating double encounter six-crossing link products, two-dimensional gel electrophoresis was used, where it turns out (fortunately) that the knots migrate slower than the links. Upon excising the six-crossing knot band, eleven knots were found, and <u>all</u> of them were 6_2! Figure 17 shows an incredible EM picture of one of these DNA knots.

Counting composites and enantiomers separately, there are eight 6-crossing knots. Eleven knots were found, all of which were 6_2. If Tn3 produces 6-crossing knots at random (no stereoselectivity), the probability of this occurring is $(1/8)^{11} \approx 1.2 \times 10^{-10}$! Since the geometric mechanism model explains all observations, the authors of [20] conclude that it is correct.

Recent knot theory arguments have emerged which strengthen the case for the recombination model put forth in [20]. Consider an unoriented

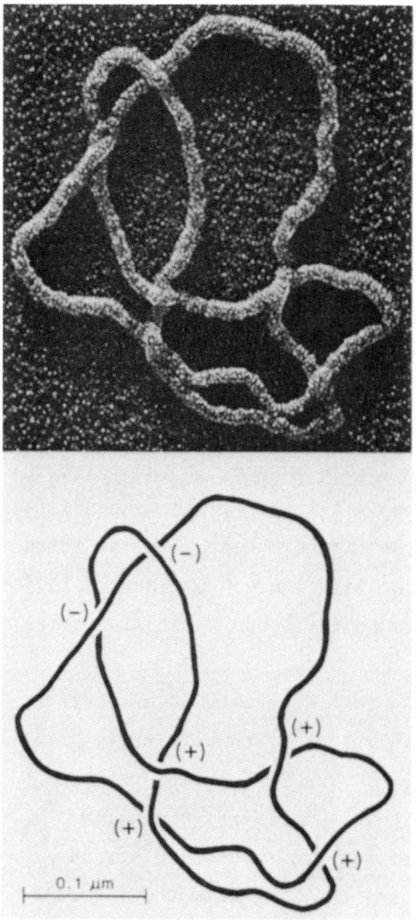

FIGURE 17. The DNA Knot 6_2 (from Ref. 19)

knot (2 component link) which has two resolution sites (one on each component). A single recombinant event (or strand exchange) on this configuration consists of two moves. First, a global move in which the resolution sites are aligned in parallel within a small 3-cell by an ambient isotopy of 3-space. The small 3-cell intersects the configuration in only the two parallel strands. Second, a local move inside the 3-cell, in which the parallel strands are broken at the resolution sites, and recombined (as in Figure 12). In the local move, the rest of the knot (link) outside the 3-cell remains unchanged.

CONJECTURE [21], [16]. It is impossible to go from the unknot to the (+) figure 8 cantenane via a single recombination event. In other words, the reaction pathway postulated in Figure 15 cannot be short-circuited.

The conjecture is true. One argument goes as follows: J. White [23] learned in Geneva from M. Boileau that if the figure 8 catenane can be obtained from the unknot via a single strand exchange, then it would be possible to obtain the 2-fold cover of S^3 (branched along the figure 8 catenane) by integral Dehn surgery on a knot in S^3. But the 2-fold branched cyclic cover of S^3 (branched along the figure 8 catenane) is the lens space L(8, -5). However, if L(p, q) is obtained via integral Dehn surgery on a knot, then $\pm q$ is congruent to t^2 mod p (for some t), which is not true for ± 5 mod 8.

Another result was relayed to me in a letter of S. Kinoshita [5]. This is an unpublished theorem of Y. Shinohara. This result appears to be of utility in deciding which configurations are related by a single strand exchange.

THEOREM (Shinohara). Let L' be the oriented link (knot) obtained from the oriented knot (link) L via a single recombinant strand exchange (as in Figure 12). If $\sigma(L)$ denotes the signature of L, and n(L) denotes the nullity of L, then

$$\sigma(L) = \sigma(L') \text{ and } |\, n(L) - n(L') \,| \leq 1 \qquad (1)$$

or

$$n(L) = n(L') \text{ and } |\, \sigma(L) - \sigma(L') \,| \leq 1. \qquad (2)$$

ACKNOWLEDGEMENT

The author was partially supported by ONR Grant N00014-84-K-0761

REFERENCES

1. Bauer, W. R., Crick, F.H.C., and White, J. H., Supercoiled DNA, Scientific American 243 (1980), 100-113.
2. Balaban, A. T., ed., Chemical Applications of Graph Theory, Academic Press, (1976).
3. Dean, F. B., Stasiak, A., Koller, T., and Cozzarelli, N. R., Duplex DNA Knots Produced by Escherichia Coli Topoisomerase I, J. Biol. Chem. 260 (1985), 4795-4983.
4. Fuller, F. B., The Writhing of a Space Curve, Proc. Nat. Acad. Sci. USA 68, No. 4 (1971), 815-819.
5. Kinoshita, S., Letter of 5 September, 1985.

6. Knop, J. V., Muller, W. R., Szymanski, K., and Trinajstic, N., Computer Generation of Certain Classes of Molecules, Editions Kemija U Industriji, Zagreb, Yugoslavia (1985).

7. Little, C. N., Non-alternate ± Knots, Trans. Roy. Soc. Edinburgh 39 (1900), 771-778.

8. Michels, J.P.J., and Wiegel, F. W., Probability of Knots in a Polymer Ring, Phys. Letts. 90A (1982), 381-384.

9. Rolfsen, D., Knots and Links, Publish or Perish, Inc., (1976).

10. Schill, G., Catenanes, Rotaxanes, and Knots, Academic Press (1971).

11. Simon, J., Topological Chirality of Certain Molecules, Topology 25 (1986), to appear.

12. Strogatz, S., Estimating the Torsional Rigidity of DNA from Supercoiling Data, J. Chem. Phys. 77 (1) (1982), 580-581.

13. Strogatz, S., Topology of Zigzag Chromatin, J. Theor. Biol. 103 (1983), 601-607.

14. Thistlethwaite, M. B., Knot Tabulations and Related Topics, Aspects of Topology in Memory of Hugh Dowker, 1912-1982, London Math. Soc. Lecture Notes #93 (1985), Cambridge Univ. Press.

15. Vologodskii, A. V., Lukashin, A. V., Frank-Kamenetskii, M. D., and Anshelevich, V. V., The Knot Problem in Statistical Mechanics of Polymer Chains, Sov. Phys. JETP 39 (1974), 1059-1063.

16. Walba, D., Topological Stereochemistry, Tetrahedron (1985), to appear.

17. Wang, J. C., DNA Topoisomerases, Scientific American 247 (1982), 94-109.

18. Wang, J. C., DNA Topoisomerases, Ann. Rev. Biochem. 54 (1985), 665-697.

19. Wasserman, E., Chemical Topology, Scientific American 207 (1962), 94-102.

20. Wasserman, S. A., Dungan, J. M., and Cozzarelli, N. R., Discovery of a Predicted DNA Knot Substantiates a Model for Site-specific Recombination, Science 229 (12 July 1985), 171-174.

21. Wasserman, S. A., and Cozzarelli, N. R., Determination of the Stereostructure of the Product of Tn3 Resolvase by a General Method, Proc. N.A.S. (USA) 82 (1985), 1079-1083.

22. White, J. H., Self-Linking and the Gauss Integral, Am. J. Math. 91, No. 3 (1969), 693-728.

23. White, J. H., Letter of 12 July, 1985.

Continuous Versus Discrete Symmetry

SHMUEL WEINBERGER / Department of Mathematics, The University of Chicago, Chicago, Illinois

The character of the theory of group actions on manifolds differs greatly in the case of finite groups from the case of (positive dimensional) compact Lie groups. Yet, of course, there is a connection; every Lie group has many finite subgroups so one can restrict and try to (re)construct the original action. This is not a new idea. For linear actions on the sphere, for instance, the finite subgroups detect since characters do, and characters are continuous functions while the elements of finite order are dense. Many nonlinear settings have also been studied; some rather simple examples where the finite subgroups do not detect will be given shortly. Yet, the study has been rather *ad hoc*. One examines the theory of G-actions on M as H ranges over the subgroups of G, and *computes* the forgetful map!

In this note I shall describe some results that are true quite generally and available even in cases where one does not have such a good grasp on the actions of either the Lie groups or the finite groups. I shall not, however, strive for the greatest generality, but rather merely state some results that give the flavor and some counterexamples indicating certain of the phenomena encountered in formulating the more general principles.

The theorems and examples below answer several questions in Schultz's problem list [6].

THEOREM 1. Any two homotopy equivalent free absolute neighborhood retract (ANR) actions of a compact connected Lie group on a simply connected manifold conjugate for each finite subgroup can be connected by a one-parameter family of such actions.

Notice that for finite groups, or more generally (since we are dealing with free actions), for locally linear actions, all deformations are essentially constant, so that Theorem 1 gives a characterization of equideformable actions as well as uniqueness theorem for the locally linear setting. For ANR actions, nonconstant families can be obtained by shrinking decompositions (see [3]). Before formulating smooth results, let us examine each of the hypotheses.

Homotopy equivalence will be discussed in more detail below. Freeness does not seem to be necessary, but I have only verified this in cases. The following example emphasizes the importance of connectedness:

EXAMPLE. Let $S^1 \times \mathbb{Z}_7$, parameterized by (θ,u), act freely on $S^5 \times S^3 = \{(x_1,x_2,x_3,y_1,y_2) \in \mathbb{C}^5 \mid \Sigma|x_i|^2 = \Sigma|y_j|^2 = 1\}$ in the following two ways:

$$(\theta,u)(\vec{x},y_1,y_2) = (\theta\vec{x},uy_1,uy_2)$$

and

$$(\theta,u)(\vec{x},y_1,y_2) = (\theta\vec{x},uy_1,u^2y_2).$$

One can use the G-signature formula to distinguish these actions. On the other hand, surgery theory can be used to identify the restrictions of the actions to arbitrary finite subgroups.

Despite this example, one can isolate and analyze the contribution of disconnectedness of the Lie group. It is rather related to the problem of nonsimple-connectivity of the ambient manifold. For instance, one can study the two actions of S^1 on $(S^5 \times S^3)/\mathbb{Z}_7$: note that the space $(S^5 \times S^3)/\mathbb{Z}_7$ is the same for both actions. A more elementary example follows.

EXAMPLE. Let N_1 and N_2 be nonisomorphic h-cobordant manifolds and consider $S^1 \times N_1$ and $S^1 \times N_2$, with the obvious circle actions.

In the smooth case, deformations are constant so Theorem 1 provides a topological equivalence. Smoothly though, there is no equivalence in general but one can describe a construction that totally describes this failure.

EXAMPLE. (Modification about an orbit): For any $x \in M$ consider an open equivariant neighborhood of the form $G^k \times D^{n-k}$. Replace this by $G \times \mathring{\Sigma}^{n-k}$ where Σ^{n-k} is a homotopy sphere. If Σ does not bound a parallelizable manifold then some finite subgroup will detect the difference in the action. For $n-k \equiv 3 \bmod 4$ there are always Σ's that do not change the action on finite subgroups, and often these do change the G-action. The $n-k \equiv 1 \bmod 4$ case is more subtle. In any case, this is a way of producing examples.

THEOREM 1'. For $n-k \neq 4$, the smooth version of Theorem 1 holds with modification about an orbit replacing the statement on one-parameter families of actions.

SPECIAL CASE. Free actions of S^1 and S^3 on homotopy spheres, except in some low dimensions, are determined by restriction to finite subgroups.

The homotopy problem ignored in the above theorem(s) is serious and interesting. I have only made progress for toral groups, where the slogan is "finite subgroups control finite primes--the rationals are uncontrolled" (I refer here to Sullivan's "Genetics of Homotopy Theory" [8]). This principle is more clearly understood in the context of trying to construct actions. For the remainder of this announcement we restrict to S^1-actions.

We approach the problem of S^1 actions in two steps: passing from $\mathbb{Z}_n$ actions for all n to $\mathbb{Q}/\mathbb{Z}$ actions, and from $\mathbb{Q}/\mathbb{Z}$ actions to S^1 actions. First we eliminate a prima facie ambiguity.

THEOREM 2. Let $\mathcal{X}$ be a simply connected finite-dimensional homotopy type. If for each n there is an $X_n \in \mathcal{X}$ having a free $\mathbb{Z}_n$ action, then there is an $X \in \mathcal{X}$ having a free $\mathbb{Z}_n$-actions for all n. If the X_n's are finite complexes, X can also be chosen finite.

It is important to note that the final $\mathbb{Z}_n$ actions might have nothing to do with the original ones. Returning now to the construction of $\mathbb{Q}/\mathbb{Z}$ actions:

THEOREM 3. Under the hypotheses of Theorem 2 there is always an $X \in \mathcal{X}$ having a "free" $\mathbb{Q}/\mathbb{Z}$ action. If all the X_n's are finite, the obstruction to finitising such a $\mathbb{Q}/\mathbb{Z}$-complex X lies in $\lim^1 Wh(\mathbb{Z}_n)$.

By a free $\mathbb{Q}/\mathbb{Z}$ action, all we mean is that all elements other than the identity act without fixed points. We assume noting like proper discontinuity.

For the second statement interpret the $\lim^1$ as referring to the sequence obtained from the transfer maps on Whitehead groups induced by inclusions $\mathbb{Z}_k \subset \mathbb{Z}_\ell$ for $k|\ell$. A consequence of these theorems and some surgery is:

COROLLARY. If M^m is a closed even dimensional simply connected manifold, or a noncompact simply connected manifold with simply connected ends, $m \geq 5$, and if M has free $\mathbb{Z}_n$ actions for all n then M has a free $\mathbb{Q}/\mathbb{Z}$ action.

Some homotopy theory leads to a characterization:

COROLLARY. $\mathcal{X}$ has free $\mathbb{Z}_{p^n}$ actions for all n if and only if $\mathcal{X}$ has the $\hat{\mathbb{Z}}_p$-homotopy type of a space with a free S^1-action.

This last corollary shows the difficulty in obtaining an S^1 action on $\mathcal{X}$ are well understood.) One has all the p-adic information necessary for finite primes but lacks the rational information.

PRETHEOREM 4. One does not need rational data to build $Q/\mathbb{Z}$ actions.

The idea is that $\mathbb{Z}_{(p)}$ - and $\hat{\mathbb{Z}}_p$-homotopy theory differ only in the rationals, and that while [2] uses $\mathbb{Z}_{(p)}$ homotopy theory, much goes through with the weaker $\hat{\mathbb{Z}}_p$ version. I shall not make this pretheorem precise here.

COROLLARY. There are manifolds with free $\mathbb{Q}/\mathbb{Z}$ actions but no S^1 actions.

This is much stronger than the assertion that there are $\mathbb{Q}/\mathbb{Z}$ actions that do not extend to S^1 actions. That is true for any odd dimensional simply connected manifold that has free $\mathbb{Q}/\mathbb{Z}$ actions!

The key to all the theorems is a study of $\varinjlim X/\mathbb{Z}_n$, assuming the $\mathbb{Z}_n$'s form part of an S^1 action. This is achieved using [1]. Theorem 1 is proven using [10] and [15] to reduce to homotopy theory, and an argument analogous to one in [4] to do that homotopy theory. One needs [3,7] to interpret the result in terms of deformations. Theorems 2 and 3 are proven simultaneously by examining the combinatorics of a tree of homotopy types and making use of [9]. The $\lim^1$ obstruction is inevitable once one realizes that finiteness in the present context is more a question of homeomorphism than of homotopy equivalence. Finally, "Pretheorem 4" comes out of [2] and [8] and the ideas going into Theorem 1.

ACKNOWLEDGEMENTS.

I would like to thank the Courant Institute of the Mathematical Sciences, Cal. Tech., and MSRI for their hospitality while I worked on parts of what is reported here. I also thank Sylvain Chappell, Peter May,

Mel Rothenberg, and Julius Shaneson for their help.

The author was partially funded by an NSF postdoctoral fellowship.

REFERENCES

1. Bredon, G., Sheaf Theory, McGraw Hill 1967.
2. Chappel, S., and Weinberger, S., Homology Propogation of Group Actions (preprint).
3. Edwards, R., The topology of manifolds and cell-like maps, Proc ICM Helsinki 1978.
4. Milnor, J., The homology of Lie groups made discrete, Comm. Math. Helv. 58 (1983), 72-85.
5. Quinn, F., Ends of Maps I, Annals of Math. 110 (1979), 275-331.
6. Schultz, R., editor. Proc. AMS Summer Meetings on Transformation Groups 1983 (to appear).
7. Siebenmann, L., Approximating Cellular Maps by Homeomorphisms, Topology 11 (1973), 271-294.
8. Sullivan, D., The Genetics of Homotopy Theory and the Adams Conjecture, Annals of Math, 100 (1975), 1-79.
9. Sullivan, D., Infinitesimal Computations in Topology, Pub. Math. d'I.H.E.S. 47 (1977), 269-331.
10. Wall, C.T.C., Surgery on Compact Manifolds, Academic Press, 1970.

The Knotting of Theta Curves and Other Graphs in S^3

KEITH WOLCOTT / Department of Mathematics, University of Iowa, Iowa City, Iowa

INTRODUCTION

One overall approach to the knot theory of graphs is to seek knots or links associated with a graph G so that questions about G can be translated into questions about knots and links. The simplest attempt would be to associate with G the set of knots and links in G. For example, a theta-curve contains three simple closed curves and of course if two theta curves contain different knots, the theta-curves are different. But an example of Kinoshita [9] showed this approach is too simple to classify graphs since there exist nonplanar theta-curves for which all three constituent knots are trivial. (See sections 1 and 2 for another proof that Kinoshita's theta-curve is nonplanar and infinitely many more such examples.)

In section 3 we consider two ways of associating knots and links with a theta-curve. First put the graph in a "normal" position that realizes the graph as a certain combination of a tangle and a planar graph. The numerator link and denominator knot [4] of that tangle become useful invariants of the theta-curve and we are able to prove (corollary 3.11) that a theta-curve is planar if and only if its three constitutent knots and associated numerator link are planar. Thus (by an effective and simple procedure) we can reduce the problem of triviality of theta-curves to the problem of triviality for knots and links. In addition, the invariant consisting of the numerator link provides a proof that Kinoshita's theta-curve is chiral.

Our second way of associating knots with a theta-curve applies only to theta-curves that are "locally unknotted", that is, the three constitutent knots are trivial. For each of the three trivial simple closed curves J_i in θ, the arc $\overline{\theta \setminus J_i}$ lifts to a knot K_i in the two-fold cyclic branched cover of S^3 branched over J_i. It is clear from basic facts about branched covers [6] that the unordered triple (K_1, K_2, K_3) of knots in S^3 is an isotopy invariant of θ. For example, (section 1) for Kinoshita's theta-curve, the K_i are all (5,3) torus knots. The fact that these are nontrivial knots provides a new proof that Kinoshita's theta-curve is nonplanar, and furthermore, the fact that these knots are chiral shows that Kinoshita's theta-curve is chiral. (We have learned that Kinoshita recently obtained the same chirality result by a similar argument.) By analyzing a theta-curve in terms of tangles we actually can show that the three knots K_i completely determine planarity, that is (theorem 3.15), a locally unknotted theta-curve is planar if and only if the three knots K_i are trivial. (In fact, it suffices to know that one of the three knots K_i is trivial!) Thus the problem of deciding whether or not a theta-curve is planar reduces to determining whether or not its three constitutent knots and one knot in a branched cover are trivial.

In section 4 we discuss a way to construct more complex graphs from simpler graphs. This method is to remove three-ball neighborhoods of an order-n vertex in each of two graphs and then glue the remaining two three-balls together in such a way that the n points of each graph in the boundary of each ball are identified. (Of course, matching different points usually produces different sum graphs.) We prove (theorem 4.2) that any order-three connect sum of two nonplanar theta-curves is nonplanar. For vertices of order $n \geq 3$ there is a question of whether the sum operation is well-defined, as the attaching map can introduce a braid into the graph. For $n = 3$, aside from deciding which points are matched with which, there actually is no problem. For $n \geq 4$ there is.

DEFINITIONS. By a <u>graph in S^3</u> we mean a one dimensional finite polyhedron in S^3. We will say G_1 is <u>equivalent</u> to G_2 (denoted $G_1 \simeq G_2$) if and only if there exists an orientation-preserving homeomorphism of S^3 that maps G_1 to G_2. A graph G is <u>locally unknotted</u> if every knot contained in G is the trivial knot. A graph G is <u>locally unlinked</u> if every link contained in G is a planar link. Thus locally unlinked implies locally unknotted. A graph G is <u>chiral</u> if there is no isotopy of S^3 that maps G to its mirror image. Equivalently, by [5], G is <u>chiral</u> if there does not exist an orientation-preserving homeomorphism of S^3 that maps G to its mirror image.

1. KINOSHITA'S THETA-CURVE

Kinoshita's theta-curve, shown in figure 1, is an example of a locally unknotted graph that is knotted [9]. We make the following stronger statement.

THEOREM 1.1. Kinoshita's theta-curve is chiral. (S. Kinoshita has communicated to us that he has recently obtained the same result using similar techniques.)

Proof. Assume Kinoshita's theta-curve G is achiral. Thus there exists an orientation-reversing homeomorphism h of S^3 that fixes G. Choose one of the three simple closed curves C contained in G and construct the two-fold branched cover of S^3 branched over C. (see figure 2). Since G is locally unknotted, C is the trivial knot and thus this branched cover is just S^3. The lifting of the arc $\overline{G \setminus C}$ in the double cover is a (5,3) torus knot. We can assume h maps C to itself because of the symmetries of G illustrated in the second presentation in figure 1 (i.e., compose h with an orientation-preserving homeomorphism of S^3 that fixes G but exchanges the simple closed curves of G.) Now there are two cases. The homeomorphism h may preserve the orientation of C or it may reverse it. In both cases, h induces an orientation-reversing map of the double covers [6], that is, of S^3, that fixes a (5,3) torus knot. This is a contradiction since torus knots are chiral. Therefore Kinoshita's theta-curve is chiral.

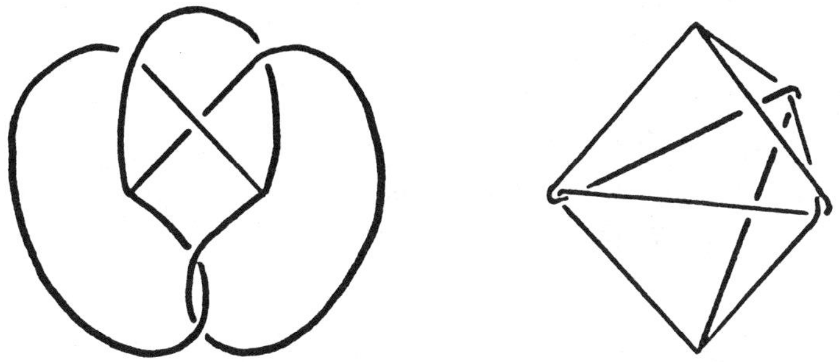

FIGURE 1. Two Presentations of Kinoshita's Theta-curve

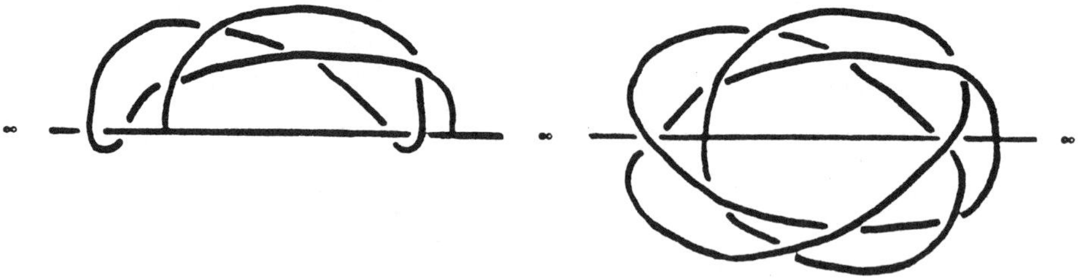

FIGURE 2. G and its Lifting in a Two-fold Branched Cover

REMARK. In the above proof we allowed for the possibility that h could induce any of twelve possible mappings of G to itself, that is, permuting the three arcs and exchanging the two vertices. Six of these can be seen to exist because of the symmetries of G illustrated in the second presentation of figure 2. P. Gilmer and R. Litherland [8] have shown using tangles and infinite cyclic covers that the other six do not exist, in particular there is no isotopy of S^3 that fixes each vertex and one edge of G while exchanging the other two edges.

2. OTHER EXAMPLES OF LOCALLY UNKNOTTED CHIRAL THETA-CURVES

For integers i, j, and k define $\theta(i,j,k)$ to be the locally unknotted theta-curve shown in figure 3. If one of the three integers is zero then the theta-curve is planar and if $i = j = k = 1$ we have $\theta(1,1,1)$ which is Kinoshita's theta-curve. Also we note that $\theta(i,j,k)$ is the mirror image of $\theta(-i,-j,-k)$. Next we observe that revolving $\theta(i,j,k)$ 120° about a vertical axis intersecting the two vertices cyclicly permutes the three integers, while turning $\theta(i,j,k)$ upside down exchanges two of the integers. Thus, permuting i, j, and k does not result in a different graph. Therefore we assume $i \leq j \leq k$.

THEOREM 2.1. The theta-curves in the collection $\theta(i,j,k)$, $i \leq j \leq k$ with $i, j, k \neq 0$, are chiral and mutually distinct.

Proof. Since $\theta(i,j,k)$ is the mirror image of $\theta(-i,-j,-k)$ we need only prove the chirality of $\theta(i,j,k)$ where at most one of i, j, or k is negative. Another presentation of $\theta(i,j,k)$, with one of its simple closed curves C drawn as a trivial knot projection with no crossings, is shown

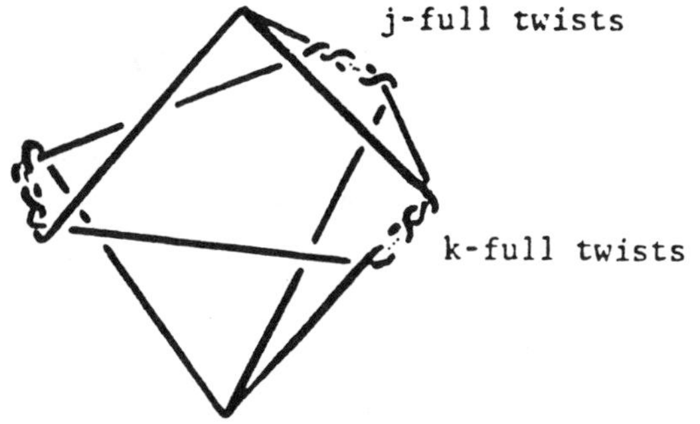

FIGURE 3. $\theta(i,j,k)$

in figure 4. In S^3, the n-fold branched cover of S^3 branched over C, the arc $\ell = \overline{\theta(i,j,k)\backslash C}$ lifts to n arcs, $\tilde{\ell}$ joining two points. One of these arcs, along with the lifting of C, forms a theta-curve that does not depend on the choice of the arc of $\tilde{\ell}$. This theta-curve, which we will call $\theta(i,j,k)(C,n)$, contains the trivial knot C and two other knots K(C,n,1) and K(C,n,2). If A is another simple closed curve of $\theta(i,j,k)$, a presentation of $\theta(i,j,k)$ with A drawn as a trivial knot projection with no crossings looks like figure 4 with i, j, and k permuted. To see that

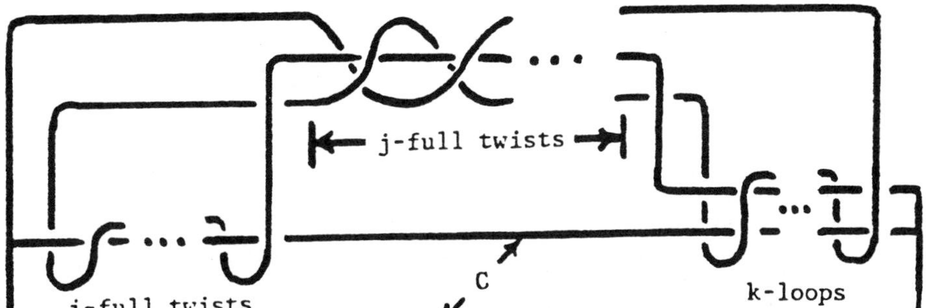

FIGURE 4. Another Presentation of $\theta(i,j,k)$

TABLE 1

m=any positive integer   0=trivial knot   θ=planar theta-curve
(p,q)=(p,q)-torus knot   (p)=twist knot with p half-twists

<table>
<tr><th>n-fold cover</th><th></th><th>n=|k|</th><th>n=|i|</th><th>n>|k|</th><th>n>|i|</th></tr>
<tr><td>i<0</td><td>k>|i|</td><td>0
(2j-1)</td><td>k=m|i|
θ(j,-1,m)
k≠m|i|
(2j)
0</td><td>All θ but
(k-i) fold
cover[1]</td><td>All θ but
(k-i),
(k-i)/m,and
|i|/k fold
covers[1,3]</td></tr>
<tr><td></td><td>k<|i|</td><td>|i|=mk
θ(j,-m,1)
|i|≠mk
0
(2j-1)</td><td>(2j)
0</td><td>All θ but
(k-i),
(k-i)/m,and
|i|/m fold
covers[1,3]</td><td>All θ but
(k-i) fold
cover[1]</td></tr>
<tr><td></td><td>k=|i|</td><td>θ(j,-1,1)</td><td>θ(j,-1,1)</td><td></td><td></td></tr>
<tr><td>k<0</td><td>|k|>i</td><td>0
(2j)</td><td>|k|=mi
θ(j,1,-m)
|k|≠mi
(2j)
0</td><td>All θ but
|k-i| fold
cover[2]</td><td>All θ but
|k-i|,
|k-i|/m,and
|k|/m fold
covers[2,3]</td></tr>
<tr><td></td><td>|k|<i</td><td>i=m|k|
θ(j,m,-1)
i≠m|k|
0
(2j)</td><td>(2j-1)
0</td><td>All θ but
|k-i|,
|k-i|/m,and
i/m fold
covers[2,3]</td><td>All θ but
|k-i| fold
cover[2]</td></tr>
<tr><td></td><td>|k|=i</td><td>θ(j,1,-1)</td><td>θ(j,1,-1)</td><td></td><td></td></tr>
<tr><td>j<0</td><td>k>i</td><td>0
(-2j)</td><td>k=mi
θ(j,1,m)
k≠mi
(-2j)
0</td><td>θ</td><td>All θ but
k/m and
(k-i)/m
fold
cover[3]</td></tr>
<tr><td></td><td>k<i</td><td>i=mk
θ(j,m,1)
i≠mk
0
(-2j)</td><td>(-2j)
0</td><td>All θ but
i/m and
|k-i|/m
fold
covers[3]</td><td>θ</td></tr>
<tr><td></td><td>k=i</td><td>θ(j,1,1)</td><td>θ(j,1,1)</td><td>(-2j+1,2)
(-2j+1,2)</td><td>(-2j+1,2)
(-2j+1,2)</td></tr>
<tr><td>i>0
j>0
k>0</td><td>k>i</td><td>0
(2j-1)</td><td>k=mi
θ(j,1,m)
k≠mi
(2j-1)
0</td><td>θ</td><td>All θ but
k/m and
(k-i)/m
fold
covers[3]</td></tr>
<tr><td></td><td>k<i</td><td>i=mk
θ(j,m,1)
i≠mk
0
(2j-1)</td><td>(2j-1)
0</td><td>All θ but
i/m and
|k-i|/m
fold
covers[3]</td><td>θ</td></tr>
<tr><td></td><td>k=i</td><td>θ(j,1,1)</td><td>θ(j,1,1)</td><td>(2j-1,2)
(2j-1,2)</td><td>(2j-1,2)
(2j-1,2)</td></tr>
</table>

$\theta(i,j,k)$ is chiral we assume not and thus have an orientation-reversing homeomorphism h of S^3 that fixes $\theta(i,j,k)$. If $h(C) = A$, the induced orientation-reversing homeomorphism of the n-fold branched cover of S^3 branched over C [6] maps $\theta(i,j,k)(C,n)$ to $\theta(i,j,k)(A,n)$, $K(C,n,1)$ to $K(A,n,1)$, and $K(C,n,2)$ to $K(A,n,2)$. If $K(C,n,i)$ and $K(A,n,i)$ are different knots for either $i = 1$ or 2, then h cannot map C to A. If $K(C,n,i)$ and $K(A,n,i)$ are the same chiral knots for either $i = 1$ or 2, then h could not have been an orientation-reversing homeomorphism. Therefore if we allow A to range over the simple closed curves of $\theta(i,j,k)$ and each case results in one of the above situations for some n-fold branched cover then $\theta(i,j,k)$ is chiral. Table 1 lists, for various n-fold branched covers of $\theta(i,j,k)$ branched over C, either $\theta(i,j,k)(C,n)$ or both knots $K(C,n,1)$ and $K(C,n,2)$. From this table it can be seen that every theta-curve in the given collection is chiral.

Similarly, Table 1 can be used to distinguish any two theta-curves $\theta(i,j,k)$, $i \leq j \leq k$. Since $\theta(i,j,k)$ is the mirror image of $\theta(-i,-j,-k)$ we need only consider theta-curves with at most one negative parameter.

3. DETERMINING PLANARITY OF THETA-CURVES AND HANDCUFF GRAPHS

Corollaries 3.11 and 3.13 of this section show that the planarity of a locally unlinked theta-curve or handcuff graph is entirely dependent on the planarity of an associated link. It may be interesting to note that our proof of theorem 3.10 uses the facts that nontrivial knots in S^3 have property R ([7] or [11]) and that nontrivial strongly invertable knots have property P [3]. First we state some preliminary definitions and results about tangles which can (except for the idea of normal form) be found in [4] and [10].

A (2-string) <u>tangle</u> is a pair (B,t) where B is a three-ball and t is a pair of disjoint arcs of B with $t \cap \partial B = \partial t$.

We need to distinguish two notions of equivalence:

Two tangles (B,t) and (B,r) are <u>equal</u> if there is a homeomorphism [equivalently isotopy] of pairs from (B,t) to (B,r) that is the identity on ∂B.

Two tangles (B_1,t_1) and (B_2,t_2) are <u>equivalent</u> if there is a homeomorphism of pairs from (B_1,t_1) to (B_2,t_2). (This is from [10] which differs from [4].)

As in [4], a tangle (B,t) is <u>integral</u> if it consists of n horizontal half-twists of t ($n \in Z$). If i, j, k, ⋯ , ℓ, m are integral tangles, then the <u>product</u> ((⋯(ij)k⋯ℓ)m) (see [4]) of these integral tangles is called a <u>p/q rational tangle</u> where p/q (including $1/0 = \infty$) is the continued fraction $m + 1/\ell + 1/\cdots/k + 1/j + 1/i$. We will refer to a p/q rational tangle as (B,p/q).

Thus the 0-tangle (B,0) consists of two horizontal arcs in B and the ∞-tangle, (B,∞) consists of two vertical arcs in B. A result of [4] is the following. If (B,p/q) and (B,m/n) are two rational tangles, then (B,p/q) is equal to (B,m/n) if and only if p/q = m/n.

A tangle (B,t) is <u>untangled</u> if it is equivalent to (B,0), the 0-tangle.

The <u>numerator</u> [resp. <u>denominator</u>] of a tangle (B,t) will be the knot or link

$$(S^3, t \cup 0) = (B,t) \cup_{id} (B,0) \quad [(S^3, t \cup \infty) = (B,t) \cup_{id} (B,\infty)].$$

We will use the definition of prime tangle given in [10]. That is, the tangle (B,t) is <u>prime</u> if it has the following properties:

(1) Any two-sphere in B which meets t transversely in two points bounds in B a ball meeting t in an unknotted spanning arc;

(2) (B,t) is not untangled.

A tangle (B,t) with numerator equal to a two-component link with linking number zero will be said to be <u>normalized or in normal form</u>. Note that normalizing a tangle (B,t) only involves adding an integral tangle to it or multiplying by an integral tangle. Thus a normal form of (B,t) is equivalent to (B,t).

There exist two planar graphs having two vertices each of order three, a theta-curve and a "handcuff" graph (see figure 5). Let θ(B,t) and ϕ(B,t) be the theta-curve and handcuff obtained from a tangle (B,t) as in figure 6.

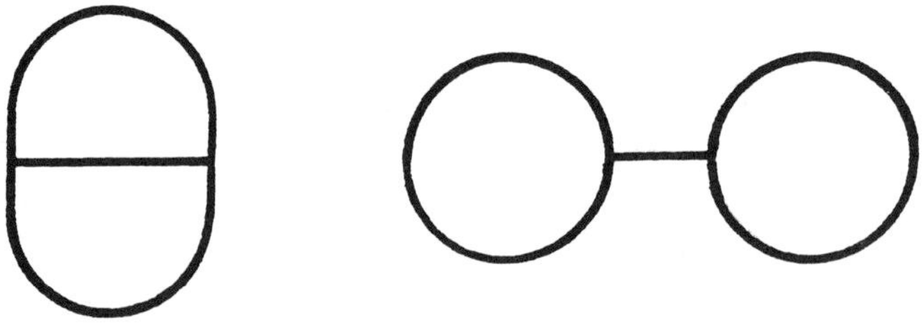

FIGURE 5. Planar Theta-curve and Planar Handcuff Graph

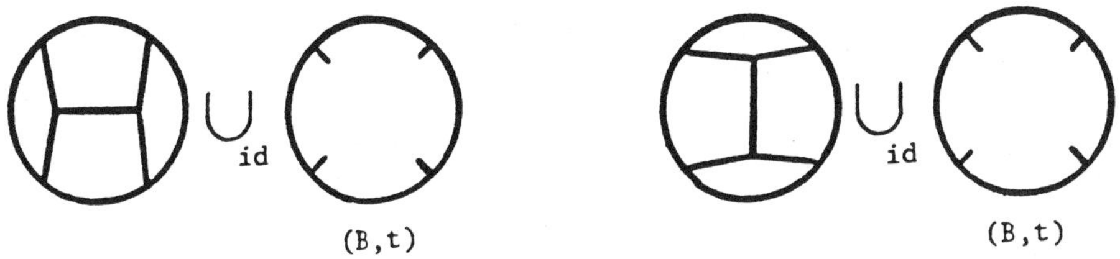

FIGURE 6. $\theta(B,t)$ and $\Phi(B,t)$

LEMMA 3.1. Let (B,g) be a pair consisting of the three-ball B and the graph g shown in figure 7. Let ℓ be the arc of g that connects the two vertices of g and let D be a spanning disk in B that is transverse to ℓ. If h : (B,g) --> (B,g) is a homeomorphism of pairs that fixes the vertex x and the vertex y of g and preserves the orientation of B, then h is isotopic, by an isotopy that keeps the graph g fixed setwise, to n half-twists about the disk D, and possibly 180° revolution about a horizontal axis containing ℓ. (By "half-twist" we mean a map such as

$$(\psi,\phi,s) \to (\psi,\phi+s\pi,s),\ 0\leq s\leq 1,\ (\psi,\phi,s) \to (\psi,\phi+\pi,s),\ 1\leq s\leq 2,$$

where (ψ,ϕ) are polar coordinates on D and we view B as $D\times[-2,2]$.)

Proof. By composing h with half-twists about the disk D and possibly 180° revolution about a horizontal axis containing ℓ we can get a new map, which we will also call h, which is the identity on $\partial B \cap \partial g$. Then since $h|_g$ is isotopic to the identity on g, we can deform h, holding g setwise invariant, so that h is the identity on g. The map h restricted to ∂B is an element of the pure mapping class group of S^2 with four points fixed.

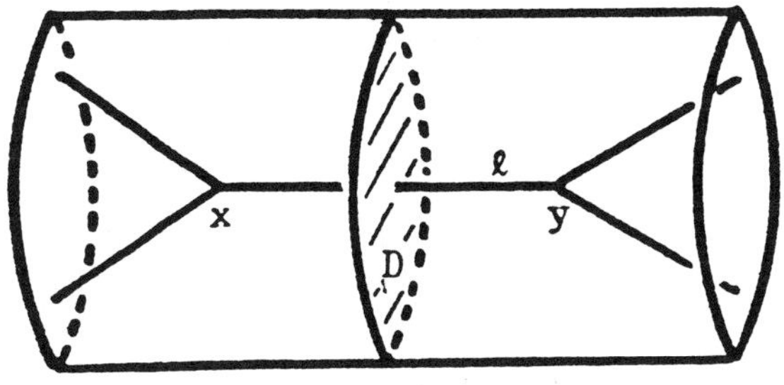

FIGURE 7. (B,g)

By theorem 4.5 of [2], $h|_{\partial B}$ is isotopic to compositions of Dehn twists about disks on ∂B that contain two of the points in the set of four points in ∂g. Such Dehn twists can be extended uniquely up to isotopy to B [1] by coning over ∂B. This can be viewed as braiding the four arcs of $g \cap$ (collar of ∂B) with a pure combed braid. Thus we can deform h to be the identity on ∂B at the expense of braiding the four arcs of $g \cap$ (collar of ∂B) with a pure combed braid. This same braiding can be accomplished by multiplying (B,g) by a sequence i,j, ··· ,k,l,m of even integral tangles. Since g is fixed by h and $(B,\overline{g\setminus\ell})$ is the ∞-tangle, $(B,\overline{g\setminus\ell})$ times the sequence of even integral tangles must equal the ∞-tangle. Thus the continued fraction obtained from i,j,k, ··· ,l,m must equal ∞ [4]. The only sequences of even integers that result in continued fractions that equal ∞ are of the form 0,i,0,j,0,k, ··· ,0,l,0,m. Therefore the only braiding that can occur when deforming h to the identity on ∂B can be accomplished with half-twists about the disk D. Now since h is the identity on ∂B, by [1] h is isotopic to the identity by an isotopy that is the identity on ∂B. We need this isotopy to be the identity on g. To achieve this we first deform h, holding g pointwise invariant, so that h is the identity on the disk D. (Use an innermost disk argument.) D then splits B into two three-balls B_1 and B_2 and h is the identity on ∂B_1 and on ∂B_2. The graph $g \cap B_i$ consists of three radial lines of B_i for i = 1,2. Alexander's proof in [1], which shows that h restricted to B_i is isotopic to the identity by an isotopy that is the identity on ∂B_i, also has the property that if h is the identity on a radial line of B_i then the isotopy is also the identity on the radial line. Thus h is isotopic to the identity by an isotopy that is the identity on ∂B and on g. The conclusion of the lemma follows.

Let G be any theta-curve, or any handcuff graph whose consituent link has linking number zero. We restrict handcuff graphs to constituent linking number zero since these are the only ones that can be described by a normalized tangle. Let T(G) be the set of normalized tangles that can be used to describe G.

THEOREM 3.2. Let G_1 and G_2 be either two theta-curves or two handcuff graphs that have nonvacuous normalized tangle descriptions $T(G_1)$ and $T(G_2)$. Then the following are equivalent:

i) $T(G_1) = T(G_2)$

ii) $T(G_1) \cap T(G_2) \neq \emptyset$

iii) G_1 is equivalent to G_2.

Also, if G_1 is planar then $T(G_1)$ contains only the 0-tangle. In general, T(θ-curve) contains at most twelve tangles and T(handcuff) at most four, modulo vertical twisting.

Proof. i) => ii) Clear.

ii) => iii) Let (B,t) be an element of $T(G_1) \cap T(G_2)$. If G_1 and G_2 are theta-curves, then $G_1 \simeq \theta(B,t) \simeq G_2$. If G_1 and G_2 are handcuff graphs, then $G_1 \simeq \Phi(B,t) \simeq G_2$.

iii) => i) Let (B,t) be an element of $T(G_1)$. Let $G_1(B,t)$ denote either $\theta(B,t)$ or $\Phi(B,t)$ depending on whether G_1 is a theta-curve or a handcuff graph. Thus $G_1(B,t)$ is equivalent to G_1 which is equivalent to G_2. Let h be an orientation-preserving homeomorphism of S^3 with $h(G_1(B,t)) = G_2$. Let $A = \overline{S^3 \backslash B}$. Let ℓ be the arc of G_1 that is contained in A. We can isotope G_2 so that h(A) = A and ℓ and $h(\ell)$ have the same orientation. (Figure 8 illustrates this situation for when G_1 and G_2 are theta-curves.) By lemma 3.1 we can assume that h restricted to A consists of n half-twists about a spanning disk transverse to ℓ and possibly 180° revolution about a horizontal axis containing ℓ. This allows us to deform h, without changing $h(G_1 \cap A)$, to the identity on A at the expense of "twisting" the ball B. Now we have that h : (B,t) --> $(B, B \cap G_2)$ is a map of pairs that is the identity on ∂B. Thus (B,t) equals $(B, B \cap G_2)$, which is an element of $T(G_2)$. (Since numerators of equal tangles are equal, $(B, B \cap G_2)$ is normalized.) Therefore $T(G_1)$ is a subset of $T(G_2)$. The above argument is symmetric with respect to G_1 and G_2, so $T(G_2)$ is also a subset of $T(G_1)$. Thus $T(G_1)$ equals $T(G_2)$.

Note that the above argument indicates that there are at most twelve [resp. four, modulo vertical twisting] elements in the set T(G) for G equal to a theta-curve [handcuff]. That is, there are three [one] choices for the arc of G that is excluded from B and for each of these choices the

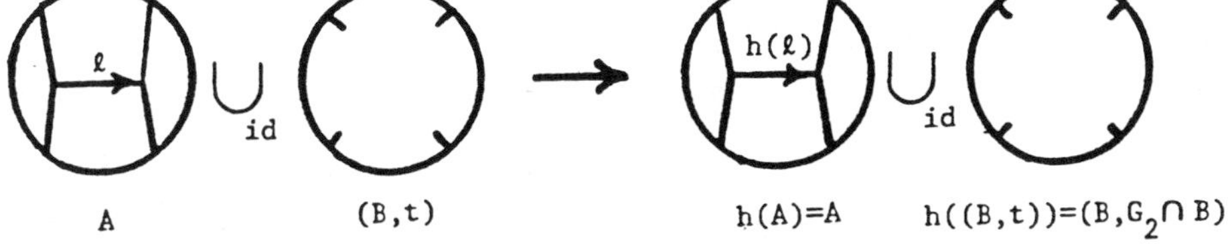

FIGURE 8. $G_1(B,t)$ and G_2

normalized tangle (B,t) can then be revolved 180° about a vertical axis, revolved 180° about a horizontal axis or both. [For handcuff graphs, normalizing the tangle does not control twisting about the arc excluded from the ball B, so any vertical twisting may occur.] Thus for instance if G is planar then T(G) contains only the 0-tangle.

COROLLARY 3.3. Let (B,t) be a normalized tangle. Then $\theta(B,t)$ is planar if and only if $\Phi(B,t)$ is planar.

Proof. If $\theta(B,t)$ is planar then by theorem 3.2 $T(\theta(B,t))$ equals the set of normalized tangles describing the planar theta-curve which only contains the 0-tangle. Thus (B,t) equals the 0-tangle. Therefore $\Phi(B,t)$ is planar.

If $\Phi(B,t)$ is planar then again by theorem 3.2 (B,t) is equal to the 0-tangle. Therefore $\theta(B,t)$ is planar.

Now suppose (B,t) and (B,r) are tangles and we add a tangle (B,s) to both (B,t) and (B,r) via the identity on ∂B. If the knot or link $t \cup s$ is different from the knot or link $r \cup s$, then (B,t) is not equal to (B,r). The simplest application of this method of distinguishing tangles is to compare the numerators and the denominators of tangles. If G is a theta-curve [resp. handcuff] and (B,t) a normalized tangle that describes G, the denominator [numerator] of (B,t) is a knot [link] contained in G. Thus, to obtain more information than contained in the constituent knots or links of G, it seems useful to look at the numerator links of tangles describing theta-curves and the denominator knots of tangles describing handcuffs.

Let NT(G) be the set of unoriented link types of numerators of tangles in T(G), and let DT(G) be the set of denominators. Because of the symmetries of the tangles in T(G), (see the end of the proof of theorem 3.2), if G is a theta-curve there are at most three elements in NT(G) and at most three elements in DT(G). If G is a handcuff graph there is at most one element in NT(G) and there can be infinitely many elements in DT(G). Using this terminology we have the following corollary to theorem 3.2.

COROLLARY 3.4. If G_1 and G_2 are two theta-curves [resp. handcuffs with constituent links having linking number zero] and $NT(G_1) \neq NT(G_2)$ $[DT(G_1) \neq DT(G_2)]$, then G_1 is not equivalent to G_2.

Given a theta-curve and a normalized tangle description $\theta(B,t)$ of it, we will call the two-component link consisting of the numerator of (B,t) the <u>associated link</u> of $\theta(B,t)$. For a given handcuff graph and a normalized

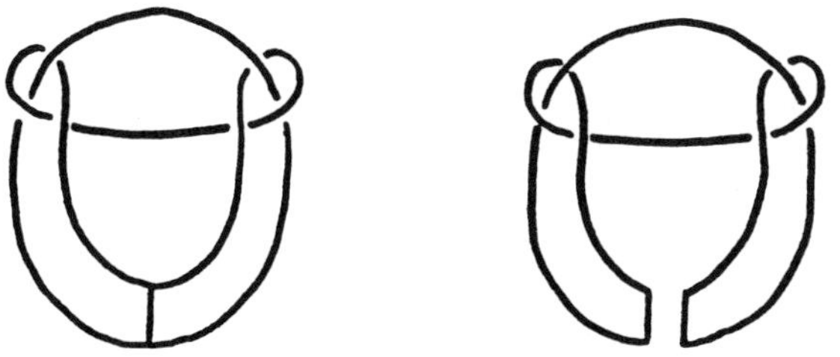

FIGURE 9. A Locally Unlinked Handcuff and its Nontrivial Associated Knot

tangle description of it, $\Phi(B,t)$, we will call the knot consisting of the denominator of (B,t) the associated knot of $\Phi(B,t)$.

COROLLARY 3.5. Let (B,t) be a normalized tangle.

1) If $\theta(B,t)$ is planar, then the associated link of $\theta(B,t)$ is planar.
2) If $\Phi(B,t)$ is planar, then the associated knot of $\Phi(B,t)$ is planar.

EXAMPLES. The handcuff graph shown in figure 9 is locally unlinked but the associated knot (figure 9) is the square knot. Thus the handcuff, by corollary 3.5, is nonplanar.

One of the three associated links of Kinoshita's theta-curve (they are actually all equal because of the symmetries of Kinoshita's theta-curve) is shown in figure 10. Depending on how we orient the components, its two variable polynomial is $p(\ell,m) = \mu + m\ell + m\ell^{-1} + m^2\ell^{-4}\mu + m^3\ell^{-3} - m^3\ell^{-1}$ or $p(\ell,m) = \mu\ell^{-2} + 2m\ell^{-1} + m^2\ell^{-4}\mu + m^3\ell^{-3} - m^3\ell^{-1}$, where $\mu = -m^{-1}(\ell + \ell^{-1})$ The polynomials are not equal to μ, the polynomial of the trivial two component link, therefore the link is nonplanar and corollary 3.5 implies that Kinoshita's theta-curve is nonplanar. Also the above polynomials are not symmetric with respect to ℓ and ℓ^{-1} so the associated link is chiral. Corollary 3.4 then implies that Kinoshita's theta-curve is chiral.

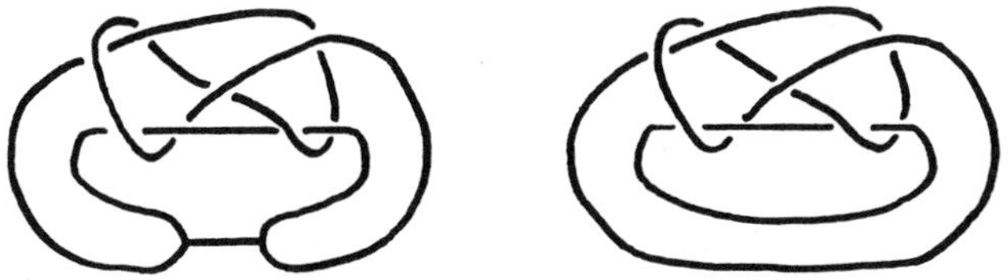

FIGURE 10. Kinoshita's Theta-curve and its Nonplanar Associated Link

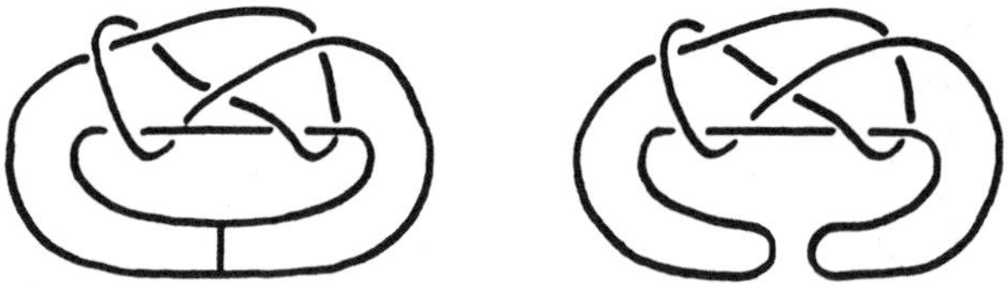

FIGURE 11. A Nonplanar Theta-curve with Planar Associated Link

Examples of nonplanar theta-curves that have planar associated link and nonplanar handcuffs that have planar associated knot are shown in figures 11 and 12. The associated links and knots may not detect the nonplanarity of theta-curves and handcuff graphs in the case where the graphs themselves contain nontrivial knots or links. Outside this case, they are (corollaries 3.11 and 3.13) perfect detectors. First we need a few preliminary results.

LEMMA 3.6. A tangle (B,t) is rational if and only if (B,t) is untangled.

Proof. (This must be well-known, but we have not found an explicit statement in the literature.) Assume (B,t) is untangled. Let h be a homeomorphism of (B,t) to (B,0), the 0-tangle. By theorem 4.5 of [10] $h|_{\partial B}$ is isotopic to a composition of Dehn twists about disks in ∂B that contain two of the points in ∂t. Thus we can isotope h to the identity on ∂B at the expense of braiding the four arcs of the 0-tangle in a collar of ∂B. Since this braiding can be accomplished as products of integral tangles, (B,t) is a rational tangle.

The other direction follows from the definitions since products of twists can be undone by a homeomorphism of B.

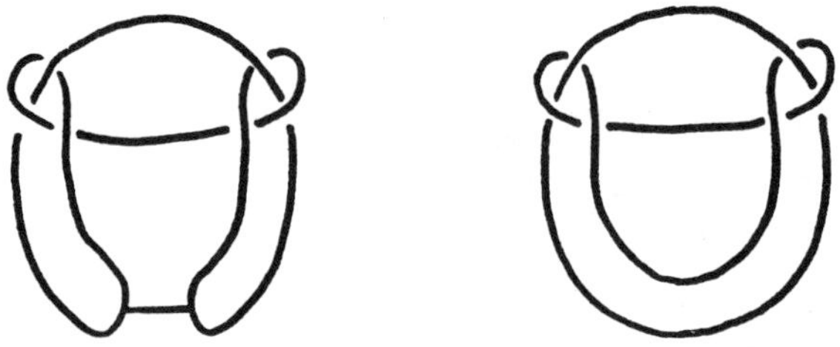

FIGURE 12. A Nonplanar Handcuff with Trivial Associated Knot

LEMMA 3.7. Let (B,p/q) be a rational tangle with denominator equal to the trivial knot. Then (B,p/q) equals (B,p), an integral tangle.

Proof. The double cover of S^3 branched over the denominator is the lens space L(q,p) [12]. The branching knot is trivial, so $L(q,p) = S^3$. Thus q = 1.

LEMMA 3.8. Let (B,p/q) be a rational tangle with numerator equal to the planar two-component link. Then (B,p/q) equals (B,0) the 0-tangle.

Proof. The double cover of S^3 branched over the numerator is the lens space L(p,q) [12] which must be $S^2 \times S^1 = L(0,1)$. Thus p/q = 0.

LEMMA 3.9. If (B,t) is a normalized tangle with denominator equal to the trivial knot and (B,t) is not equal to the 0-tangle, then the double cover of B branched over t is a nontrivial knot exterior in S^3 of a strongly invertible knot.

Proof. First we will prove that (B,t) is a prime tangle. Suppose (B,t) is an untangle that is not equal to the 0-tangle. Lemmas 3.6 and 3.7 imply that (B,t) equals an integral tangle. But since (B,t) is a normalized tangle, (B,t) equals the 0-tangle, which contradicts the assumption (B,t) is not equal to (B,0). Next suppose there is a two-sphere in B which meets t transversely in two points and bounds a ball in B meeting t in a knotted spanning arc. Then the denominator of (B,t) would not be the trivial knot. Thus (B,t) is a prime tangle.

The double cover M of B branched over t is an irreducible, boundary-irreducible three-manifold with torus boundary [10, Theorem 5]. By gluing a solid torus T, which is the double cover of the ∞-tangle branched over its two arcs, to M, we can construct the double cover $M \cup T$ of S^3 branched over the denominator of (B,t). The manifold $M \cup T$ is just S^3 since the branching set is the trivial knot. Thus M is a knot exterior. M is not a solid torus since M is boundary-irreducible. To see that M is the exterior of a strongly invertible knot, we observe that M is the exterior of the knot in $M \cup T$ consisting of the core of T and that the involution of $M \cup T$ obtained from the covering map inverts the core of T. ([12] constructs the two-fold branched cover of the ∞-tangle.) Thus M is the exterior of a strongly invertible knot. (See the proof of theorem 3.15 for a way to construct the knot whose exterior is M.)

THEOREM 3.10. Let θ(B,t) be a locally unknotted theta-curve described by the normalized tangle (B,t). Let the knot or link K equal $t \cup s$ in $(B,t) \cup_{id} (B,s)$ where (B,s) equals (B,0) or (B,1/q) for any $q \in Z \setminus \{0\}$. Then θ(B,t) is planar if and only if K is planar.

COROLLARY 3.11. Let $\theta(B,t)$ be a locally unknotted theta-curve described by the normalized tangle (B,t). Then $\theta(B,t)$ is planar if and only if its associated link is planar.

Proof of corollary. Let the tangle (B,s) in theorem 3.10 be equal to $(B,0)$. Then K is the associated link.

Proof of theorem 3.10. If $\theta(B,t)$ is planar, theorem 3.2 implies that (B,t) equals the 0-tangle. Thus by construction K is planar.

Assume $\theta(B,t)$ is nonplanar and we will show that K is nonplanar. Since $\theta(B,t)$ is nonplanar, (B,t) is not equal to $(B,0)$. Since $\theta(B,t)$ is locally unknotted, the denominator of (B,t) is the trivial knot (the denominator is one of the knots in $\theta(B,t)$). Lemma 3.9 implies that M, the double cover of B branched over t, is a nontrivial knot exterior in S^3 of a strongly invertible knot J. Let $K_{p/q}$ be the knot of link $t \cup p/q$ in $(B,t) \cup_{id} (B,p/q)$. The double cover of B branched over p/q is a solid torus T (since $(B,p/q)$ is equivalent to $(B,0)$). The double cover of S^3 branched over $K_{p/q}$ is then $M \cup T$ where the attaching map corresponds to p/q Dehn filling on the knot manifold M [12]. Thus the three-manifold $M \cup T$, which we will call $M(p/q)$, is the double cover of S^3 branched over $K_{p/q}$. If $K_{p/q}$ is the planar two-component link, then $M(p/q)$ is homeomorphic to $S^2 \times S^1$. The knot J is nontrivial and therefore has property R ([7] or [11]) so $M(p/q)$ is not $S^2 \times S^1$. Thus $K_{p/q}$ is not a planar two component link. If $K_{p/q}$ is the trivial knot, then $M(p/q)$ is homeomorphic to S^3. Strongly invertible knots have property P [3], so $p/q = \infty$. Thus $K_{p/q}$ is only planar for $p/q = \infty$. In particular, each of the possible knots K given in the theorem is nonplanar.

REMARK. The above proof actually shows that if $\theta(B,t)$ is locally unknotted but not planar, then the knot or link obtained by adding any rational tangle (except ∞) to (B,t) is nontrivial.

We have the analogous results for handcuff graphs.

THEOREM 3.12. Let $\Phi(B,t)$ be a locally unlinked handcuff graph described by the normalized tangle (B,t). Let the knot K equal $t \cup s$ in $(B,t) \cup_{id} (B,s)$ where (B,s) equals $(B,1/q)$ for any $q \in Z$. Then $\Phi(B,t)$ is planar if and only if K is trivial.

COROLLARY 3.13. Let $\Phi(B,t)$ be a locally unlinked handcuff graph described by the normalized tangle (B,t). Then $\Phi(B,t)$ is planar if and only if its associated knot is trivial.

Proof of corollary. Let tangle (B,s) in theorem 3.12 be equal to (B,∞). Then K is the associated knot.

Proof of theorem 3.12. If $\Phi(B,t)$ is planar, then theorem 3.2 implies that (B,t) equals the 0-tangle. Thus by construction K is trivial.

Assume $\Phi(B,t)$ is nonplanar and we will show that K is nontrivial. Since $\Phi(B,t)$ is nonplanar, (B,t) is not equal to (B,0). Since $\Phi(B,t)$ is locally unlinked, the numerator of (B,t) is the planar two-component link and thus (B,t) is either prime or untangled. If (B,t) is untangled, lemma 3.6 implies (B,t) equals (B,p/q) for some p/q. Lemma 3.8 implies that (B,p/q) equals (B,0) which is a contradiction. Thus (B,t) must be prime.

The double cover M of B branched over t is an irreducible, boundary-irreducible three-manifold with torus boundary [10]. Suppose K is the trivial knot. The double cover of S^3 branched over K is then just S^3. Since this branched double cover can be constructed by gluing the branched double cover of (B,1/q) to M, S^3 contains M. Thus M is a nontrivial (M is boundary-irreducible) knot space. Since the numerator of (B,t) is the planar two-component link, the double cover of S^3 branched over the numerator is $S^2 \times S^1$ and can be constructed by gluing a solid torus T, which is the branched double cover of (B,0), to M. But nontrivial knots have property R ([7] or [11]), so $M \cup T$ cannot be $S^2 \times S^1$. This is a contradiction thus K is nontrivial.

Now we will look at a second method for associating knots with a theta-curve. This method will apply only to locally unknotted theta-curves. First we prove the following lemma.

LEMMA 3.14. Let (B,t) be a normalized tangle such that $\theta(B,t)$ is locally unknotted. Then (B,t) is prime if and only if $\theta(B,t)$ is nonplanar.

Proof. Suppose $\theta(B,t)$ is planar. Theorem 3.2 implies that (B,t) equals the 0-tangle and thus is not prime.

Suppose (B,t) is not prime. Then since $\theta(B,t)$ is locally unknotted, (B,t) is untangled. By lemma 3.6 (B,t) is rational. Since the denominator of (B,t) is a knot in $\theta(B,t)$, it is the trivial knot, so lemma 3.7 implies that (B,t) equals (B,p) for some integer p. Since (B,t) is in normal form, p = 0 and thus (B,t) equals the 0-tangle. Therefore $\theta(B,t)$ is planar.

Let $\alpha_i (i = 1,2,3)$ be the three arcs comprising G and let J_i be the (trivial) simple closed curve $G \setminus \mathrm{int}\,\alpha_i$. The arc α_i lifts to a knot K_i in the two-fold cyclic branched cover of S^3 branched over J_i (which is again S^3).

THEOREM 3.15. Let G be a locally unknotted theta-curve with K_1, K_2 and K_3 obtained as above. Then the following are equivalent:

i) G is planar.

ii) All three of K_1, K_2, and K_3 are trivial.

iii) At least one of K_1, K_2, or K_3 is trivial.

Proof. i) $\Rightarrow$ ii) $\Rightarrow$ iii) are clear so we just need to show iii) $\Rightarrow$ i).

Assume G is nonplanar. Since G is locally unknotted, lemma 3.14 implies that the set T(G) of normalized tangles describing G contains only prime tangles. Let α_i be one of the arcs of G and (B,t) be one of these prime tangles that excludes the arc α_i. The denominator of (B,t) is the trivial knot since it is one of the knots in G. By lemma 3.9 the double cover M of B branched over t is a knot exterior of a nontrivial knot K. We now show that the knot K equals K_i.

The exterior of the knot K in S^3 is the double cover of B branched over t. Let A be a copy of B and (A,s) be the ∞-tangle. The knot s $\cup$ t in $S^3 = (A,s) \cup_{id} (B,t)$ is the denominator of (B,t) and thus is the trivial knot. S^3 is the double cover of S^3 branched over s $\cup$ t and $\overline{S^3 \setminus M}$ is the double cover of A branched over s. Since K is the core of the torus $\overline{S^3 \setminus M}$ that is the double cover of A branched over s, to construct the knot K we need only construct S^3 as the double cover of S^3 branched over s $\cup$ t and keep track of the three-ball A and $\overline{S^3 \setminus M}$, its double cover branched over s. Therefore we redraw s $\cup$ t as a straight line with endpoints identified at infinity. (See figure 13 for a particular example.) As in figure 13,

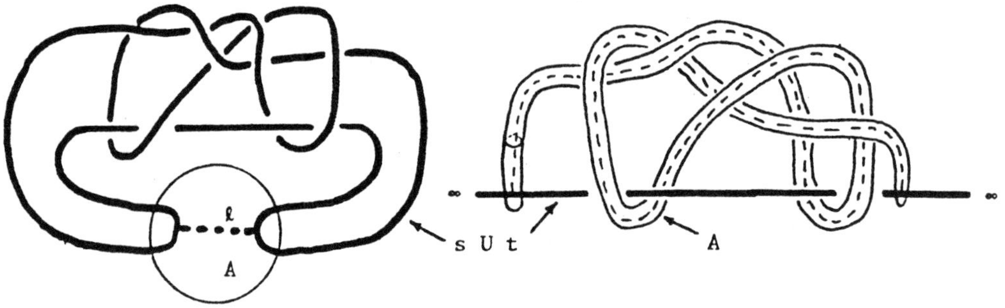

FIGURE 13. Two Presentations of θ(-1,1,1)

if we add the dotted line ℓ in A so that $s \cup t \cup \ell$ equals $\theta(B,t)$, then the lifting of ℓ in the double cover of S^3 branched over $s \cup t$ is K_i and also equals K.

4. "CONNECT SUMS" OF GRAPHS

In this section we will look at ways to construct more complex graphs from simple ones.

One such method is the classical connect sum which is used for knots (see figure 14). S. Suzuki has studied this type of connect sum for graphs and has proven a unique prime decomposition theorem [14].

Another method of combining graphs is what we will call order-n vertex connect sum, which is defined as follows. If graphs G_1 and G_2 each have a vertex of order n, remove three-ball neighborhoods of each of these vertices and glue the remaining three-balls together in such a way that the n points of G_1 and the n points of G_2 in the boundary of each three-ball are identified (see figure 15). This is just classical connect sum if n equals two. The element of the mapping class group of S^2 with n points fixed setwise that is chosen to do the gluing determines a braid b of n strings that "appears" in the new graph $G_1 \#_b G_2$. Now we ask, for which braids b is it true that for any graphs G_1 and G_2 that $G_1 \#_b G_2$ is equivalent to $G_1 \#_e G_2$ (where e is the trivial braid)? Such braids form a subgroup of the braid group of the disk on n strings. We have been calling this subgroup M_n the marionette subgroup since it consists of braids that can occur in the tangled strings of a marionette. It is easy to see that if b is not a pure braid then b is not an element of M_n. (Look at order-n connect sums of θ_n-curves that contain knots.) Thus M_n is a subgroup of the pure braid group P_n on n strings.

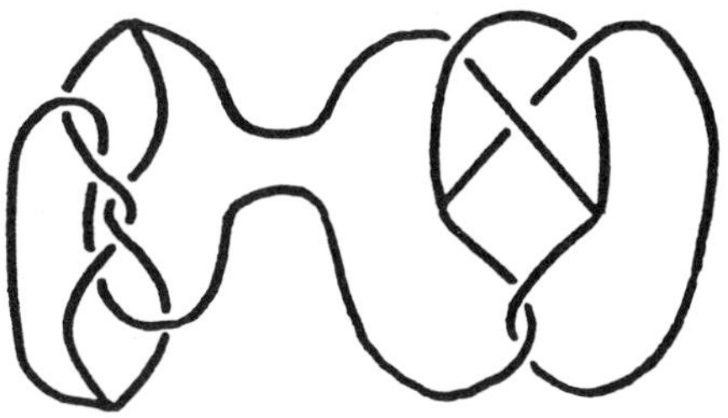

FIGURE 14. A Connect Sum of $\theta(-1,-1,1)$ with $\theta(1,1,1)$

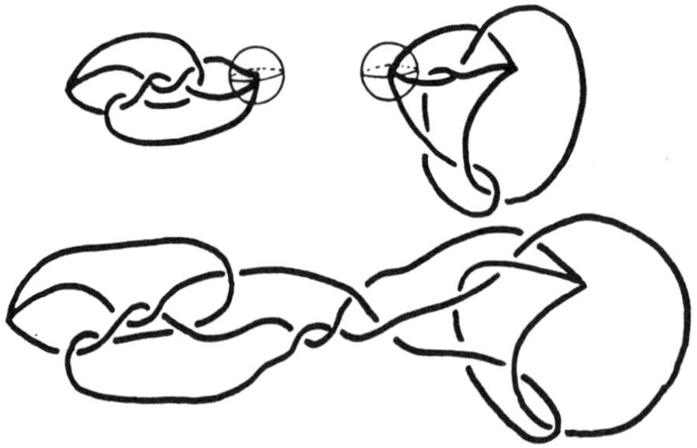

FIGURE 15. An Order-three Connect Sum of $\theta(-1,-1,1)$ with $\theta(1,1,1)$

In a different setting, the group M_n has been studied in [13] which shows that M_n is a normal subgroup of P_n and that M_3 is all of P_3. Thus for theta-curves and handcuff graphs we need only consider six different connect sums for any two given vertices. For $n \geq 4$, $M_n \neq P_n$ [13] so order-n connect sum always involves the ambiguity of an included braid, more precisely, the coset of that braid in P_n/M_n. For $n = 4$, P_4/M_4 is free of rank two.

LEMMA 4.1. Any order-three vertex connect sum of two locally unknotted theta-curves is locally unknotted.

Proof. This is true since the connect sum of two trivial knots is trivial.

THEOREM 4.2. Let G_1 and G_2 be two theta-curves with G_1 nonplanar. Then any order-three vertex connect sum of G_1 and G_2 is a nonplanar theta-curve.

Proof. If G_2 is planar then any order-three connect sum with G_1 equals G_1 and thus is nonplanar. Thus we can assume G_2 is nonplanar. If either G_1 or G_2 contains any nontrivial knots then the order-three vertex connect sum contains a nontrivial knot. Thus we need only consider the locally unknotted case. If G_1 and G_2 are locally unknotted, lemma 3.14 implies that any normalized tangles that describe G_1 or G_2 are prime tangles. Any order-three vertex connect sum $G_1\#G_2$ has a normalized tangle description consisting of the addition of two normalized tangles describing G_1 and G_2. By [10] the addition of two prime tangles is prime. Thus any $G_1\#G_2$ has a normalized prime tangle description. By lemma 4.1 $G_1\#G_2$ is locally unknotted so lemma 3.14 implies $G_1\#G_2$ is nonplanar.

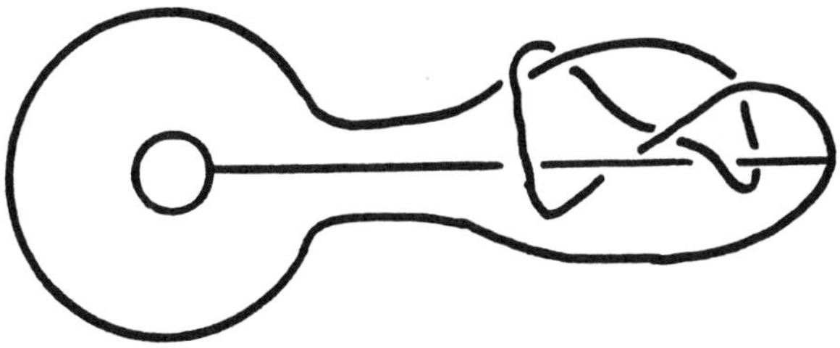

FIGURE 16. A Planar Order-three Connect Sum of a Planar Handcuff with Kinoshita's Nonplanar Theta-curve

The above result is not true in general. For example, an order-three vertex connect sum between a planar handcuff and any locally unknotted theta-curve results in a planar handcuff. Figure 16 shows this for Kinoshita's theta-curve.

ACKNOWLEDGEMENTS

This work was supported by the University of Iowa Stereochemical Topology project, ONR contract #N00014-85-K-0099. This paper will be part of my Ph.D thesis and I wish to thank Jonathan Simon for his great patience, many helpful comments, and enthusiasm as my thesis advisor.

REFERENCES

1. Alexander, J. W., On the Deformation of an n-cell, Proc. National Academy of Science 9 (1923), 406-407.
2. Birman, J. S., Braids, Links, and Mapping Class Groups, Annals of Math. Studies number 82. Princeton (1974).
3. Bleiler, S., and Scharleman, M., Tangles, Property P and a Problem of Martin, MSRI preprint.
4. Conway, J. H., An Enumeration of Knots and Links, and some of their Algebraic Properties, Computational Problems in Abstract Algebra, Proc. Conf. Oxford, 1967, Ed. John Leech, Pergamon Press, 1970, 329-358.
5. Fisher, G. M., On the Group of all Homeomorphisms of a Manifold, Trans. Amer. Math. Soc. 97 (1960), 193-212.

6. Fox, R. H., Covering Spaces with Singularities, Algebraic Geometry and Topology: A Symposium in Honor of S. Lefschetz, Fox et al eds., Princeton (1957), 243-257.

7. Gabai, D., 3-manifolds and Foliations, Abstracts from Workshop on 3-manifolds, January 9-19, 1985.

8. Gilmer, P., Litherland, R., Personal letter to J. Simon (1985).

9. Kinoshita, S., On Elementary Ideals of Polyhedra in the 3-sphere, Pacific J. Math. 42 (1972), 89-98.

10. Lickorish, W.B.R., Prime Knots and Tangles, Trans. Amer. Math. Soc. 267 (1981), 321-332.

11. Menasco, W. M., Branched Surface and Property-R, Abstracts from Workshop on 3-manifolds, January 9-19, 1985.

12. Montesinos Amilibia, J. M., Seifert Manifolds that are Ramified Two-sheeted Cyclic Coverings, (Spanish) Bol. Soc. Mat. Mexicana (2) 18 (1973), 1-32.

13. Shepperd, J.A.H., Braids Which can be Plaited with their Threads Tied Together at each End, Proc. Roy. Soc. Ser. A 265 (1961/62), 229-244.

14. Suzuki, S., A Prime Decomposition Theorem for a Graph in the 3-sphere, Preprint, Waseda University (1985).

Index